HERPETOLOGICAL EXPLORATIONS
OF
THE GREAT AMERICAN WEST

This is a volume in the Arno Press collection

BIOLOGISTS AND THEIR WORLD

See last pages of this volume
for a complete list of titles

HERPETOLOGICAL EXPLORATIONS OF THE GREAT AMERICAN WEST

Edited with an Introduction by
Kraig Adler

Volume I

ARNO PRESS
A New York Times Company
New York • 1978

Editorial Supervision: MARIE STARECK

Reprint Edition 1978 by Arno Press Inc.

Special thanks are given to the American Museum of Natural History Library and to the State Historical Society of Wisconsin Library for lending many of the selections in this volume.

BIOLOGISTS AND THEIR WORLD
ISBN for complete set: 0-405-10641-6
See last pages of this volume for titles.

Manufactured in the United States of America

Publisher's Note: The selections which were originally published by the U. S. Government Printing Office are not under copyright protection. Page 235 appears as page 255 in "Description de Quelques Espèces de Reptiles de la Californie," in all available copies. The text is correct. All color plates have been reproduced in black and white in this edition.

Library of Congress Cataloging in Publication Data
Main entry under title:

Herpetological explorations of the great American West.

(Biologists and their world)
Reprint of articles published 1835-1905.
1. Reptiles--The West--Addresses, essays, lectures.
2. Amphibians--The West--Addresses, essays, lectures.
I. Adler, Kraig. II. Series.
QL653.W4H47 598.1'0978 77-81100
ISBN 0-405-10682-3

Herpetology in Western North America During The Nineteenth Century

WITHIN sixty years after the American Revolution the fauna and flora of the original colonies were well known. Like other groups of organisms, the reptiles and amphibians had at first been described and named by Europeans but by 1800 this task had fallen largely to a small but highly productive group of American naturalists primarily centered in Philadelphia. Their work led to the first comprehensive, illustrated survey of the reptiles of the United States, John Edwards Holbrook's *North American Herpetology* . . . (1842). This treatise, exhaustive though it was, covered only a small part of the new nation, then primarily an area from New England south to the peninsula of Florida ceded by Spain in 1821. The few species then known from west of the Appalachian frontier were included by Holbrook but gave a limited and, as we now know, grossly incomplete picture of the rich herpetofauna of that part of America — but this void was to be rapidly filled.

The Lewis and Clark expedition in 1803-1806 yielded the first reasonably accurate measure of the true width of the continent and the vastness of the unexplored territory. It was not until after the War of 1812 that Americans began looking westward, particularly at the immense Louisiana Territory which had been purchased from France in 1803. In 1819-1820 a scientific party under Major Stephen H. Long surveyed the Great Plains and reached as far west as the Rocky Mountains of Colorado. This group included Thomas Say, the prominent Philadelphia naturalist, who described some of the commonest species of amphibians and reptiles from the plains region. Long's party was attached to the "Yellowstone Expedition", a military detachment sent up the Missouri River to pacify the Indians and prevent incursions by British traders into American territory. This combination of military and scientific purposes became the pattern for other government surveys in the West, but there was not to be another major one for some twenty years, due in part to Major Long's exaggerated reports of the unsuitability of the West, which he and others referred to as the "Great American Desert", for settlement and agriculture.

During this period there was relatively little biological exploration west of the Mississippi River except along the Pacific coast where small but historically important collections were made. During 1827–1829 P. E. Botta, a naturalist aboard the French merchant ship *Héros*, collected in California and the specimens were later described by Henri Marie Ducrotay de Blainville (1777–1850), an anatomist and one of the successors to Baron Cuvier at the Muséum National d'Histoire Naturelle. Perhaps the most peculiar novelty taken by Botta was a small stump-tailed snake belonging to the boa family which was named in the collector's honor by de Blainville. During the period 1838–1842 there was also an important American survey of the West coast, that of the U. S. Exploring Expedition under the command of Charles Wilkes. The herpetological collections from this extensive survey of the whole Pacific region were studied and written up by Charles Girard. But the only significant herpetological work in the interior of the West during the period from Long's expedition until the topographical surveys in the 1840s was that of a Rhenish prince who devoted most of his life to natural history and ethnology.

Alexander Philipp Maximilian (1782–1867) of the Prussian principality of Wied, located at Neuwied along the Rhine in what is now West Germany, was easily the most titled and best trained scientific observer to explore the West during the first half of the nineteenth century. After serving as a major-general in the Napoleonic wars, Maximilian resigned from the Prussian Army in 1814 and promptly embarked on an expedition to Brasil that was to establish his reputation as an explorer, ethnologist and a naturalist of the first rank. The results of the prince's exploration in the coastal forests of Brasil were published in his extensive book *Reise nach Brasilien in den Jahren 1815 bis 1817*, issued in 1820–1821, and also in an atlas illustrating the animals and plants (*Abbildungen zur Naturgeschichte Brasiliens*, 1822–1831). Later, at the age of 50, he left the comforts of his palace once again to make his second (and last) celebrated exploration, the ascent of the great Missouri River in the northern plains of North America. Maximilian's written accounts of this trip are second in importance only to those of Lewis and Clark as a record of the biota and people of the trans-Mississippian region. Maximilian's report was published in parts over the period 1838–41 under the title *Reise in das innere Nord-Amerika . . . in den Jahren 1832 bis 1834*; there were also several foreign language translations of this scientifically-important and widely-read work. At that time Illinois and Missouri were the frontier states and beyond them lay Indian country into which relatively few whites had gone before. It was indeed fortunate that a person of such ability witnessed this Indian civilization in its final days of glory, because three years after Maximilian's return to Europe a smallpox epidemic wiped out approximately one quarter of the entire Indian population along the Missouri. And those who were not killed by disease later fell victim to excessive whiskey consumption, military encounters and finally to the onslaught of settlers.

Maximilian arrived at Boston in July 1832, and wintered at New Harmony, Indiana, where a socialistic commune had been established along the Wabash River. There the prince befriended Thomas Say, whose official duties in the community were described in his title, Superintendent of Literature, Science and Education. Together with Say, Maximilian collected local amphibians and reptiles and those that were new to science the prince subsequently named. Maximilian eventually reached St. Louis where he boarded the American Fur Company's small steamboat *Yellowstone* to ascend the Missouri, the longest river in North America. In 1833 the Missouri was considered the most treacherous of Western rivers, although now long-since tamed by a system of dams. Today it is difficult to imagine that the dangers to Maximilian and his small party in this now-civilized region were so great, but his journal records bear testimony to the great risks that he took. Maximilian belonged to that breed of romantic explorers to whom the rigors associated with discovery were not only endured but, apparently, even necessary for the experience to be savored to the fullest possible extent.

FIGURE 1. Maximilian being introduced to Indian chiefs at Fort Clark where the prince spent the bitter winter of 1833-34. Karl Bodmer, the artist who captured the visual record of Maximilian's expedition, is at the prince's left.

Accompanying Maximilian was the young Swiss painter Karl Bodmer (1809-1893) whose watercolor landscapes and portraits of Indians and their dwellings comprise the most extensive pictorial representation of the time, as photographic techniques had not yet been developed. Early in August, after a journey of nearly 2700 miles from St. Louis, Maximilian reached the westernmost point of the trip, Fort McKenzie, near the present-day city of Great Falls, Montana. But Maximilian did not remain long and left on September 14th to ship down the Missouri to Fort Clark and there emulate Lewis and Clark by spending the winter among the Mandan and Minnetaree Indians. It was a tribute that very nearly cost him his life.

Maximilian's party, numbering only seven, utilized a small, heavily-laden boat and thus were completely exposed to depredations by the various tribes along the Missouri and to the physical dangers posed by the river itself. But to Maximilian this sort of travel was preferable to that on the steamboat, as he could stop on command wherever he wished to collect specimens or reconnoiter. Finally, on November 8, 1833, they reached Fort Clark in what is now central North Dakota. This was a typical frontier fort, small (Maximilian recorded that the stockade was only 45 paces square), and having a total white population averaging about a dozen. The forts along the Missouri were operated by the trading companies which at that time were primarily interested in animal pelts. Although St. Louis was only five weeks away by boat, Maximilian chose to spend the next five months at Fort Clark during which time he made undoubtedly the most important observations of the entire trip.

Life in the fort was spartan, but the extended stay allowed time for detailed observations of the local Indians and extensive collections of artifacts and natural history specimens. Because there was no military protection at the fort, Maximilian and the other whites were at the mercy of Indians in the nearby encampments who made frequent trips to the fort to trade goods but also to help themselves to virtually anything else they wanted, thus producing confrontations that required all the diplomatic skills Maximilian and the other whites could muster. The winter was a particularly brutal one, with temperatures often below -20°F and sometimes even below -40°; on one occasion the mercury sank into the ball and froze! The buckets of water in Maximilian's own living quarters froze solid as did Bodmer's paints. These were no royal quarters, yet Maximilian persevered and continued to interview Indians and trappers and even spent some nights in nearby Mandan huts. Eventually the fort went on starvation rations when the last dried meat was consumed at Christmas time. Thereafter, except for occasional buffalo kills, the food consisted of weak coffee and maize boiled in water. Like others at the fort, the prince eventually contracted scurvy and nearly died but the long winter finally broke and fresh fruits could be brought upriver to the fort. Maximilian's party reached St. Louis late in May and he left the United States shortly after, his great adventure over and his contribution to American natural history and ethnology forever established.

This was the last major expedition by a foreigner for in 1842 the American government sent the first of what was to become a steady flow of surveying parties into the West. The earlier expeditions, together with the scattered reports of trappers and scouts — the legendary "mountain men" — provided some knowledge of the West, but the practical inadequacy of that knowledge was abruptly recognized in 1845 when México refused to acknowledge the legality of Texas's annexation by the United States and war seemed inevitable. Accurate maps of the area along the Mexican border did not exist and President Polk promptly dispatched a team of surveyors under William Hemsley Emory (1811-1887). Lieutenant Emory belonged to that small band of mostly West Point-trained army officers who made up the Corps of Topographical Engineers, a group whose exploits are collectively as colorful and heroic as any other explorers in history. Although never at any time having more than 36 officers, this tiny group more than any other succeeded in laying the foundation for uniting the continental nation. Their primary tasks were to produce maps, lay out boundaries and gather information relevant for settlements; in short, they served as a "Department of Public Works" for the Old West. But they performed valuable scientific work as well, taking meteorological measurements and, where possible, amassing collections of specimens. Some of them, like Emory, were in close contact with the leading scientists in the East and thus played a vital role in describing the biota of the West. Emory himself was a flamboyant character with red whiskers and picturesque language who looked more like a cavalryman than the engineer he was. His ruggedness, an essential requisite for survival in the Old West, was first demonstrated in 1846 during the Battle of San Pasqual when Mexican troops attempted to recapture southern California. With Kit Carson as their guide, Emory's detachment of dragoons surprised the Mexican squadron of lancers and they mounted a classic all-out cavalry charge. Emory joined the commanding general at the head of the column and eventually became the hero of the battle when he killed a Mexican lancer who was about to finish off the general. Immediately after the Mexican war ended in 1848, a commission was established to survey the new boundary and Emory was placed in charge but because of political squabbles the resulting reports, including the one by Baird on reptiles here reprinted, were not published until a decade later.

FIGURE 2 (clockwise from top left): Maximilian, W. H. Emory, Charles Girard and Spencer F. Baird.

When these Western collections were sent to Washington the government assigned them to the Smithsonian Institution where they fell into the able hands of Spencer Fullerton Baird (1823–1887). Baird had joined the Smithsonian shortly after it was founded in 1846 and was given the primary task of building up a natural history collection. He was precocious as a student and following graduation from college at the age of 17 he spent a year on his own studying nature; on one occasion he covered 400 miles on foot in 21 days studying birds in the Pennsylvania mountains. By age 22 he had returned to his alma mater as Professor of Natural History. Baird's own preserved collection, together with that of the Wilkes Expedition mentioned earlier, formed the nucleus of the Smithsonian's to which he added assiduously, both from government-sponsored surveys and through exchanges with other museums. Baird himself did highly competent studies on these collections and published major treatises on all the vertebrate groups. Besides being a fine naturalist Baird was an extraordinarily accomplished administrator and organizer, becoming the first Commissioner of Fish and Fisheries in 1874 and later the head of the Smithsonian Institution itself in 1878. He was also instrumental in setting up the famed oceanographic institute at Wood's Hole. But in spite of his many official duties Baird never forgot the influence of his own teachers who included Audubon and devoted much of his time to encouraging aspiring young naturalists.

Baird's herpetological work often was in collaboration with his chief assistant, Charles Frédéric Girard (1822–1895), a Frenchman by birth who had accompanied his teacher Louis Agassiz to America in 1847. Girard participated briefly in some of the Western surveys, but most of his work was done at the museum in Washington. Shortly before the Civil War began Girard went to France and later supplied the Confederate army with drugs and surgical supplies. Although he returned to the United States briefly in 1863 to tour the Southern states, he spent the remainder of his life practicing medicine in France. Another of Baird's protégés, Robert Kennicott (1835–1866), performed much of the actual laboratory work. Kennicott was later named director of the Chicago Academy of Sciences but shortly thereafter died, at the age of 30, on an expedition to Alaska. Together, Baird and his associates described and named a large proportion of the reptiles and amphibians of the West, particularly the lizards and snakes which are so numerous in the Southwest. Baird's published work in herpetology, unlike his on birds and mammals, consisted of little more than lists or descriptions of new species and had few critical comments on classification or biology although in fairness it should be mentioned that Baird's extensive manuscript review of North American reptiles and amphibians was never published and instead was incorporated into the major reviews of those groups by E. D. Cope. Nevertheless, Baird's papers included in this anthology form the foundation of our knowledge of many Western species of amphibians and reptiles.

During the 1850s government-sponsored expeditions penetrated into many different parts of the West. Captain Howard Stansbury was ordered to study the basin of the Great Salt Lake including the indigenous Indian tribes and, perhaps more importantly, the area that had been settled in 1847 by the Mormons. Later, friction between the "Desert Saints" and Federal officials precipitated the brief Mormon War of 1857-1858 and the maps and extensive observations obtained earlier by Stansbury confirmed the continuing need for such surveys. The reptiles collected on this expedition were written up by Baird and Girard, including a new genus of lizards named in honor of the Ute Indians. Further to the south, Captain Lorenzo Sitgreaves departed from Albuquerque in 1851 in order to explore the Zuñi and Colorado rivers; he eventually reached San Diego, exploring a route which was later to be appropriated by the Santa Fe Railroad. This survey included a naturalist, Samuel Washington Woodhouse (1821-1904), whose collections of reptiles and amphibians were later studied by the prominent Philadelphia physician and naturalist Edward Hallowell (1809-1860). Woodhouse had developed an interest in

FIGURE 3. Survey party of the Corps of Topographical Engineers camped on the prairie at the Red Fork of the Arkansas River, June 1850. Standing in front of the horse is S. W. Woodhouse, the party's surgeon and naturalist, his pistol, knives and rifle displayed at his side.

natural history through a childhood association with the Academy of Natural Sciences in Philadelphia. Officially he was the surgeon for Sitgreaves's detachment, but in the course of his allied zoological activities got a Yavapai arrow through his leg and was bitten by a rattlesnake, costing him the use of his hand for months. Small recompense it was that Baird named a toad in his honor. In 1852, captains Randolph Barnes Marcy and George B. McClellan explored in Arkansas, Oklahoma and northern Texas, successfully discovering the source of the Red River, and their herpetological collections were also published by Baird and Girard. Goetzmann describes in detail these and other army surveys in the American West, providing maps of the major routes.

The all-consuming political controversy during this period was the struggle of expansionists to secure year-round lines of communication with the Pacific coast, a debate that was heightened when California joined the Union in 1850. Like the Mexican dispute before it, this controversy became enmeshed in the mounting political and sectional differences in the country that eventually led to civil war. The focal point for these and many other differences was the debate in Congress over the proper location for a transcontinental railroad. In desperation the government attempted to break the political deadlock by appealing for an objective scientific judgment that held out the promise of an impartial solution. Accordingly, on 2 March 1853 Congress authorized a railroad survey which called for simultaneous reconnaissances along about a dozen major routes. The final published report included a separate account for each route, usually referred to by the name of the chief engineer or by the parallel or both, but the whole survey was so complicated and the political aspect so confusing that the public documents are a bibliographer's nightmare, although Meisel gives the best accounting.

The attempt to decide the issue on strictly scientific grounds was an abject failure and a Southern route was chosen largely on political and economic grounds. Despite the fact that the railroad surveys failed in their primary purpose, they provided a great scientific opportunity to explore the West and at government expense. Baird fully recognized this tremendous potential and quickly saw to it that naturalists were attached to most of the surveys and that their collections were returned East for study and publication. These were truly the harvest days for the Eastern museums. A few of the naturalists recommended by Baird, like James Graham Cooper (1830–1902), wrote up their own accounts and also provided the earliest first-hand life history data. Cooper had served in the Civil War and was later associated with the California Academy of Sciences. He is better known for his

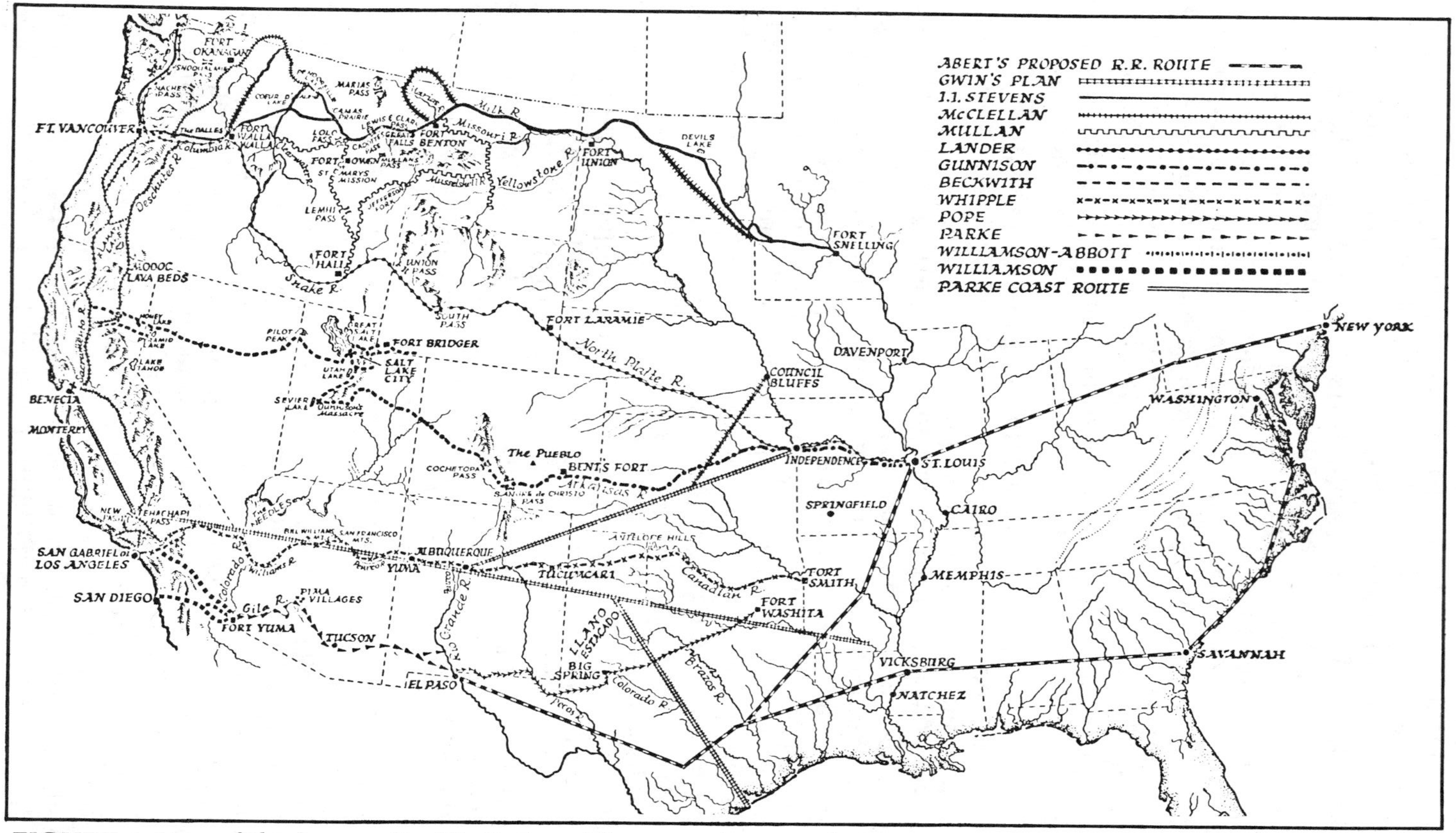

FIGURE 4. Map of the famous Pacific Railroad Surveys showing the routes explored by the several engineers of the Topographical Corps in 1853–54, as well as routes suggested by Senator Gwin of California and the chief of the Corps, Colonel Abert.

ornithological work; indeed, the Cooper Ornithological Club was named in his honor in 1893.

During the Civil War the railroad surveys were suspended but were eventually succeeded by the various explorations of the territories and geological surveys in the 1870s. These studies were undertaken by a new generation of naturalists including Henry Crècy Yarrow (1840–1929), a Union Army surgeon during the War and a participant in Wheeler's Survey in the West; Elliott Coues (1842–1899), the preeminent ornithologist and, like Yarrow, also an Army doctor later associated with the Smithsonian Institution; and Henry Wetherbee Henshaw (1850–1930), an explorer-naturalist of the West and a noted ethnologist. The dominant herpetologist of the day, however, was Edward Drinker Cope (1840–1897). Cope had been a protégé of Baird and also came under the influence of Joseph Leidy, professor at the University of Pennsylvania and once America's leading paleontologist. But Cope was quite unlike Baird and Leidy in temperament and style. While they preferred the quiet solitude of academic work, Cope abhorred routine and only accepted a university professorship when he could no longer support himself through private means. Publishing his first paper while only nineteen, he became the dominant figure in American zoology during his professional lifetime and his influence in anatomy, paleontology and biology was truly international in scope. During the 1870s Cope participated in several of the Western surveys and it was during this period that his famous feud with O. C. Marsh of Yale College began. At stake were the enormous collections of vertebrate fossils, especially those of dinosaurs, being unearthed mainly in the West and, ultimately, dominance of the field of paleontology. Cope was the better scientist but Marsh had available massive personal resources and also influential friends, so the result was a virtual standoff. Before the dispute ended when Cope died, the rivalry had engulfed the National Academy of Sciences, the Geological Survey, the Smithsonian Institution and many of America's leading biologists, paleontologists and geologists.

Cope's most epochal contributions were in paleontology but he also published extensively on living vertebrates including reptiles and amphibians. The scientific journal *Copeia*, dealing with lower vertebrates, was named in his honor. Although Cope described hundreds of new genera and species in his more than 1400 published works, his descriptive research was erratic and often annoyingly sketchy and inattentive to detail. He was clearly more at home in generalizing his work, in erecting higher classifications based on morphology, and in explaining animal distributions on a continental basis. Though a confirmed neo-Lamarckian, he was Darwinian in approach: his mind could easily sweep across vast quantities of data to produce grand generalizations. But the state

FIGURE 5 (clockwise from top left): J. G. Cooper, H. C. Yarrow, Alfredo Dugès and Edward D. Cope.

of biology in America, particularly that in the West, was then more in need of describers and namers and Cope's modern approach to many topics was generally ignored until the early part of the twentieth century.

During this same period, similar surveys were also being conducted in México which had been visited by many European naturalists during the nineteenth century, but the most extensive collections were those of the French and British. The French effort was an adjunct to the attempt by Napoleon III to establish a puppet Mexican empire under Archduke Maximilian of Austria, but after the latter's execution in 1867 in the peasant uprising led by Benito Juárez, the legitimate president of México, the scientific field party had to return to France. The biological results were published in the series *Mission Scientifique au Mexique et dans l'Amérique Centrale,* during the period 1868–1909. The massive volumes on reptiles and amphibians took nearly forty years to complete. The parallel British work was undertaken in a completely different way, being the joint project of two Cambridge University classmates, Frederick Du Cane Godman and Osbert Salvin, who privately sponsored an extensive expedition that resulted in the famous series *Biologia Centrali-Americana,* published during the period 1879–1915. One of the first collections in the Baja California peninsula was made by the French naturalist Léon Diguet whose specimens were reported on by François Mocquard (1831–1917) of the Muséum National d'Histoire Naturelle in Paris. But after these foreign visits México was left practically without any naturalists except for an immigrant Frenchman, Alfredo Augusto Delsescautz Dugès (1826–1910). Dugès travelled about the country making collections of all animal groups and provided the first comprehensive lists of the herpetofauna of his adopted country. The contribution reprinted here is perhaps his most important one to Mexican herpetology and may seem modest even by the scientific standards of the day. However, one must recall that unlike his contemporaries in America and Europe Dugès worked in isolation from major museum reference collections and he also suffered from chronic illnesses including deafness. Moreover, throughout his residence in Guanajuato he supported himself through a private practice as a gynecologist and his services as professor and director of the museum at the local college were purely voluntary.

To this point in time the Darwinian Revolution had had very little impact in herpetology, even on the brilliant Cope. Species descriptions were invariably typological and the importance of population and geographic variability within species was not appreciated. The exploration of the West had yielded descriptions and lists of species but little more. It was a logical extension of the descriptive phase of natural history, the "Naturalist s Period", which had ended about 1850 in the East. It was another generation of biologists who were to use essentially modern concepts and procedures to survey and classify the reptiles and amphibians of the American West and to usher in the modern era in herpetological studies. This transition has usually been credited, above all, to Leonhard Hess Stejneger (1851-1943) who had left his native Norway in 1881 after a brief career as a lawyer. He came to the Smithsonian Institution and for a number of years worked with birds and mammals but at the age of 38 he turned his attention to herpetology which was surely more in need of his considerable talents. During a long career at the Smithsonian Institution Stejneger produced a steady stream of careful taxonomic studies on American reptiles and amphibians and was the principal author of a series of important herpetological checklists that standardized nomenclature. Stejneger also participated in the final stages of that earlier era of Western exploration and, among other contributions, wrote accounts on collections made by the Biological Survey in Arizona and California under the supervision of Vernon Orlando Bailey (1864-1942), one of the Survey's collectors in its early days and later himself a prominent naturalist.

FIGURE 6 (left to right): Leonhard Stejneger and John Van Denburgh.

All of these Western studies were finally consolidated in the work of John Van Denburgh (1872–1924) who from 1894 was associated with the California Academy of Sciences in San Francisco. As shown in his short paper on specimens he collected in the Columbia River basin in 1892, even as a student at Stanford University Van Denburgh was interested in the habits and distribution of reptiles and amphibians. Van Denburgh wrote extensive reviews of many groups of Western amphibians and reptiles and did many state surveys, leading eventually to his first monograph, *The Reptiles of the Pacific Coast and Great Basin*, published in 1897, which covered all of the coastal states plus Idaho and Nevada. Like Holbrook's work some fifty years before, Van Denburgh synthesized the exploration and work done by naturalists over the preceding century; this work was later greatly expanded in Van Denburgh's two-volume work, *The Reptiles of Western North America* (1922).

At last the bulk of America's kinds of reptiles and amphibians had been described and classified and the general limits of their ranges defined. Biologists thus had a suitable foundation on which to build. The ensuing work of the twentieth century has therefore emphasized the reevaluation of species as groups of geographically-varying populations, studies on the ecology and behavior of amphibians and reptiles and of their physiology, topics which continue to occupy us to this day. Yet the names of those hardy men whose explorations provided the foundation of our present knowledge have not been forgotten, emblazoned as they are in the names of Western mountain peaks and passes, rivers and towns. Here too is a happy instance of the often-descried practice of giving people's names to newly discovered organisms, persons whose sacrifices and toil might otherwise be forgotten. Their names are thus immediately familiar to students of American herpetology: Lindheimer, Berlandier, Wislizenus, Gambel, Marnoch and other primarily freelance collectors; Stansbury, Emory, and Graham, all topographical engineers; and the army surgeons Kennerly, Mearns, Woodhouse, Hammond and Yarrow (whose name, when Latinized, forms the specific epithet *jarrovi*). There were also other military personnel who collected specimens, including commissioned officers (Couch, Marcy), army surveyors (Schott, Clark), and even a one-time Minister to Mexico and Secretary of War, Poinsett. One genus of lizards was named by Baird for John Xántus (alias Louis de Vésey in several of the survey reports), a strange character who was once an army medical assistant in Kansas and briefly served as United States consul at Manzanillo in México. Other scientific names honor naturalists such as Say, LeConte and Henshaw, the geologist Hayden and Biological Survey collectors (Merriam, Bailey). The patronyms in themselves are perhaps uninteresting and sometimes may seem to be an impediment to the proper use of scientific terminology. But these names and the lives of the frontier naturalists they honor provide the key to a fuller, even inspiring knowledge of our rich herpetological heritage, a history not unworthy of our attention yet today.

KRAIG ADLER

Cornell University
Ithaca, New York
1 June 1977

References and Further Reading

Dall, W. H. 1915. Spencer Fullerton Baird, a biography . . . J. B. Lippincott Co., Philadelphia. xvi, 462 p.

Ewan, J. 1950. Rocky Mountain naturalists. Univ. Denver Press, Denver. xvi, 358 p.

Geiser, S. W. 1948. Naturalists of the frontier (second edition). Southern Methodist Univ., Dallas. 296 p.

Goetzmann, W. H. 1959. Army exploration in the American West 1803-1863. Yale Univ. Press, New Haven, Connecticut. xx, 509 p.

Hume, E. E. 1942. Ornithologists of the United States Army Medical Corps thirty-six biographies. Inst. Hist. Med., Johns Hopkins Univ., Monogr. (I). xxvi, 583 p.

Meisel, M. 1924-1929. A bibliography of American natural history the pioneer century, 1769-1865. Premier Publ. Co., Brooklyn, New York. 3 vols.

Osborn, H. F. 1931. Cope: master naturalist . . . Princeton Univ. Press, Princeton. xvi, 749 p.

Schmidt, K. P. 1955. Herpetology, p. 591-627. *In* E. L. Kessel (ed.), A century of progress in the natural sciences — 1853-1953. Calif. Acad. Sci., San Francisco.

Shor, E. N. 1974. The fossil feud between E. D. Cope and O. C. Marsh. Exposition Press, Hicksville, New York. xii, 340 p.

Thomas, D. and K. Ronnefeldt. (eds.) 1976. People of the first man. E. P. Dutton & Co., New York. 256 p. [Account of Prince Maximilian's expedition along the Missouri River, 1833-1834].

Acknowledgments and Credits

I thank Keir B. Sterling for comments on the manuscript and my assistant, Arlyn Sharpe, for library research. The photographs and map have been kindly supplied by the following institutions and organizations:

Fig. 1. Maximilian and Indian chiefs (copperplate engraving from Maximilian's *Reise in das innere Nord-Amerika . . .*, 1838–41; courtesy Cornell University).

Fig. 2. Maximilian (Hunt Institute, Pittsburgh); Emory (from F. Moore's *The Rebellion Record . . .*, vol. 10, 1867); Girard (from *Bulletin of the United States National Museum*, vol. 41, 1891); Baird (from *Annual Report of the Smithsonian Institution for 1897*).

Fig. 3. Survey party (from E. Hume's *Ornithologists of the United States Army Medical Corps . . .*, 1942).

Fig. 4. Map of Pacific Railroad Surveys (from W. Goetzmann's *Army exploration in the American West . . .*, 1959; courtesy Yale University Press).

Fig. 5. Cooper and Yarrow (both from E. Hume's *Ornithologists of the United States Army Medical Corps . . .*, 1942); Dugès (Hunt Institute, Pittsburgh); Cope (Historical Society of Pennsylvania, Philadelphia).

Fig. 6. Stejneger (Academy of Natural Sciences, Philadelphia); Van Denburgh (California Academy of Sciences, San Francisco).

CONTENTS

Volume 1:

Blainville, H[enri Marie] D[ucrotay] De
Description de Quelques Espèces de Reptiles de la Californie (Description of Some Species of Reptiles from California) (Reprinted from *Nouvelles Annales du Muséum d'Histoire Naturelle*), Vol. IV, Paris, 1835

Cooper, J[ames] G[raham]
Report Upon the Reptiles Collected on the Survey (Reprinted from *Reports of Explorations and Surveys, to Ascertain the Most Practicable and Economical Route for a Railroad from the Mississippi River to the Pacific Ocean*, Vol. X), Washington, D. C., 1859

Cope, Edward D[rinker]
Report on the Recent Reptiles and Fishes of the Survey, Collected by Campbell Carrington and C. M. Dawes (Reprinted from *Preliminary Report of the United States Geological Survey of Montana* by F. V. Hayden), Washington, D. C., 1872

Cope, E[dward] D[rinker]
Notes on the Geographical Distribution of Batrachia and Reptilia in Western North America (Reprinted from *Proceedings of the Academy of Natural Sciences*, [Vol. XXXV], 1883), Philadelphia, 1884

Coues, Elliott
Synopsis of the Reptiles and Batrachians of Arizona (Reprinted from *Report Upon Geographical and Geological Explorations and Surveys West of the One Hundredth Meridian, in Charge of First Lieut. Geo. M. Wheeler*, Vol. V, Engineer Department, United States Army), Washington, D. C., 1875

Coues, Elliot and H[enry] C[rècy] Yarrow
Notes on the Herpetology of Dakota and Montana (Reprinted from *Bulletin of the United States Geological and Geographical Survey*, Vol. IV, Department of the Interior), Washington, D. C., 1878

REPORT
ON THE BIOLOGICAL SURVEY COLLECTION
OF LIZARDS AND SNAKES FROM TEXAS

Vernon Bailey

REPORT ON THE BIOLOGICAL SURVEY COLLECTION OF LIZARDS AND SNAKES FROM TEXAS.

The Biological Survey collection of reptiles in the United States National Museum contains 353 specimens from Texas, including 102 specimens and 31 species of snakes from 81 localities, 252 specimens and 32 species of lizards from 167 localities. No attempt has been made to identify or include the turtles and batrachians. The material has been gathered by the field assistants of the Survey as opportunity offered in connection with other work, but none of the collectors has made a specialty of reptiles. Until a systematic field study of these groups is taken up, we can not expect to know much of the distribution and habits of the species; but so much that is vague, erroneous, and misleading has been published, especially in regard to the Texas region, that it seems doubly important to put on record all definite localities from which specimens have been positively identified. Every specimen in the present collection is fully labeled with exact locality, including altitude in the case of most specimens from the mountains, date, name of collector, and in many cases notes on habitat. Only a knowledge of the country is requisite to enable each specimen to be referred to its proper zone. Of some species there are specimens from enough localities to determine with considerable accuracy their range and zonal position in the State, but of many others the few records will be useful mainly in future works of broader scope.

But few of the collectors' field notes on reptiles, when unaccompanied by specimens, have been made use of owing to the danger of confusing closely related species, and when such notes are used they are carefully distinguished. In a few cases the records of specimens in the United States National Museum collection from localities of peculiar importance are given separately. References to published records usually are avoided.

To Dr. Leonhard Stejneger I am greatly indebted for identification of the specimens. With the aid of his assistant, Mr. Richard G. Paine, I was able to simplify his task in many cases by making preliminary determinations.[a]

Anolis carolinensis Cuvier. Carolina Chameleon.

This little chameleon-like lizard is represented in the collection by specimens from Waskom, Joaquin, Sour Lake, and Columbia, and I have seen it in abundance at Jefferson and Timpson, but never in the western half of Texa

Crotaphytus collaris (Say). Ring-necked Lizard.

Specimens of this beautiful lizard from Wichita Falls, Henrietta, Miami, Gail, Castle Mountains, Fort Lancaster, Fredericksburg, and Rock Springs, Tex., and from Roswell and Santa Rosa, N. Mex., carry the range of the species over the middle plains region of Texas, the region lying between the Pecos River and the eastern timbered

---

[a] Additional specimens identified by Mr. Paine and myself during Doctor Stejneger's absence are as follows: From Sour Lake, *Crotalus horridus*, *Callopeltis obsoletus*, *Agkistrodon contortrix*, *Storeria dekayi*, *Sceloporus consobrinus*, *Anolis carolinensis*; from Hempstead, *Tropidoclonium lineatum*; from Seguin, *Eutainia elegans marciana* and *Sceloporus spinosus floridanus*; from Washburn, *Heterodon nasicus* and *Liopeltis vernalis*, and from Cleveland, *Elaps fulvius*.

country. The above localities lie near the junction of Upper and Lower Sonoran zones, but this lizard inhabits at least a part of both zones. Fourteen out of the 15 specimens referred to *collaris* have the single row of interorbital plates. One specimen from Miami has two full rows of interorbitals, but with the large supraoculars and blunt nose of *collaris*.

Crotaphytus collaris baileyi Stejneger.

Nine specimens of *Crotaphytus* from eight localities are referred by Doctor Stejneger to *baileyi*. They are from Comstock, Alpine, Paisano, Chisos Mountains (west base), Davis Mountains (east base), Toyah, 70 miles north of Toyah, Tex., and one from the east base of Guadalupe Mountains, west of Carlsbad, N. Mex. These localities are near the junction of Upper and Lower Sonoran zones. Five of the nine specimens from Comstock, Paisano, and Chisos Mountains and two from the Davis Mountains are typical *baileyi*, with two full rows of interocular plates, small supraoculars, and relatively narrow muzzle. The specimen from Alpine has the interoculars joined in a single row, but otherwise possesses the characters of *baileyi*. The specimen from the Guadalupe Mountains and the two from Toyah and 70 miles north of Toyah have the interoculars joined in a single row and other characters intermediate between *baileyi* and *collaris*. Considering the close relationship and evident intergradation of the two forms, it seems best to follow Witmer Stone in placing *baileyi* as a subspecies of *collaris*.[a]

Crotaphytus reticulatus Baird.

Lloyd collected a specimen of this rare and apparently very locally distributed lizard at Rio Grande City, Tex., May 28, 1891.

Crotaphytus wislizenii Baird & Girard. Leopard Lizard.

A fine, large individual of this big, spotted lizard was shot near Boquillas, in the Great Bend of the Rio Grande, by McClure Surber, and Cary and Hollister each collected a specimen near Toyahvale, in the Pecos Valley. The species is not common and occurs only in the low, hot valleys of extreme arid Lower Sonoran zone.

Holbrookia texana Troschel.

This most brilliantly colored of the Texas lizards is represented by 11 specimens from the following nine localities: Fort Stockton, Adams, Toyahvale, Pecos River (5 miles west of Sheffield), Davis Mountains (east base), Boquillas, McKinney Spring (60 miles south of Marathon), Comstock, and Benbrook. It is a common and conspicuous species over all the hot, bare Lower Sonoran desert of western Texas and as far up the Pecos Valley as Santa Rosa, N. Mex.

---

[a] Proc. Acad. Nat. Sci. Phila., 1903, 30.

This species is so similar in both general appearance and habits to *Callisaurus draconoides*, the 'gridiron-tailed lizard' of the Death Valley country, that I never noticed the difference between them until Doctor Merriam pointed it out in a beautiful colored study of *Holbrookia texana* made by Fuertes in western Texas. Few animals possess more wonderful protective markings than these bar-tailed lizards. As they dash away, well up on their legs, with tail curled over the back, exposing their brightly colored sides and the black and white barred lower surface of the tail, they are strikingly conspicuous, until, stopping suddenly, they flatten themselves on the ground, when the speckled back blends into the earth colors and the lizards are lost to view.

Holbrookia propinqua B. & G. Long-tailed Holbrookia.

There are 15 specimens of this slender-tailed Holbrookia from five localities in southern Texas: Brownsville, Sauz Ranch, Santa Rosa Ranch and Padre Island in Cameron County, and King's Ranch in Nueces County.

Holbrookia maculata Girard. Spotted-sided Holbrookia.

The collection contains 19 specimens of this little, short-tailed lizard from the following eight localities in Texas: Mouth of Devils River, Fort Stockton, Fort Davis, Alpine, Paisano, Dimmitt (45 miles south), Henrietta, and Amarillo, places lying some in Upper and some in Lower Sonoran zones of the arid and the semiarid regions. Apparently the species has not been taken south of Devils River and San Antonio.

Holbrookia maculata lacerata Cope. Spotted-tailed Holbrookia.

One specimen of the spotted-tailed lizard from Cotulla, two from 15 miles west of Japonica, and one from 25 miles southwest of Sherwood considerably extend the southern and western range of the species. Apparently it belongs to Lower Sonoran zone.

Uta stansburiana B. & G.

Specimens of this little lizard from El Paso, Pecos City, and Fort Stockton carry its range across the extremely arid part of Lower Sonoran zone in western Texas and mark the eastern limit of a widely distributed species.

Uta ornata B. & G.

Twelve specimens of this little lizard are from the following localities in western Texas: Mouth of Pecos, Langtry, Ingram, Chisos Mountains, Altuda, Paisano, and Fort Davis. The Chisos Mountain specimen was taken at 6,000 feet, and the four Fort Davis specimens at approximately 5,700 feet, in the midst of Upper Sonoran zone. Altuda, Paisano, and Ingram are at the lower edge of the zone, while

Langtry and mouth of Pecos, as well as the type locality of the species, Devils River, are just below the edge in Lower Sonoran. However, enough Upper Sonoran species of plants cling to the cold walls of side canyons in the Langtry, Pecos, and Devils River country to account for the presence of such rock-dwelling species, and I am inclined to consider the range of this lizard as strictly Upper Sonoran, at least in Texas. If all the southern Arizona and California records of the species are correct, it is certainly Lower Sonoran in that region.

Sceloporus torquatus poinsettii (B. & G.).

This splendid, big, scaly rock lizard is represented in the collection by 14 specimens from the following localities in western Texas: Japonica, East Painted Cave, Marathon (50 miles south), Chisos Mountains (6,000 feet), Paisano, Davis Mountains (5,700 feet), Fort Stockton, Castle Mountains, near Tóyah, and Guadalupe Mountains (south end of Dog Canyon, at about 6,700 feet). The species ranges throughout the width of Upper Sonoran zone, but in many places comes well into Lower Sonoran, as at Toyah and Fort Stockton.

Sceloporus clarkii B. & G.

The collection contains but 4 Texas specimens of this big scaly lizard—3 from Boquillas and 1 from Langtry. Both localities are on the Rio Grande, in extremely arid Lower Sonoran zone.

Sceloporus spinosus floridanus (Baird).

Eighteen specimens of this medium-sized *Sceloporus* from southern Texas come from the following localities: Seguin, Ingram, Brownsville, Rio Grande City, Lomita Ranch (6 miles north of Hidalgo), Devils River, Langtry, and Pecos River (50 miles from mouth), which carry its range across the State in Lower Sonoran zone.

Sceloporus consobrinus B. & G.

Six localities in Texas are represented by the 8 specimens of this medium-sized *Sceloporus:* Joaquin, Sour Lake, Kerrville (Lacey's Ranch), Santa Rosa (Cameron County), Langtry, and Fort Davis. The range of the species apparently covers nearly the whole State where there are trees, in both Upper and Lower Sonoran zones.

Sceloporus dispar B. & G.

Lloyd collected 5 specimens of this little slender *Sceloporus* at Lomita Ranch, 6 miles north of Hidalgo, in June, 1891.

Sceloporus merriami Stejneger.

This beautiful little *Sceloporus*, which Doctor Stejneger has just described as new,[a] is represented by 5 specimens, 2 from the East

[a] Proc. Biol. Soc. Washington, XVIII, p. 17, Feb. 5, 1904.

Painted Cave, near the Rio Grande, a mile below the mouth of the Pecos; 1 from Comstock; 1 from the Pecos River Canyon, 55 miles northwest of Comstock, and 1 from Boquillas, in the Great Bend of the Rio Grande. Apparently the species is confined to the rocky walls of the canyons of the Rio Grande and Pecos rivers.

Phrynosoma cornutum (Harlan). Horned Toad.

This commonest and longest-horned species of the Texas horned toads is represented by specimens from El Paso, Grand Canyon of Rio Grande, Alpine, Altuda, Valentine, Davis Mountains (east base), Toyahvale (20 miles southeast), Fort Stockton, Fort Lancaster, Painted Caves, Carrizo, Roma, Rio Grande City, Sauz Ranch (Cameron County), King's Ranch (Nueces County), Corpus Christi, Center Point, Llano, Dimmitt, Henrietta, and Tascosa, and from Antioch and Virginia Point in eastern Texas, a series of localities covering at least the whole arid Lower Sonoran zone of the State and extending irregularly into the humid eastern division. Considerable variation however, appears within this range. The three specimens from Virginia Point and one from Antioch are much darker than any of the others, with sharply marked face bars and gray, thickly spotted bellies, while those of the upper Rio Grande Valley, El Paso, and Rio Grande Canyon are the lightest and brightest colored of all, with narrow face bars and white bellies. The transition from these extremes is gradual across the State, but the dark individuals from the eastern localities, the gray ones from the grassy plains, the strongly marked ones from the Pecos Valley, and the paler specimens from the Rio Grande Valley, indicate color forms comparable with the subspecies of horned larks found breeding in regions of corresponding differentiation.

Phrynosoma hernandesi (Girard).

This rusty-brown horned toad with short, stubby horns is represented in the collection by a single specimen from Texas, collected at about 7,000 feet altitude in the southern end of the Guadalupe Mountains. Farther north it is abundant in Upper Sonoran and Transition zones.

Phrynosoma modestum Girard.

Anota modesta of Cope and others.

Specimens of this little gray, short-horned horned toad from the west base of the Davis Mountains, 40 miles south of Alpine, 20 miles northwest of Toyah, Salt Valley, at west base of Guadalupe Mountains, and Big Springs carry the range of the species well over the arid region of western Texas in a series of localities where Upper and Lower Sonoran species are more or less mixed. Apparently the species belongs to Lower Sonoran zone and extends to its extreme upper limit.

A specimen from the west base of the Davis Mountains is rusty brown instead of ashy gray, like those from other localities, a peculiarity of coloration agreeing with the brown *Crotalus lepidus* from this lava soil region.

Coleonyx brevis Stejneger. Gecko.

A single specimen of this odd little brown and yellow lizard was collected by Merritt Cary at Sheffield, August 9, 1902.

Ophisaurus ventralis Linn. Glass Snake.

The glass snake is represented by two specimens from Texas, one collected by Lloyd near Santa Rosa, Cameron County, and the other by H. P. Attwater and W. H. Rawson, 3 miles north of Kerrville. These localities help to fix a western limit for this legless lizard of the Lower Sonoran zone of the Southeastern States.

Gerrhonotus liocephalus infernalis (Baird).

A single specimen of this glassy smooth lizard was collected in the Chisos Mountains at approximately 6,000 feet altitude in Upper Sonoran zone. It was nosing about in the dry leaves under scrub oaks on the mountain side in the manner peculiar to the individuals of the genus.

Cnemidophorus tessellatus (Say). Whip-tailed Lizard.

This species is represented by specimens from Boquillas, Langtry, Fort Stockton (35 and 45 miles west), Castle Mountains, Monahans, and Van Horn, all of which localities lie in the extremely arid Lower Sonoran zone of western Texas. They mark the eastern limit of the known range of the species.

Cnemidophorus perplexus B. & G.

Two specimens of this species were collected by Cary, one at Pecos and one 4 miles west of Adams, in Pecos County. Both localities are in Lower Sonoran zone.

Cnemidophorus sexlineatus (Linn).

The collection contains specimens of this eastern species of the whip-tailed lizard from Waskom, Long Lake, Nacogdoches, Henrietta, Canadian, and Padre Island.

Cnemidophorus gularis B. & G.

This seems to be the commonest and most widely distributed of the whip-tailed lizards in Texas. There are specimens from Brownsville, Lomita Ranch (Hidalgo County), Rio Grande City, Roma, Carrizo, Cotulla, San Diego, near Alice, Corpus Christi, Cuero, Kerrville, Fort Lancaster, Fort Stockton, Devils River (near mouth), Comstock, Painted Caves, Paisano, and Marfa, and up the Pecos Valley in New Mexico to Santa Rosa and Ribera. Among these localities Ribera is the only one fairly out of Lower Sonoran zone.

Leiolopisma laterale (Say).

Specimens of this little slender Eumeces-like lizard from Waskom, in northeastern Texas, and from Velasco, near Columbia and near the mouth of Navidad River in southeastern Texas, do not extend the range of the species, which apparently covers all the eastern part of the State.

Eumeces quinquelineatus (Linn).

Hollister obtained a specimen of this species at Joaquin, near the eastern boundary of Texas.

Eumeces guttulatus (Hallowell). Skink.

Cary collected a specimen of this little skink at the east base of the Davis Mountains, 20 miles southwest of Toyahvale, at about 5,000 feet, and I took one at 6,800 feet in the south end of the Guadalupe Mountains, well toward the upper edge of Upper Sonoran zone.

Eumeces obsoletus (B. & G.). Skink.

A specimen of this larger skink was collected in the southern end of the Guadalupe Mountains, at 6,800 feet, in the same locality with the little *guttulatus*.

Eumeces brevilineatus Cope.

Lloyd collected a single specimen at Paisano " in a damp fernery at 5,300 feet." Cope records it from Helotes Creek (the type locality), on the front line of hills 20 miles northwest of San Antonio, and from near Fort Concho (across the river from San Angelo).[a] While these localities lie at the upper edge of Lower Sonoran zone the species, if an inhabitant of damp gulches, may well belong to Upper Sonoran.

Diadophis regalis B. & G. Ring Snake.

A specimen of this little spotted-bellied snake was collected in the Chisos Mountains at 5,000 feet, June 3, 1901. The two previously recorded localities for Texas—Fort Davis and Eagle Springs[b]—are both close to 5,000 feet and, like the Chisos Mountain locality, at the edge of Upper and Lower Sonoran zones.

Heterodon nasicus B. & G. Hog-nosed Snake.

Specimens of the hog-nosed snake from Cameron County (El Haboncillo), Sycamore Creek, North Llano River, Washburn, and Amarillo carry its range over the whole north and south length of the State through Lower and Upper Sonoran zones in the semiarid region. It is not a common or conspicuous species, and with the exception of the specimens collected, I have not seen it in Texas.

[a] Ann. Rep. U. S. Nat. Mus. for 1898 (1900), p. 665.
[b] Ibid., p. 745.

Heterodon platirhinos Latreille. Blow Snake.

A single immature specimen collected by Lloyd, at Matagorda, does not add much to our knowledge of the range of the species, which apparently covers eastern Texas and reaches west to the Pecos and Devils rivers. All the Texas records are from Lower Sonoran zone, but farther east the species also ranges across at least Upper Sonoran.

Opheodrys æstivus (Linn.). Rough Green Snake.

The little green snake is represented by 3 specimens from Corpus Christi, Kerrville, and Rock Springs. From a range over the whole eastern or humid division of Lower Sonoran zone the species apparently reaches in Texas its western limit. It does not enter the plains region nor the arid region except where brushy gulches enable it to cross the rough country south of the lower arm of the Staked Plains. The specimens from near Kerrville and Rock Springs were taken in gulches with such vegetation as pecan, sycamore, elm, black cherry, oak, and abundant underbrush.

Liopeltis vernalis De Kay. Smooth Green Snake.

A specimen of the little smooth green snake was collected at Washburn by Gaut in July, 1904.

Bascanion flagellum Shaw. Whip Snake.

The coach whip or whip snake is represented in the collection by specimens from Matagorda, Matagorda Peninsula, Padre Island, and the Chisos Mountains, and by a flat skin from Devils River. While apparently distributed over the whole State, the species is most abundant, or at least most frequently seen, in the half brushy, half open arid region, where it is one of the commonest snakes.

Bascanion ornatum B. & G.

Two specimens of this rare species were collected in Texas, one at the head of Devils River, by Cary, the other on the Rio Grande, 8 miles south of Comstock, by Hollister, both in July, 1902. Hollister wrote on the back of the label of his specimen, "Up in bushes."

Cope records the species from two localities only, "Western Texas" and "Howard Springs, Texas."[a]

Drymobius margaritiferus (Schlegel).

Lloyd collected a single specimen of this species at Brownsville on July 17, 1891.

Callopeltis obsoletus (Say). Pilot Snake; Mountain Black Snake.

A specimen was collected by Lloyd, near the mouth of the Nueces, November 21, 1891, and another by Gaut, near Sour Lake, April 1, 1905.

[a] Ann. Rep. U. S. Nat. Mus. for 1898 (1900), p. 814.

Drymarchon corais melanurus Dum. & Bibr.

Lloyd collected a specimen of this Mexican black snake at Brownsville, July 6, 1891.

Pituophis sayi (Schlegel). Prairie Bull Snake.

The prairie bull snake is represented in the collection by 3 specimens from Gail, Comstock (20 miles north), and Paisano, and by a flat skin from the head of Dog Canyon, in the southern Guadalupe Mountains. It is common over at least middle and northern Texas in Lower and Upper Sonoran zones. In a prairie-dog town near Gail I killed an unusually large individual, measuring 7 feet 8 inches in length. It was about 3 inches in diameter—large enough to have readily swallowed a full-grown prairie dog. Near Rock Springs a smaller individual was found in the act of swallowing a freshly killed squirrel (*Citellus m. parvidens*).

Lampropeltis getula holbrooki Stejneger. King Snake.

Lloyd collected two specimens at Matagorda in January and February, 1892, and I collected one at Arthur in June of the same year.

Rhinocheilus lecontei B. & G.

A single specimen of this beautiful yellow and black ringed snake was collected about 30 miles west of Rock Springs, where a tongue of Lower Sonoran runs up into the Upper Sonoran plains. If the published records can be trusted, the species ranges over the whole arid Lower Sonoran zone of Texas.

Chionactis episcopus isozonus Cope.

A single specimen of this little, bright-colored, pink and black ringed snake was collected in the Chisos Mountains at about 6,000 feet altitude, in Upper Sonoran zone.

Natrix fasciata transversa (Hallowell). Water Snake.

Specimens from Lipscomb, the Nueces River (near mouth), mouth of Devils River, the Pecos River (55 miles northwest of Comstock), and from Carlsbad, N. Mex., together with previously published records, indicate for the species a range in most of the rivers of the western half of Texas, excepting the Rio Grande. Lipscomb is apparently the only locality where the species has been found outside of pure Lower Sonoran, and this is just at the edge of the zone.

Natrix clarkii (B. & G.). Striped Water Snake.

A specimen of the striped water snake was collected by J. D. Mitchel at Carancahua Bay, Calhoun County, Tex., in January, 1892, and the National Museum collection contains 5 specimens from Indianola, including the type, and one specimen from Galveston.

The following letter from Mr. Mitchel accompanied the specimen and 8 of the well-developed embryos:

This snake was captured with 3 others in a salt marsh on Carrancahua Bay, Calhoun County. It took to the salt water freely. One had 4 small mullets in its stomach, another some fiddler crabs. The other inclosure is part of the womb of one of the snakes, showing embryos. She had 4 on one side and 10 on the other, 14 in all.

Storeria dekayi Holbrook.

Lloyd collected a specimen of this little brown snake at Barnard Creek, west of Columbia, March 4, 1892, and Gaut collected one at Sour Lake, March 14, 1905.

Eutainia elegans marciana (B. & G.). Garter Snake.

There are specimens of this plain little striped snake from Brownsville, Santa Rosa Ranch (Cameron County), Corpus Christi, Victoria, Seguin, Sycamore Creek, Devils River, Paisano, and Boquillas. It is the common garter snake of the whole arid Lower Sonoran zone of western Texas, apparently reaching its eastern limit at Victoria.

Eutainia proxima (Say). Spotted Garter Snake.

The collection contains 12 specimens of this garter snake from Brownsville, Lomita Ranch (Hidalgo County), Sycamore Creek, Corpus Christi, and San Antonio River, near San Antonio. The species apparently has a wide range, including most of Texas.

Eutainia cyrtopsis Kennicott.

One specimen of this beautiful garter snake with black nuchal spots was taken in the Davis Mountains, July 12, 1901, at about 5,700 feet, in Upper Sonoran zone.

Tropidoclonium lineatum Hallowell.

Gaut collected a specimen of this little striped snake at Hempstead, February 28, 1905.

Tantilla gracilis B. & G.

I collected a specimen of this tiny brown snake at Lacey's Ranch, near Kerrville, May 5, 1899. The species has been recorded from various localities in eastern and southern Texas in Lower Sonoran zone.

Elaps fulvius (Linn.). Coral Snake.

Four specimens of the coral snake from Brownsville, Corpus Christi, Cleveland, and Kerrville nearly double the list of definite localities for the State. Cope records the species from Indianola, San Diego, Fort Clark, and Hempstead, but his records for Cameron County, Rio Grande, and Rio Pecos are too indefinite for practical purposes in outlining distribution. A live specimen sent to Doctor Stejneger from Beaumont, December 4, 1903, carries the range of the species nearly across the State in both humid and arid Lower Sonoran zones, but the Cleveland and Beaumont specimens are much darker and richer in coloration than those from farther west.

Fortunately this is not a common species in Texas. I have found it but once in the State. From its beautiful colors and harmless appearance it is likely to be handled carelessly, and its bite is dangerous.

Agkistrodon piscivorus (Lacépède). Cottonmouth; Water Moccasin.

A single specimen of the water moccasin, collected by William Lloyd at the mouth of Devils River, September 24, 1890, slightly extends the known range of the species. (Indianola and Eagle Pass are the westernmost records given by Cope.[a]) A specimen collected in the Big Thicket of Liberty County was lost, but I could not have been mistaken in its identity.

Agkistrodon contortrix (Linn.) Copperhead.

A specimen of the copperhead collected by H. P. Attwater near San Antonio and three others collected near Kerrville, Arthur, and Sour Lake help to fill out the range of the species in the State. Cope records it from Cook County, Sabinal, and between Indianola and San Antonio.[b] All of these localities are in Lower Austral zone.

Sistrurus catenatus consors (B. & G.). Massasauga.

There is a single specimen of the massasauga in the Biological Survey collection, from Santa Rosa, Cameron County, Tex., collected by Lloyd in 1891.

Crotalus horridus Linn. Eastern Rattlesnake.

Gaut collected a specimen of this rattlesnake in the Big Thicket, 8 miles northeast of Sour Lake, April 1, 1905.

Crotalus atrox B. & G. Western Diamond Rattler.

This, the largest of the Texas rattlesnakes, ranges throughout at least the arid part of Lower Sonoran zone of Texas. In the collection there are specimens from Corpus Christi, Japonica, Devils River, Comstock, Sycamore Creek, Eagle Pass, Langtry, and Boquillas, and from as far up the Pecos Valley as Santa Rosa, N. Mex. Mr. Cary also saved a flat skin from Pecos, Tex., and Doctor Fisher one from Colorado, Tex. Specimens recorded by Cope[c] from Indianola, San Antonio, and Brazos River apparently mark the eastern border of the known range of the species. I have never found it in eastern Texas.

Throughout its Texas range this is the commonest rattlesnake.

---

[a] Ann. Rep. U. S. Nat. Mus. for 1898 (1900), p. 1135.
[b] Ibid., pp. 1137 and 1138.
[c] Ibid., p. 1166.

While it is often reported as excessively abundant, I have never found more than a dozen individuals in a season's field work of four or five months in its favorite haunts. Over most of the range we do not see more than an average of one or two in a month's field work. On an 18-days' camping and collecting trip—April 24 to May 11, 1900—from Corpus Christi to Brownsville and return, we saw only five rattlesnakes, where we had been led to expect hundreds; and in this region of dense cactus beds and thorny thickets they find perfect protection and probably reach their maximum abundance. Specimens from extreme southern Texas are reported by the residents as reaching a length of 11 to 13 feet; but the largest specimen I have seen alive measured only 50 inches in length.

Crotalus confluentus Say. Plains Rattlesnake.

This small, dull-colored rattlesnake is represented in the Biological Survey collection from Texas by one specimen from Amarillo, on the Staked Plains, in Upper Sonoran zone. I am familiar with the species on Upper Sonoran plains of New Mexico and Nebraska, but have not found it elsewhere in Texas. The specimen collected by Captain Pope (No. 4962, U. S. N. M.), and labeled "Pecos River, Texas," has not even a date by which to locate the place where obtained; but as it is well known that Captain Pope's specimens from the top of the Staked Plains on the east and from the Guadalupe Mountains on the west were labeled "Pecos River," this record can have no zonal significance. The specimen recorded by Cope [a] from San Antonio, Tex., is entered in the Museum catalogue as collected "between San Antonio and El Paso," and the one recorded from Rio San Pedro (Devils River) is catalogued from "between Ash Creek and Rio San Pedro." Mr. Brown's record for Pecos,[b] based on a specimen collected by Meyenberg, can not, as Mr. Brown admits, be used for "minute zone work," as Meyenberg's specimens, while said to have been collected somewhere "within a day's journey by team of Pecos," may have come from much farther away. In 1902, at a place 75 miles northwest of Pecos, I met one of his men bringing in a wagonload of live animals, among them numerous snakes and lizards which had been collected along the base of the Guadalupe Mountains, mainly in Upper Sonoran zone, but probably also in Transition. As these specimens apparently were sent out as collected at Pecos, in Lower Sonoran zone, the difficulty of using Meyenberg's material for zonal work is apparent.

Crotalus molossus B. & G.

This is the common rattlesnake of the Guadalupe Mountains in Upper Sonoran zone on both sides of the Texas and New Mexico line.

---

[a] Ann. Rep. U. S. Nat. Mus. for 1898 (1900), p. 1172.

[b] Proc. Acad. Nat. Sci. Phila., 1903, p. 551.

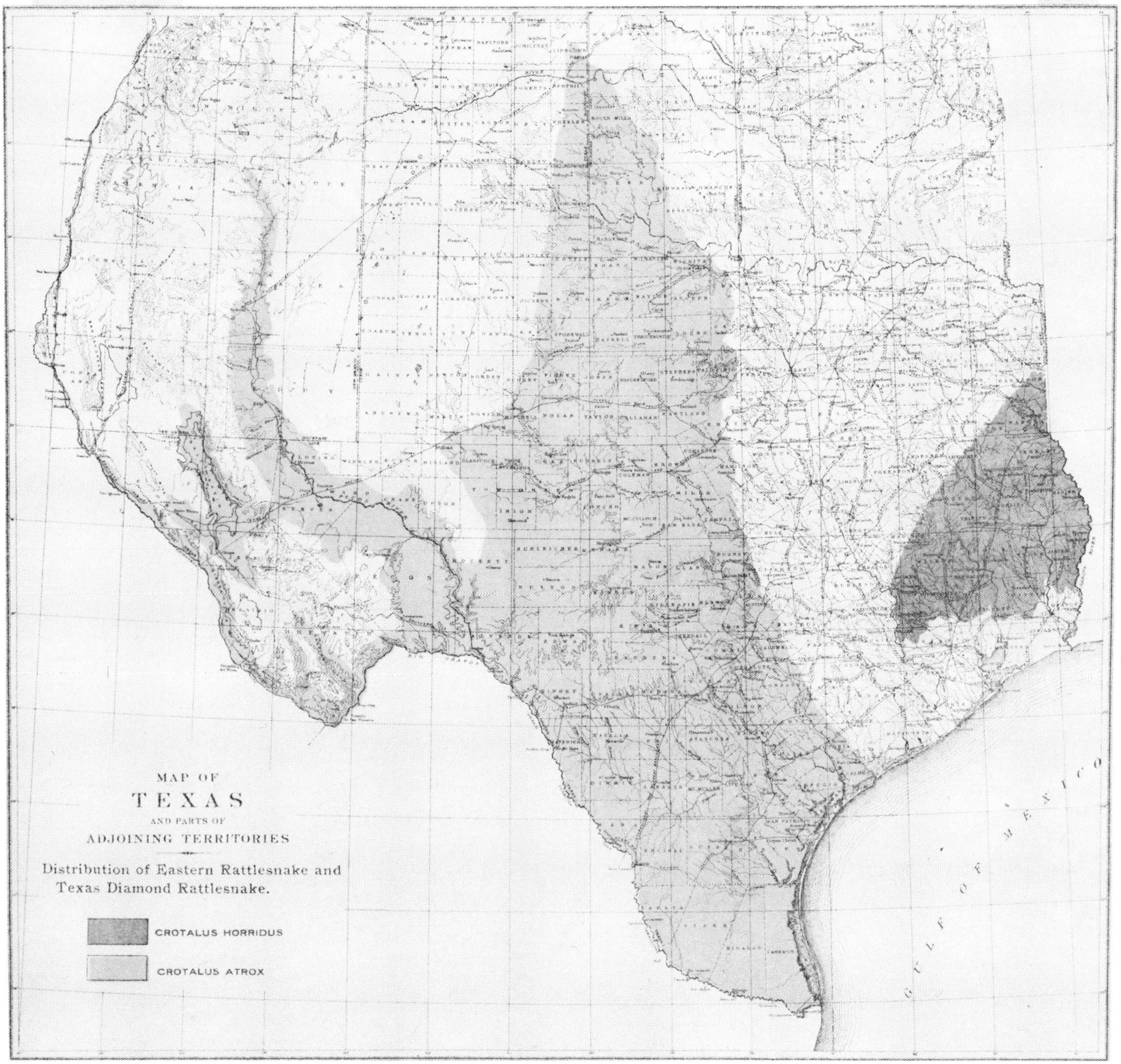
MAP OF
TEXAS
AND PARTS OF
ADJOINING TERRITORIES
Distribution of Eastern Rattlesnake and Texas Diamond Rattlesnake.
CROTALUS HORRIDUS
CROTALUS ATROX
GULF OF MEXICO

Specimens were collected near the edge of Transition zone on the east and west slopes of the mountains at 6,300 and 6,800 feet, but I assume that the species belongs to Upper Sonoran. A flat skin collected by Cary at a point 25 miles west of Sheffield is apparently this species.

We found this snake in August, 1901, in the gulches high up on the range. It is pugnacious, quick to sound its rattle and throw itself on the defensive. Because of its prevailing color of olive green, we always referred to it in the field as the " green rattlesnake."

Crotalus lepidus Kennicott.

Five specimens of this little rattlesnake from western Texas, the mouth of the Pecos, Paisano, Chisos Mountains, and Davis Mountains, indicate a range confined to the most arid and rocky part of the State in both Upper and Lower Sonoran zones. The specimen from the mouth of the Pecos was in Lower Sonoran zone on the gray limestone ledges at the east end of the High Bridge. In color it is pale ashen gray, with the pattern faintly indicated in dusky lines. The color has not changed materially in the alcoholic specimen from that of the live animal collected May 20, 1900, when I called it the ' white rattlesnake ' to distinguish it from anything previously known to me. Mr. Cary reports several seen on limestone ledges along the Pecos Canyon at Howard Creek and Sheffield, all of which were whitish in color like the lime rock on which they were found. One specimen from the Chisos Mountains at about 6,000 feet, one from Paisano, and two from 5,700 feet in the Davis Mountains, were all taken in Upper Sonoran zone. Others were seen in both the Chisos and Davis mountains at similar altitudes, but none lower. All of these specimens in life or when freshly killed, and the others seen but not collected, were dark rusty brown with a pinkish tinge, heavily marked with velvety black crossbars. The brown has faded considerably in the alcoholic specimens, as shown by comparison with a careful color study of the fresh snake made in the field by L. A. Fuertes.

These little brown rattlers were fairly common about our camp near the head of Limpia Creek, in the Davis Mountains. On the dark brown lava soil they were very inconspicuous, and they had a way of suddenly springing up between our feet that made us slightly nervous.

REPTILES

Spencer F[ullerton] Baird
and
Charles Girard

REPTILES.

By SPENCER F. BAIRD and CHARLES GIRARD.

Siredon lichenoides, Baird.

Pl. I.

Spec. char.—Body uniform blackish brown, covered all over with licheniform patches of grayish yellow; snout rounded; tail compressed and lanceolated; toes broad and short.

The addition of an authentic new species to the genus *Siredon* will justly be considered as of great interest to herpetologists. Two species are now clearly ascertained to exist; perhaps a third, if the one mentioned by Prof. Owen* be really such. It is not improbable that many more exist, as we have accounts of many localities of "fish with legs," in various parts of Mexico, New Mexico, and Texas, although, as yet, we have been able to procure only these two species. The possession of these, however, allows a comparison of characters by which the absolute generic features of the group can be better ascertained.

The figures hitherto published of *S. mexicanus*, and the imperfect sketch of *S. maculatus*, are far from being satisfactory, and do not allow any accurate comparison to be made of their specific features. As these will have to be critically redrawn in order to meet the wants of science, we have endeavoured to obtain, and we hope with success, figures of *S. lichenoides* that will enable future comparisons to be satisfactorily made. Our specimen is a little more than six inches and a-half in length, the figures being all of natural size. Whether this be the absolute size of the species which it represents, we are unprepared to state. If such was the case, it is considerably smaller than *S. mexicanus*. The tail forms nearly the half of the entire length, and the head a little less than the fifth of the same.

The head is ovoidal, much broader than deep, and the snout rounded, a character which at once will distinguish our species

---

* Annals and Magazine of Natural History, xiv. 1844, 23.

from *S. maculatus*, the chief character of which consists in the truncated snout. The eyes are of medium size, situated behind the angle of the mouth. Their position and size in *S. maculatus* we do not know accurately, but in *S. mexicanus* these organs would be considerably smaller and more approximated to the end of the snout, judging of these peculiarities from the various sketches given of that species, and proportionally much smaller than in a specimen of *Siredon* in our possession from the city of Mexico, and of about the same size as our *S. lichenoides*. If these proportions are correct in the drawings of *S. mexicanus*, we would undoubtedly have a species which could not be accurately characterized until further information should be obtained. The nostrils are very small and near the end of the snout. The body is subcylindrical, subfusiform, broader and deeper at its origin than on any point backward. The tail is very much compressed, elongated, and tapering into a point. The dorsal membrane commences at the occiput, rising gradually until the middle of the tail, whence it diminishes again toward its pointed tip. The membrane under the tail is lower than that above, extending from behind the vent to the tip of the tail, and reaching its greatest height on its anterior third, but diminishing more rapidly forward than backward. The anal opening is very large, elongated, and rendered very conspicuous by the great development of the fleshy masses which constitute its margin.

The presence of four external flaps, provided with respiratory fringes, is a generic character, belonging to all the species hitherto known. Their real appearance has been misrepresented in many sketches, as we could satisfy ourselves by the examination of two species preserved in alcohol. The branchial fringes do not extend all along the upper edge of the branchial flap. They occupy densely the lower edge of that cutaneous appendage from its origin to its tip, and thence for a short space above, but much less developed here than below, as we have endeavoured to show in the profile of fig. 2. The fringes themselves are very much flattened, tapering, and disposed upon a double row, so that each of them appears as if double; but it is easy to ascertain that the row on either side does not combine with the other.

The fore and hind legs have nearly the same length when measured from their bases to their extremities; the hind ones, however, are much thicker, and the toes of both pairs are neither so slender nor so elongated as in *S. mexicanus* and *S. maculatus*.

The ground colour is blackish brown; there are irregular patches of grayish yellow spread all over the body, head, and tail, reminding us of surfaces over which lichens grow, whence the specific name by which we designate this species.

It was caught by R. H. Kern, Esq., in Spring Lake, at the head of Santa Fé Creek, in New Mexico, accordingly a member of the fauna of the basin of the Rio Grande del Norte.

In a revision of the *North American Tailed Batrachia*, published in the Journal of the Academy of Natural Sciences, 2d series, vol. i., p. 281, (1849,) we intimated a doubt as to whether *Siredon* or the axolotls were adult animals. Their wonderful resemblance to the larvæ of *Ambystoma punctata* was our chief ground for this belief. Since then, however, we have seen the description and figures by Sir Everard Home of *S. mexicanus*, in which he clearly indicates, in one individual at least, ovaries distended with eggs. We have also seen specimens with the genital apparatus presenting the tumid and highly developed appearance of salamanders in general, when in the breeding season.

CNEMIDOPHORUS TIGRIS, Baird and Girard.

Pl. II.

SPEC. CHAR.—Scales on the subguttural fold small in size; four yellowish, indistinct stripes along the dorsal region.

This species, one of the most elegant of its genus, is the third hitherto described as found in the United States, for we have no doubt that the lizard referred to in Long's expedition, under the name of *Ameiva tesselata*, will come under this genus, and be closely allied to our species. No specimen of this being extant at the present time in any known collection, a direct comparison with the other species of *Cnemidophorus* was not possible. That *A. tesselata*, however, although closely allied to, is not identical with our species, we think that any one will be convinced on comparing Say's description with the figures in pl. II. These are of natural size, and exhibit most admirably the structure of the regions in their most minute details. The plates of the head above (fig. 3), below (fig. 2), and on the sides (fig. 1), need no further description. The minute scales of the back and upper portion of the legs contrast strikingly with the eight rows of large scales of the belly and those of the lower part of the hind legs, as well as with those of the inferior surface of the head and throat. On the tail again, the scales assume another character, well opposed too; they are longer than broad, arranged in annular rows, or else verticillated and slightly carinated. The tail is cylindrical, and two and a-half times the length of the body and head together: it tapers gradually and terminates in a point.

The ground colour appears to have been bluish yellow, marked with irregular patches of black. In some specimens, four longitudinal yellow stripes may be seen extending from the occiput to the base of the tail, and occasionally to a little distance on the latter. In the young state, the black patches predominate, unite, and form, as it were, the ground colour, and the yellow constitutes irregular small spots.

A series of individuals of different sizes were collected by Capt. Stansbury in the valley of the Great Salt Lake.

Genus CROTAPHYTUS, Holbrook.

GEN. CHAR.—Head covered with small and polygonal plates. The occipital proper, minute. Teeth on the jaws, pterygoids and palatines, rudimentary on the latter. Broad auditory aperture. Femoral pores present; no anal ones. Tail very long.

Of the genus *Crotaphytus*, a typical form peculiar to North America, but one single species has hitherto been discovered, the *C. collaris*, observed for the first time by Major Long's party, on their expedition to the Rocky Mountains. The discovery of a second species of this genus within the limits of the United States will be received with interest: both of these have a very close generic relationship, the differences being found in some minor details of their structure.

The generic characters are the following:—The occipital region, vertex, and front are covered with small and polygonal plates. On the superciliary region and around the nostrils the plates are scarcely larger than the scales of the back. The odd occipital plates are inconspicuous, being but very little larger than the surrounding ones. The auditory apertures are broadly open. Teeth exist on the maxillaries, palatines, and pterygoids; conical, acute, and slightly curved on the anterior part of the jaws, they are compressed and tricuspid on the posterior. The palatine teeth are rudimentary. The skin is folded under the throat. The scales of the upper part of the body are polygonal, and smaller than those of the lower part, or belly, and tail; those under the head have nearly the size of those of the back. The femoral pores are very distinct; the anal ones are wanting. The tail is cylindrical, longer than the body and head together.

The genus *Crotaphytus* differs from *Holbrookia* in having external auditory apertures, teeth on the pterygoids, and but a minute occipital plate. The shape of the head is likewise more elongated and pointed in front.

CROTAPHYTUS WISLIZENII, Baird and Girard.

PLATE III.

SPEC. CHAR.—Head proportionally narrow and elongated. Cephalic plates and scales on the back very small. Yellowish brown, spotted all over with small patches of deeper brown or black.

C. wislizenii has the same general form and appearance as *C. collaris*, exhibiting the same contracted neck and fold under the throat, the same compact body, the same cylindrical and elongated tail, and the same shape and proportions of locomotive members and terminating toes. The differences by which the two are distinguished, although of a comparatively minor character, are readily appreciable when both are directly compared. Thus the head of *C. wislizenii* is proportionally more elongated and narrower than that of *C. collaris*. The small and polygonal plates which cover its upper surface and sides are smaller, as well as those of the lip of the lower jaw. The scales of the back are likewise smaller, and those of the belly larger. The tail is somewhat longer, and its scales larger in *C. collaris;* these are subverticillated in both species, and subcarinated from the middle of the back toward its extremity. The pores of the lower surface of the thighs are more conspicuous in *C. collaris*, independently of the fact that they are generally less so in the female than in the male of the same species. Immediately behind the vent, at the origin of the tail, there exists, in the male, a row of large scales more uniform in *C. collaris* than in *C. wislizenii*. The specimen figured on our plate III. being a female, these anal plates are not to be seen in fig. 4.

In the colours of the body distinctive marks will at once be found. *C. collaris* possesses on the sides of the neck a double band of black bordered with white, which does not exist in *C. wislizenii*. The upper surface of the body of the former is scattered all over with small yellow dots, which indeed are found in the latter, but are much smaller and more numerous, having in addition, intermixed with them, irregular roundish brown spots, extending even to the head. The tail is irregularly annulated with brown and yellow.

Nothing is known of the habits of the crotaphyti. We found in the stomach of one specimen of the *C. wislizenii* the remains of a species of *Cnemidophorus* allied to *C. sexlineatus.*

The specimen which we have had figured was caught near Santa Fé, by Dr. Wislizenus, during the Mexican war. To him we take great pleasure in dedicating it, in testimonial of his zeal for science, during the arduous duties of an army surgeon, while attached to Colonel Doniphan's command. On his return, the specimen was sent to Dr. Le Conte, and by him kindly transmitted to the Smithsonian Institution. The same species was obtained by Colonel J. D. Graham, between San Antonio and El Paso del Norte, while on his journey to the boundary-line, in May last.

Genus Holbrookia, Girard.

Syn. *Cophosaurus*, Trosch. Arch. f. Naturg. (1850, I.) 1852.

Gen. char.—Head covered with small and polygonal plates. No auditory aperture. No teeth on the palatine bones. A fold of the skin on the breast. Femoral pores present, but no anal ones.

This genus was established in 1850, upon a small lizard, much less remarkable in its general aspect than in its structure. In its appearance it is so similar to certain species of the Chilian proctotreti, that at first glance no one could suppose it to be different from the latter genus. But in examining it more attentively, we find no external opening to the ears, a fact that cannot but strike the observer. There are no teeth on the palatines—another character which distinguishes *Holbrookia* from *Proctotretus*. Both of these genera have a fold under the throat, but the former has femoral pores, which are wanting in the latter. The anal pores are absent in *Holbrookia*, while they exist in *Proctrotretus*.

The genus *Holbrookia* will, no doubt, prove somewhat related to *Crotaphytus*, having, like the latter, the upper surface of the head covered with small and polygonal plates, and well-developed femoral pores. The elongated tail of the crotaphyti, although so disproportionate when compared to *Holbrookia maculata*, will no longer appear as a feature peculiar to the genus, so soon as we shall have an opportunity to give a description and a figure of another species of the same genus, and which was latêly collected by Mr. John H. Clark, zoologist to Colonel J. D. Graham, while in charge of the survey of the United States and Mexican boundary.

As these pages were passing through the press, we received the numbers iv. and v. for 1850, and ii. and iii. for 1851, of the "Archiv für Naturgeschichte." The present editor of that periodical, Dr. Troschel, describes, in the fourth number for 1850, a new saurian genus, under the name of *Cophosaurus*, a species of which was brought to Germany by the geologist, Ferd. Rœmer, who visited Texas some years ago. Having the same saurian in our possession, collected in the same locality, as a second but undescribed species of our genus *Holbrookia*, we could readily identify the genus *Cophosaurus*. It is to be regretted that the description of our genus *Holbrookia* which reached Germany in the summer of 1851, that is before the publication of the number iv. of the Archiv. für Naturgeschichte for 1850, has been overlooked by this able German zoologist. The absence of auditory apertures, "aures externæ nullæ," which is the most striking character of our genus, would have struck Dr. Troschel, had he been aware of our description in the Proceedings of the American Association for the Advance of Science, fourth meeting, held at New Haven, in August, 1850.

As it may hereafter be questioned which of these generic names has the priority, inasmuch as the volume in which *Cophosaurus* is published bears the date of 1850, we deem it proper to call attention to the fact that its publication took place in 1852. Even at the date at which we write these lines, (April, 1852), the year 1850 of the Archiv. für Naturgeschichte has not been completed. This fact shows how an author may be deprived of the fruit of his labours by an anachronism of this kind.

HOLBROOKIA MACULATA, Girard.

PL. VI. FIG. 1–3.

Syn. *Holbrookia maculata*, GIRARD, Proc. Amer. Assoc. Adv. Sc., IV. (1850), 1851, 201.

SPEC. CHAR.—Tail about the length of the trunk. Head subcircular, slightly conical in front. Pectoral fold bordered with large scales.

The general form of this species is rather thick and short than elongated, especially in the female: the young and the males are more slender. The body is subcylindrical, the tail conical and very stout at its origin, tapering however suddenly away. The entire length is between three and four inches, as shown by the figures, all of which are drawn of natural size. The tail is of the length of the body, the head excluded. The latter is oval, broader than high; its summit being convex, and its snout truncated. It is covered with small, irregular, and polygonal plates, larger on the middle line of the skull than above the eye, the nose, and the nape. There is a supraorbital carina, with small elongated plates, scarcely to be seen with the naked eye. The infraorbital plates are less numerous, but longer. The eyes occupy the middle of the length of the head. The eyelids are bordered by a row of minute and pointed plates, forming a serrated edge. The nostrils are nearer the end of the snout than the eye. The angle of the mouth extends to the posterior rim of the orbit. The upper jaw is bordered with a row of small, very elongated plates, obliquely imbricated. Margining the lower jaw there are two rows of small angular plates, the larger ones being at the angle of the mouth. There is a single row of small conical teeth on both jaws; those in front are acute and slightly recurved; those behind stouter and erect, with a carina separating the rounded crown from the body of the tooth. The posterior extremity of the tongue has a semilunar notch. As observed in the generic paragraph, there are no external auditory apertures; the tympanum is covered by scales altogether similar to those of the neck. On both sides of the neck and immediately behind the angle of the mouth, is a fold of the skin, which vanishes in a depression under the head. Farther backward, and on the breast, is situated another fold, constituting an elegant neck ring, which, however, does not extend higher than the shoulders.

The anterior legs are shorter and more slender than the posterior. There are five toes, similar in each pair of limbs, elongated, slender, terminated by a compressed and recurved nail. The toes and nails of the posterior limb, however, are a little longer in proportion to the limbs themselves. The fourth toe is the longest. the two external ones the shortest, the second and third nearly equal. There are eleven femoral pores on each thigh.

The scales are slighty imbricated, subcarinated on the back and sides, smooth underneath. They are smaller on the neck and at the base of the limbs than on the sides and back. Those on the tail are indistinctly verticillated. The smallest ones are found under the head, in the region of the groins and behind the vent; they are larger on the abdomen than under the tail. The toes are entirely covered with scales.

The colour, according to a drawing made from life by an artist of great merit, Mr. William H. Tappan, while on the River Platte in 1848, is olivaceous brown, slightly violaceous on the sides of the head. On each side of the body there are two, sometimes three oblong spots, of a deep black; and on the body two rows of quite large, irregular blackish-brown patches, with a band of a lighter colour between each row. There is an indication of a third row of these patches, less apparent, however, in the male, in which again we find two orange-red lines, the uppermost extending to the end of the snout in passing over the eye; the other follows the lower jaw. In advance of the eyes the orange hue of these lines passes into brighter yellow.

This species inhabits the valley of the Platte River, as collected there by W. H. Tappan. We have received several specimens from Texas, collected by General Churchill, one of which is the original from which the accompanying drawings have been made. The individual represented is a female.

Genus Uta, Baird and Girard.

Gen. char.—Upper part of body covered with minute scales; a pectoral fold; auditory apertures; femoral pores, but no anal ones.

The genus which we now establish will not fail to attract the attention of herpetologists, having a relation to both *Sceloporus* and *Holbrookia*. The former genus it resembles in having the upper surface of the head covered with similar scales, and in being provided with auditory apertures. On the other hand, the body is covered with scales, like those in *Holbrookia*, while on the tail they are much large than in any of the above genera, thus contrasting greatly with those of the back. The genus *Uta*, moreover, has a subgular fold of skin, constituting a neck-ring similar to that in *Holbrookia*. Its elongated tail would recall to mind the genus *Crotaphytus*, were not the palatine teeth absent. Femoral pores exist, while anal ones are wanting.

Besides the species described below, this genus embraces two others, one of which is entirely new to science, and was sent in by Colonel J. D. Graham from the boundary-line, and which we call *Uta ornata*; while the other was described as a *Sceloporus*, and first as a mere variety of *S. grammicus*, afterward, however, separated under the specific name of *S. microlepidotus*. The minuteness of the scales to which it makes allusion becomes now a generic character. Thus the species included in the genus *Uta* are, generally speaking, of a smaller size than the scelopori proper.

UTA STANSBURIANA, Baird and Girard.

PL. V. FIG. 4–6.

SPEC. CHAR.—Tail slender, elongated, and conical, provided with large scales arranged in verticils; a subgular fold in addition to the pectoral one.

This species is one of the handsomest and most graceful among the small saurian forms peculiar to North America. In its general aspect it recalls to mind the cnemidophori, having, like the latter, the body covered with small polygonal scales, while on the tail the scales are large, elongated, and verticillated. The proportions in length between the body and tail, and the fold under the throat, are likewise similar in both *Cnemidophorus* and *Uta*. But when we compare the shape and structure of the head in the two, we detect differences which are not merely generic, but cause them to belong to different families. The upper surface of the head is covered with small and polygonal scales, as in *Sceloporus*, *Crotaphytus*, and *Holbrookia;* its shape is rather rounded than conical or elongated, and is broad and flattened, while it is narrower and higher than broad in cnemidophori and allied genera. The body is subfusiform, the tail slender, conical, and nearly one and a-half times the length of the body and head together.

The occipital plate is polygonal and comparatively large; three or four superciliaries, the vertical and frontal plates, are a little larger than the many others by which they are surrounded. The nostrils are very conspicuous, and approximated to the end of the snout; they are situated interior to the prolongation of the superciliary ridge, and composed of elongated and narrow plates. The rostral plate is narrow, and is scarcely to be seen when viewed from above, but much more conspicuous than in *Holbrookia* when looked at in front. The plates lining the margin of the jaws are quadrangular, elongated, and much narrower on the upper than on the lower jaw. Under the chin, three or four pairs of polygonal scales are considerably the largest. Between the latter and those on the margin of the lower jaw two other rows may be seen, composed of scales a little larger than the subgular ones. The auditory aperture is moderate, surrounded by irregular folds of the skin which extend under the throat; from its anterior edge proceed three or four conical and acute scales; the posterior pectoral fold is provided along its margin with a row of large, smooth, and subquadrangular scales. The anterior and upper surface of the locomotive members are covered with scales nearly of the same size of those of the belly, but carinated and a little narrower posteriorly. Scales, similar to those of the belly, extend on the lower part of the hind legs, are somewhat larger, especially on the anterior margin. The toes are slender and terminated by short, pointed, and slightly recurved nails. The small scales which cover the middle region of the back, from the anterior limbs to the origin of the tail, are somewhat longer than those of the occiput and sides from the ear to the hind limbs. The posterior part of the thigh and groins are covered with minute scales similar to those of the sides of the body. The largest scales are seen on the tail, as already observed; they are verticillated and strongly carinated.

The colour below is uniform greenish-yellow, except under the head, where the green predominates, intermingled with brownish, narrow bands. Above it is blackish-brown, marbled with greenish-yellow, or whitish-yellow irregular dots. Sometimes a double row of dorsal patches of a deeper black may be seen along the back in some individuals, recalling to mind a similar distribution of colour in *Sceloporus scalaris*, *S. graciosus*, and *Holbrookia maculata.*

This species is from the valley of the Great Salt Lake, where it was collected by Captain Howard Stansbury; in remembrance of whose services to the country and to science we have designated it under the name which it now bears.

Sceloporus graciosus, Baird and Girard.

Pl. V. Fig. 1–3.

Spec. char.—Head subconical; scales of the back proportionally large; tail of medium size, slender and conical.

This small and graceful species has a much greater affinity with the Mexican *Sceloporus scalaris* than with *S. undulatus* of the United States. The most striking character, as compared with *S. scalaris*, consists in the marked difference in size between the scales of the back and those of the base of the tail. The latter is proportionally longer than in *S. scalaris*, and shorter than in *S. undulatus*. The body of our species is subcylindrical, and rather short; the specimen figured being a female, the abdomen is represented in a state of too great expansion. The head is ovoidal, depressed, subconcave above. The occipital plate is heptagonal, broad, and linear posteriorly, pointed and very acute anteriorly. It is surrounded by twelve smaller plates, six of which, very small, line the posterior edge, while the other six are distributed on the remaining circumference, three to the right and three to the left. There are two vertical plates, the anterior one the largest and pentagonal. In advance of the latter, seven or eight polygonal frontal plates form a conspicuous group, while near the extremity of the snout the plates are as minute as the scales on the neck. The plates which line the margin of the jaws are small, narrow, elongated, and inconspicuous. On the chin there are three pairs comparatively larger and conspicuous. The scales on the back are spade-shaped and strongly carinated from the neck to the tip of the tail. On the latter region they are a little more acute posteriorly and verticillated. On the sides of the abdomen they are smaller, and their outline less regular. The abdominal scales are smooth, irregularly lozenge-shaped, bidentated posteriorly as in *S. scalaris*. The opening of the ear is subtriangular, protected by several projecting scales arising from its anterior edge. The toes and nails are very slender; the latter are proportionally long and slightly curved. When the fore legs are stretched backward, the tip of their toes will reach to the knees of the hind ones when the latter are brought forwards.

The brilliancy of the general hue having disappeared by immersion in alcohol of the specimens collected, we are not prepared to describe this accurately. As to the markings, they differ somewhat from those of *S. scalaris*. The row of the large crescent spots along the back is more compact, and not so distinctly bordered with white. The yellowish band that runs from behind the eyes backward to the middle of the tail is much broader; and besides, there is a second similar band extending from below the snout, and passing under the eye and above the auditory aperture, to the insertion of the hind locomotive limbs; the sides, therefore, are not ornamented with vertical, slightly undulating dark stripes; the irregular patches that are seen on that region are entirely deprived of any white margin. The abdomen in the male is blue indigo, as in most species of the same genus. The neck and throat are unicolor in both sexes.

This species inhabits the valley of the Great Salt Lake, where it was collected by Captain Stansbury and Lieutenant Gunnison.

Elgaria scincicauda, Baird and Girard.

Pl. IV. Fig. 1–3.

Syn. *Tropidolepis scincicaudus*.—Skilt. Amer. Jour. of Sc. VII. 1849, 202, fig. 1–3.

Spec. char.—Dusky green above; light ash colour below. Eleven transverse black bands on the back, interrupted on the dorsal line; white dotted posteriorly, six or more on the tail. Thirteen to fourteen rows of scales, well carinated.

The individual of this species which we have had figured, although not quite full-grown, exhibits, nevertheless, all the essential characters for its identification and specific distinction. A much younger specimen was figured by Mr. Skilton, in the seventh volume of the second series of the American Journal of Sciences and Arts, where it is given as a species of *Tropidolepis*. The scales, indeed, are carinated, but this is not a character peculiar to the genus *Tropidolepis*. A closer examination soon reveals the characteristic features of the gerrhonoti, of which the genus *Elgaria* is a mere subdivision. The conical and tapering tail, which is longer than the body and head together, forms the prominent distinctive mark of the genus in which we place the present species. The longitudinal area of minute scales, which extends from behind the ears to the insertion of the hind legs, belong to gerrhonoti generally, and distinguishes them from *Tropidolepis* or *Sceloporus*. In elgaria and gerrhonoti, in general, there are neither femoral nor anal pores, while their existence in *Sceleporus* will enable any one to distinguish between them.

The hind legs of *E. scincicauda* are slightly longer and stouter than the anterior, not taking the toes into consideration, as those of the hind feet are generally much the longest. The head is flattened, and has the shape of an acute triangle, the summit of which would be rounded. The snout therefore is rather elongated, with the end conical. The plates which cover its upper surface are smooth, and faithfully represented in fig. 3. The scales of the back constitute twelve or thirteen rows, each of which is provided with a medial carina, or ridge, extending over the tail. The scales of the belly and lower surface of the tail are smooth; on the abdomen they form twelve rows, the middle one being composed of larger scales. Fig. 2 exhibits the differences of the scales under the head, throat, belly, and tail. The scales form transverse as well as longitudinal rows, not only on the tail, as is often the case, but likewise on the body itself. Those on the throat and lower surface of the head alone seem not subjected to any serial order.

Possessing only specimens that have been immersed in alcohol, and therefore have lost their general hue, we can only, in allusion to the colour, remark that the upper part of body and tail have transverse and irregular, sometimes undulating, and sometimes angularly broken bands of deep brown or black. Ten of these bands belong to the body from the occiput to the origin of the tail. On the tail itself they extend more or less toward the tip, according to the size of the specimen.

This species inhabits Oregon, about the Dalles of the Columbia River, where it has been collected by Rev. George Geary. A specimen in our possession, from California, was collected by Colonel Frémont. That which is represented on our plate we owe to the kindness of Dr. Avery J. Skilton.

PLESTIODON SKILTONIANUM, Baird and Girard.

PL. IV. FIG. 4–6.

SPEC. CHAR.—Head small, continuous with the body; tail stout, very long, and subquadrangular; olivaceous brown, with four broad bands of black.

This is a species of skink which must strike any one familiar with the general appearance of the other species of the same genus inhabiting North America. The tail, although considerably developed in all the skinks of the genus *Plestiodon*, acquires in this species much greater proportions. In the specimen figured, the tip of that organ is wanting, but when restored, the whole organ would be nearly twice the length of the rest of the body and head. Its form is rather subquadrangular than conical, and preserves a general stoutness which is not seen in the other species, in which it tapers more suddenly from its origin to its tip. The body is subcylindrical, and nearly of the same thickness from the occiput to the tail, into which it passes almost imperceptibly. The head itself is rather small, subconical, rounded on the snout; it is continuous with the body, the neck being but slightly contracted. The plates of its upper surface are represented in figure 6, which will serve as a good term for comparison with the other species. The locomotive members are very short; the fore ones rather slender, the hind ones stouter. The toes are terminated by delicate and slender nails, curved at their tip. The scales have a very smooth appearance; when examined attentively, those of the back and tail, however, exhibit four or more longitudinal furrows or stripes on each of them.

The lower part of the head and throat is whitish, the belly bluish. Above, olivaceous brown, with two broad bands of black on each side, and perhaps two narrow ones on the back extending from the head to the anterior region of the tail. A whitish or perhaps yellowish stripe extends from the opening of the ear to behind the hind legs, and a similar one from the occiput to the origin of the tail.

This species inhabits the same location in Oregon as *Elgaria scincicauda.* The specimen figured, together with several others, were collected by Rev. George Geary, and sent by him to Dr. Avery J. Skilton, to whom we are indebted for their examination, and to whom we have dedicated it, as a slight acknowledgment of gratitude.

Genus Churchillia, Baird and Girard.

Gen. char.—Three pairs of frontal plates; a very small loral, and several postorbitals. Scales carinated.

In spite of the great uniformity in the plan of structure of ophidian reptiles, and especially of the genera *Coluber* and *Tropidonotus,* we do not hesitate in separating generically from both of these, the species hereon described.

The genus *Churchillia* has three pairs of frontal plates, one more than in both *Coluber* and *Tropidonotus;* the middle pair, however, is subjected to some irregularities, by which, instead of one pair, there is only an odd plate. In either case it is a constant character proper to our genus. There is a small loral plate and several postorbitals.

The scales of the upper part of the body are carinated, as in tropidonotus, but the lateral row, which is contiguous to the ventral shields, is scarcely larger than the other—while in *Tropidonotus,* the scales composing that row are much the largest.

Churchillia bellona, Baird and Girard.

Spec. char.—Body yellowish, with a series of large subhexagonal patches of brown, bordered with black, and two or three rows of smaller patches on the sides. A brownish black band across the eyes, from top of head to the angle of the mouth.

This snake attains a considerable size. The tail, properly so called, is comparatively very short, forming only the ninth part of the length. The head is moderate, conical, detached from the body by a somewhat contracted neck. The eyes are comparatively large, occupying the middle of the length of the head. There are two anterior orbital plates, the uppermost very large, vertically elongated, reaching the upper surface of the head; the lower one, on the other hand, is very minute, situated between the third and fourth labial shields. The postorbitals are four in number, of medium, but nearly of equal size, the upper ones, however, being slightly the largest. The lower one is separated from the small anterior orbital by the fourth labial shield, which reaches the orbit.

The scales of the back and sides are keeled, elliptical in form, and a little more pointed posteriorly than anteriorly. On the occiput they are smaller, polygonal, and smooth.

The coloration of this snake resembles at first glance that of *Coluber eximius;* the ground colour is a light yellowish-brown, maculated with large patches of a deeper brown, margined with black, and much smaller patches of pure black. The dorsal row of brown patches is considerably the largest, as in *Coluber eximius.* On the abdomen there are two rows of small and semilunar black dots, the convexity of which is turned forward. A narrow band of black is seen on the upper surface of the head, in advance of the eyes, extending obliquely to the angle of the mouth, being only interrupted by the eye itself.

This species was collected by General Churchill, on his march to Mexico, on the left bank of the Rio Grande, at the crossing near Presidio del Norte, in 1846; and it is with much satisfaction that we embrace the occasion to pay a tribute of respect to one, who, during the exercise of his arduous official labours, has always found time for the advancement of natural history, by securing specimens of whatever new or interesting species of animals might fall in his way.

COLUBER MORMON, Baird and Girard.

SPEC. CHAR.—Posterior frontal plates very large; vertical plate long and very narrow on its middle; eyes very large.

The only specimen which we have seen of this snake may prove hereafter to be a young individual, as it is only one foot and a half long; but we are satisfied it will also prove to be a very distinct species. The slender and conical tail forms between a third and a fourth of the total length. The head is elongated and ovoidal, separated from the body by a contracted neck. The vertical plate is elongated and contracted on its middle. The posterior pair of frontal plates has almost twice the size of the anterior pair. The eyes are proportionally very large, protected below by the fourth and fifth labial shields, anteriorly by a large and a small orbital plate, and posteriorly by two of nearly the same size, if not of the same shape. The loral is proportionally very much developed. The nostrils are prominent. There are eight labial plates to the upper jaw and seven to the lower. The scales are smooth, elongated on the back, and posteriorly pointed, while on the sides they are broader, and the row contiguous to the ventral shield is composed of scales broader than long. On the posterior part of the head, as usual, the scales are subcircular, or polygonal, and much the smallest. On the tail they assume rather a lozenge shape. There are one hundred and seventy-five ventral shields from the throat to the vent.

The colour is brown above and yellow underneath, with a bluish tint along the sides of the abdominal region. On the back there is a row of transversely elongated patches of deep brown, bordered with black. The sides are spotted with three indistinct rows of small spots of brown and black intermixed.

This species was found by Captain Howard Stansbury's party, in the valley of the Great Salt Lake.

Hetorodon nasicus, Baird and Girard.

Spec. char.—Minute and numerous frontal plates instead of two large pairs; two brown stripes over the head; temporal patch very broad.

The essential character which distinguishes this species from the *H. platyrhinos* and *H. simus*, which it most resembles, consists in the presence of numerous small scales between the vertical and rostral plates; that is, on the space which is occupied by the two pairs of frontal plates. In *H. simus*, it is true, very minute scales may be occasionally observed on the middle line between the vertical, rostral, and frontal plates; but when this is the case, the latter are not reduced to the small size which they have in the species which we here describe.

The vertical plate in *H. nasicus* is circularly subhexagonal. The superciliaries are longer than the vertical, and thus longer than broad, while the occipital are broader than long. The nasal is proportionally large, and so is the lower loral plate, as there exists a second loral above the first, much smaller than any of those of the orbital circle. A specimen of *H. simus* examined by us has shown a second similar small loral on one side of the head, while on the other it evidently did not exist, so that it may prove either not to be constant in *H. nasicus*, or to be likewise found in other species of the same genus. The rostral plate is directed upward, as in *H. simus*, having the same general shape as in the latter.

The tail has the same proportions with regard to the body as in *H. platyrhinos*, that is about the sixth of the entire length. The eyes are considerably smaller than in *H. platyrhinos*, as we have compared specimens of both species, having the same length and size of body. The only specimen of *H. nasicus* which we have seen is not a foot long, so that we have had to compare with it the young of *H. platyrhinos*. Now, if our *H. nasicus* be immature, it is the young of a species which differs from those already known of the genus, as shown by the very great difference in size of the eyes, and those other characters which we have just described.

In coloration, our species resembles more that of *H. platyrhinos* than that of *H. simus*. The ground colour appears to be the same. There are in both species three rows of blackish-brown patches, bordered with a line of light yellow or white; but while these patches are elongated in *H. platyrhinos*, they assume a more circular shape in *H. nasicus*, although there is no regularity or symmetry about them. Two indistinct and much smaller rows exist on the sides. The lower part of abdomen and tail is almost entirely black. The throat and lower surface of head are unicolor, of the same hue as the ground colour of the body above. On the occiput, immediately behind the occipital plate, there is a lanceolated patch, and on both sides of this a much greater irregularly oval one. Two brown stripes are seen on the top of the head, one over the anterior half of the eyes, extending vertically to the mouth, and another over the posterior half of the eye, hence obliquely backward, forming over the temples an elongated but much broader patch than in *H. platyrhinos* and *H. simus*.

This species was collected in Texas by General Churchill: a specimen is preserved at the Smithsonian Institution.

SIREDON LICHENOIDES, Baird.

Pl. II.

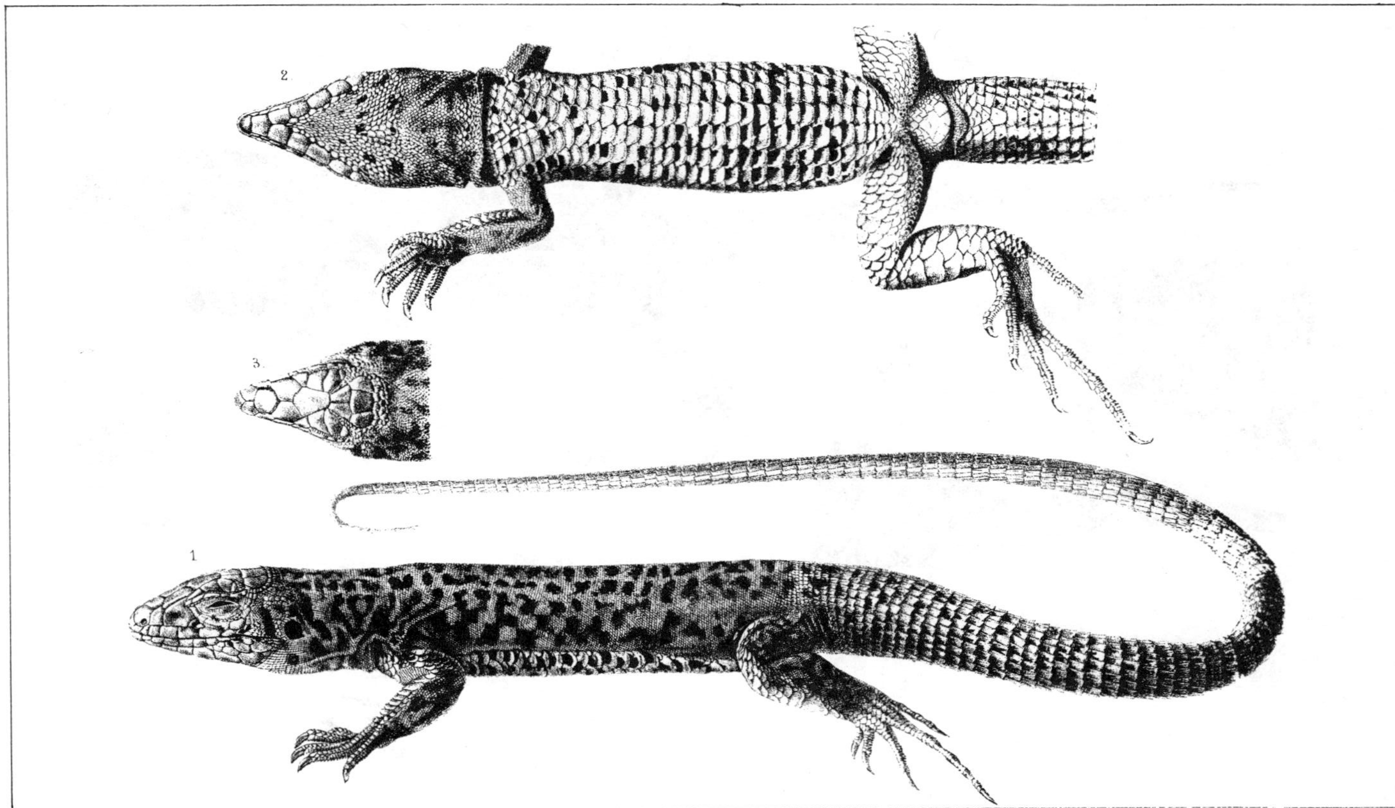

Ackerman Lith 379 Broadway N.Y.

CNEMIDOPHORUS TIGRIS, Baird & Girard

Pl. III.

CROTAPHYTUS WISLIZENII, Baird & Girard.

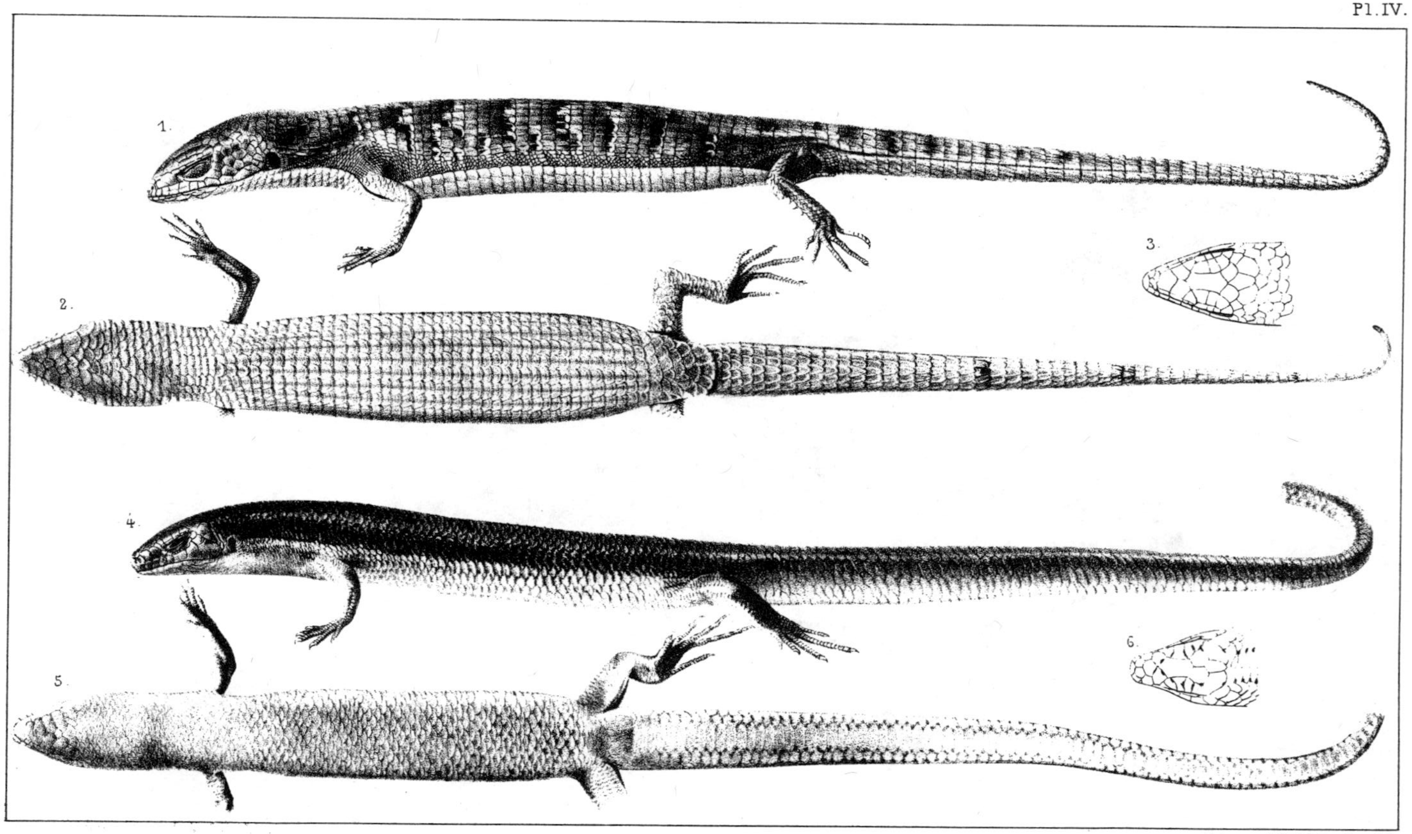

Ackerman Lith. 379 Broadway, N.Y.

Fig. 1-3. ELGARIA SCINCICAUDA, Baird & Girard. Fig. 4-6. PLESTIODON SKILTONIANUM, Baird & Girard.

Fig. 1-3. SCELOPORUS GRACIOSUS, Baird & Girard. Fig. 4-6. JTA STANSBURIANA, Baird & Girard.

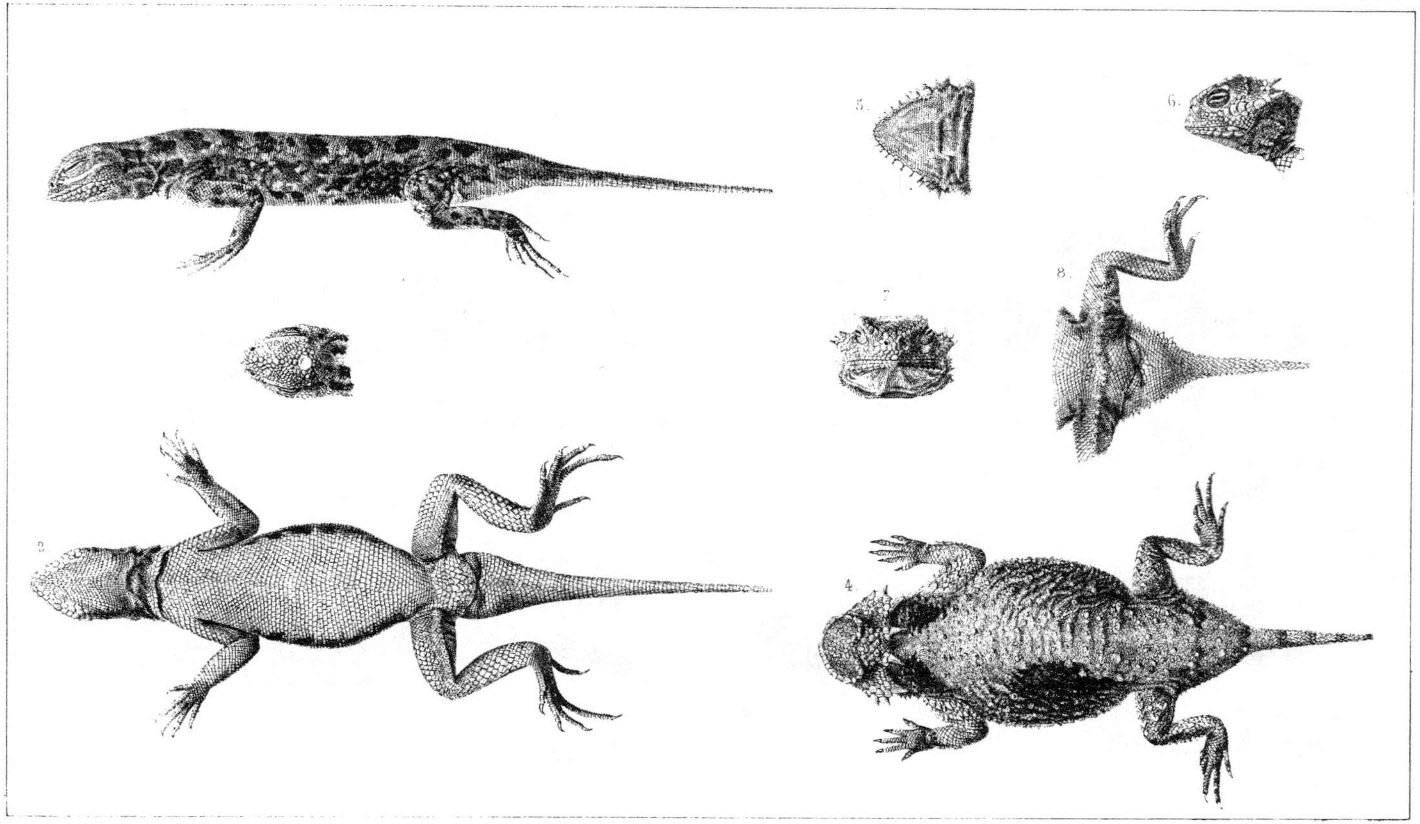

Fig. 1-3. HOLBROOKIA MACULATA, Girard   Fig. 4-8. PHRYNOSOMA MODESTA, Girard.

REPTILES

S[pencer] F[ullerton] Baird
and
C[harles] Girard

REPTILES.

BY S. F. BAIRD AND C. GIRARD.

SERPENTS.

The serpents collected by Captains Marcy and McClellan belong to ten species, distributed into eight genera. Several of these species had previously been received from other sections of the country: three, however, were first collected during the expedition. All are here figured for the first time, except *Ophibolus Sayi*, of which a hitherto undescribed variety is represented.

I. CROTALUS, Linn.

This genus is characterized by its erectile poison fangs, and by having the upper surface of the head covered with small plates resembling the scales on the body, and with only a few larger ones in front. There is a deep pit between the eyes and the nostrils. The plates under the tail are undivided, and the tail is terminated by a rattle. Scales carinated.

1. Crotalus confluentus, Say.

Zoology, Pl. I.

Spec. char.—Head subtriangular. Plates on top of head squamiform, irregular, angulated, and imbricated; scales between superciliaries small, numerous, uniform. Four rows of scales between the suborbital series (which only extends to the centre of the orbit) and the labials. Labials 15 or 18, nearly uniform. Dorsal series 27–29. Dorsal blotches quadrate, concave before and behind; intervals greater behind. Spots transversely quadrate posteriorly, ultimately becoming 10 or 12 half rings. Two transverse lines on superciliaries, enclosing about one-third. Stripe from superciliary to angle of jaws, crosses angle of the mouth on the second row above labial. Rostral margined with lighter.

SYN.—*Crotalus confluentus*, Say, in Long's Exped. Rocky Mts. II, 1823, 48. B. & G. Cat. N. Amer. Rept. I, 1853, 8.
C. Lecontei, Hallow. Proc. Acad. Nat. Sc. Philad. VI, 1851, 180.

DESCRIPTION.—This species bears a considerable resemblance to *C. atrox*, but the body is more slender and compact. Scales on the top of the head anterior to the superciliaries nearly uniform in size. Line of scales across from one nostril to the other consists of six, not four as in *C. atrox*. Superciliaries more prominent. Labial series much smaller. Upper anterior orbitals much smaller, as also is the anterior nasal. Scales on the top of the head less carinated. Scales between superciliaries smaller and more numerous, five or six in number instead of four. Two lateral rows of scales smooth; first, second, and third gradually increasing in size. Scales more linear than in *C. atrox*.

General color yellowish brown, with a series of subquadrate dark blotches, with the corners rounded and the anterior and posterior sides frequently concave, the exterior convex. These blotches are ten or eleven scales wide and four or five long, lighter in the centre, and margined for one-third of a scale with light yellowish. The intervals along the back light brown, darker than the margins of the blotches. Anteriorly the interval between the dark spots is but a single scale; posteriorly it is more, becoming sometimes two scales, where also the spots are more rhomboidal or lozenge-shaped; nearer the tail, however, they become transversely quadrate. The fundamental theory of coloration might be likened to that of *Crotalus adamanteus*, viz: of forty or fifty light lines decussating each other from opposite sides; but the angles of decussation, instead of being acute, are obtuse, and truncated or rounded off throughout. Along the third, fourth, and fifth lateral rows of scales is a series of indistinct brown blotches covering a space of about four scales, and falling opposite to the dorsal blotches: between these blotches, and opposite to the intervals of the dorsal blotches, are others less distinct. Along the fifth, sixth, seventh, and eighth rows is a second series of obsolete blotches, each covering a space of about four scales, and just opposite the intervals between the dorsal spots. The dorsal and lower series are separated by an interval of three scales, this interval light brown. Beneath, the color is dull yellowish, and ten or twelve darker half rings are visible on the tail.

In point of coloration the principal features, as compared with *C. atrox*, lie in the disposition of the dorsal blotches in subquadrate spots instead of subrhomboids; the intervals thus forming bands across the back perpendicular to the longitudinal axis. This tendency to assume the subquadrangular pattern has broken up the chain-work into isolated portions, as in *Ophibolus eximius* or *Crotalophorus tergeminus*. The intervals of the dorsal blotches are wide and darker in the middle, while in *C. atrox* they are narrow, not linear, and unicolor. The sides of the head present the usual light stripe from the posterior extremity of the superciliary; it passes, however, to the angle of the jaw on the neck, along the second row of scales above the labials. A second stripe passes in front of the eye to the labials, widening there. A small light vertical bar is seen below the pit, and another on the outer edge of the rostral. On the superciliaries are two light transverse lines enclosing a space nearly one-third of the whole surface. In *C. atrox* there is a single median line. Sometimes, as in *C. atrox*, the single blotches on the nape are replaced by two elongated ones parallel to each other.

Dorsal row of scales, 29; abdominal scutellæ, 180; subcaudal ones, 27. Total length, 34 inches; length of tail, 4 inches.

A specimen was collected the 5th of June in the Witchita mountains. Another specimen of the same species was brought home from the Cross-Timbers, Arkansas, by Dr. S. W. Woodhouse, and described by Dr. Hallowell as new, under the name of *Crotalus Lecontei*, on the ground that the anterior vertebral spots are not confluent. This we do not consider as a sufficiently distinctive character, although we have never seen a specimen with decidedly confluent markings. The notes of Dr. Leconte, quoted by Dr. Hallowell, hardly apply to the present species.

The species was first discovered by Say on Major Long's expedition to the Rocky mountains, and has not since been seen until procured first by Dr. Woodhouse, and then by Captain Marcy and the Mexican boundary commission. It was found by the latter party in Western Texas, where, however, it is rare.

Plate I represents *Crotalus confluentus* of natural size.

II. EUTÆNIA, B. & G.

This genus is composed of numerous species, some of them quite common, and known under the names of Riband, Striped, and Garter snakes: inoffensive, like most of the North American snakes. They may be recognised by three light stripes on a darker ground, the intervals between these stripes provided with alternating or tessellated blackish spots. The scales have a ridge or small keel along their middle, and are arranged in 19 or 21 longitudinal rows. The postabdominal or anal scutella is entire, like the others. There is one anterior orbital plate and three posterior. The body is either moderately stout or else slender, according to the species. Of the two described in this article, one belongs to the division with a slender body and 19 dorsal rows of scales, and the other to the second division, with a stouter body and 21 dorsal rows of scales.

2. Eutænia proxima, B. & G.

Zoology, Pl. II.

Spec. char.—Body stoutest of the division. Black above; three longitudinal stripes, the dorsal ochraceous yellow or brown, lateral greenish white or yellow. Total length about three and a half times that of the tail.

Syn.—*Coluber proximus*, Say, in Long's Exped. to Rock. Mts. I, 1823, 187.—Harl, Journ. Acad. Nat. Sc. Philad. V, 1827, 353.
Eutainia proxima, B. & G. Cat. N. Amer. Rept. I, 1853, 25.

Desc.—Deep brown, almost black, above and on the sides; beneath greenish white. Dorsal stripe on one and two half rows of scales, ochraceous yellow, lateral stripe on the 3d and 4th rows of scales, greenish yellow or white, markedly different in tint from the dorsal. Sides of abdominal scutellæ, and 1st and 2d dorsal series, of the same color as the back. On stretching the skin, numerous short white lines are visible. Occipital plates with two small approximated spots on the line of junction. Orbitals whitish. The greenish white of the abdomen becomes more yellow anteriorly.

Head more like that of *E. saurita* than of *E. Faireyi*, while the body is stouter than in either. The subcaudal scales are less numerous than in the other two allied species. Resembling *E. Faireyi* in color, it is always distinguishable by the stouter body, fewer caudal scales, and dissimilarity of color in the longitudinal stripes.

Dorsal rows of scales 19, all keeled; abdominal scutellæ 170; subcaudal ones 100. Total length 33 inches; length of tail 9 inches. Found at Camp No. 7.

The species is represented in natural size on Plate II.

3. Eutænia marciana, B. & G.

Zoology, Pl. III.

Spec. char.—Prominent color light brown; a vertebral paler line and one lateral on each side, more or less indistinct. Three series of square black spots on each side, of about 56-60 in each series, from occiput to anus. Sides of head black, with a crescentic patch of yellowish posterior to the labial plates. Three and sometimes four black vittæ radiating from the eye across the jaws. A double white spot with a black margin on the suture of occipital plates.

Syn.—*Eutainia marciana*, B. and G. Cat. N. Amer. Rept. I, 1853, 36.

Desc.—The markings about the head are generally very constant and distinct. Viewed laterally, we see first the large dark-brown patch at the back part of the head, extending as far back as the posterior extremity of the jawbones. In the anterior part of this patch is seen the crescentic patch (concave before) of yellowish white, with a more or less narrow dark-brown margin anteriorly. The next black band starts from the posterior edge of the superciliaries, and passes obliquely downwards and backwards along the posterior edge of the 6th upper labial. Similar black margins are seen on the posterior edges of the 5th and 4th labials, the intervening spaces being yellowish white, particularly on the 5th upper labial. Occasionally the posterior margins of the 7th and 3d labials have the black line as well as those mentioned, which frequently extend across to the posterior margins of the corresponding lower labials. The white spot on the anterior portion of the occipital suture is always margined with black.

The six series of black spots are arranged so as to alternate with each other. The lower or third series on each side is below the indistinct lateral stripe. The posterior edge of each abdominal scutella shows a black margined spot on each side. The dorsal line is generally a single scale in width, occasionally including portions of the lateral, and itself sometimes encroached upon by the black spots. Each spot is about a scale or a scale and a half long, and about three scales broad. The number in the dorsal series from the head to the anus varies from 56 to 60. Posterior edges of scales very slightly emarginate, if at all. All are decidedly keeled.

Dorsal scales disposed in 21 rows; abdominal scutellæ, 152; subcaudal, 75. Total length 34 inches; length of tail, 8 inches.

Collected between Camp 5 and Red river, on the open prairie.

This species is very widely distributed in the south and west. Red river forms its limit on the north, and the Gulf of Mexico on the east;

but it extends to the Rocky mountains on the west, and far into Mexico on the south. Its centre of distribution appears to be on the lower Rio Grande.

Plate III represents this species in natural size.

III. HETERODON, Pal. de B.

This genus is eminently characterized by the peculiarity of its snout, which is terminated by a triangular plate recurved upwards; hence the popular appellation of hog-nose snake. Though perfectly harmless, they exhibit a threatening appearance, when approached, in the flattening of their head and violent hissings; hence the names of blowing-viper, spreading adder, &c. Their body is short, stout, and the tail also short. The head is broad and short. The dorsal scales are carinated, and arranged in 23–27 rows. The preanal or postabdominal scutella is bifid; a chain of small plates beneath the eye, completed above by the superciliaries. There is a supplementary plate on the top of the head, behind the prominent rostral, either in contact with the frontals, or separated by smaller plates. The colors are light, with dorsal and lateral darker blotches, or else brown, with dorsal transverse light bars; sometimes entirely black.

One species only was collected on the Red River exploration. Six species are known to exist in the United States.

4. Heterodon nasicus, B. & G.

Zoology, Pl. IV.

Spec. char.—Vertical plate broader than long. Rostral excessively broad and high. Azygos plate surrounded behind and on the sides by many small plates (12–15.) A second loral. Labials short and excessively high. Dorsal rows of scales 23, exterior alone smooth. A dorsal series of about 50 blotches, with four or five other series on each side. Body beneath, black. A narrow white line across the middle of the superciliaries; a second behind the rostral. A broad dark patch from the eye to the angle of the mouth, crossing the two postlabials.

Syn.—*Heterodon nasicus*, B. & G. Reptiles in Stansbury's Expl. Valley of Great Salt Lake, 1852, 352.—B. & G. Cat. N. Amer. Rept. I, 1853, 61.

Desc.—Vertical plate very broad, subhexagonal. Occipitals short. Rostral very broad, high, more so than in the other species, outline rounded. The interval between the opposite frontals, the rostral, and the vertical occupied by a number of small plates, from 10 to 12, or more, arranged without any symmetry, on each side and behind the small azygos. The base of the rostral between the opposite prenasals is generally margined by these small plates, which sometimes, too, are seen between the vertical and the anterior portion of the superciliaries. This crowding of plates causes the anterior part of the forehead to be broader than in *H. simus*. Eye small, its centre rather posterior to the middle of the imaginary line connecting the tip of rostral with the lower angle of the postlabial, which line scarcely crosses the eyeball. Orbital plates, 10–13 in number. Loral triangular, rather longer than high, separated from the frontal by a small plate. Nasals rather short, occasionally with the lower part of the nostril bounded by a small plate. Labials 8 or 9 above, all of them higher than long; indeed, their vertical extension is much greater than in any other species: the 6th highest; centre of eye over the junction of the 5th and 6th.

Dorsal rows of scales 23, outer row smooth, rest all distinctly carinated, the keels extending to the ends of the scales; those just behind the occipital plates truncated, with obsolete carinæ. Scales on the hind part of the body rather broader and shorter than anteriorly; the inequality scarcely evident in large specimens.

Ground-color light brown or yellowish gray, with about 50 dorsal blotches from head to tip of tail; the 39th opposite the anus. These blotches are quite small, rather longer transversely, subquadrate, or rounded, indistinctly margined with black, (obsoletely on the outside;) they cover 7 to 9 scales across, are 2 to 2½ long, and separated by interspaces of 1½ scales, which are pretty constant throughout, though rather narrower on the tail. On each side of the dorsal row may be made out, under favorable circumstances, four alternating rows of blotches; the first on the contiguous edges of the scales of the first and second exterior dorsal rows; the second on the scales of the 3d row, and the adjacent edges of those in the 2d and 4th; the third on the scales of the 4th, 5th, and 6th, and the adjacent edges of the 3d and 7th; and the fourth on the scales of the 6th, 7th, and 8th rows, and the adjacent edges of those of the 5th. This last is opposite the intervals of the dorsal series; the rest alternate with it. The central inferior surface of the abdominal scutellæ is black, sharply variegated with quadrate spots of yellowish white; the portion of the scutellæ entering into the side of the body is yellowish white, with that part opposite the dorsal intervals dark brown, thus, in fact, constituting a fifth lateral series of blotches, alternating with the lowest already mentioned. The throat and chin are unspotted. The head is light brown, with a narrow whitish line finely margined before and behind with black, which crosses in front of the centre of the vertical, and through the middle of the superciliaries: a second similar but more indistinct line runs parallel to this just behind the rostral, and extending down in front of the eye. A third equally indistinct and similar line crosses the posterior angle of the vertical, and runs back on the side of the neck, behind the labials and temporal shields. There is a broad brown patch from the back part of the eye to the angle of the mouth, across the penultimate and last labial. The coloration is thus very different from that of *H. Simus*, where there is a distinct narrow black band across the forehead scarcely involving the vertical, and passing through the eye to the angle of the mouth across the last labial. Behind this a much broader yellowish band, continued without interruption into the neck behind the angle of the mouth. In *H. nasicus* the most conspicuous feature is a narrow white band, much narrower than the darker patch before and behind it. The dark patch, to the angle of the mouth, is much broader, continuous as it were, with the broad bar between the middle and anterior light lines, which corresponds with the narrow black line of *H. Simus*. The other distinguishing features are evident. The three dark patches behind the head are much as in *H. simus*.

In large specimens from Sonora and the Copper Mines of the Gila, (Fort Webster), the ground-color is yellowish gray, each scale minutely punctate with brown. The blotches are all obsolete, only one dorsal and two lateral on each side being defined by darker shades. The blotches on the sides of the abdomen are wanting, but the black in the middle is strongly marked. The other characters, however, are preserved, except that the exterior row of dorsal scales is more or less carinated.

Specimens of this species vary in the number of small postrostral plates. In some there are only three or four, in others a larger number. Sometimes, instead of a single series of median dorsal spots, there are two, in close contact, and more or less confluent. The narrow light line across the middle of the superciliaries and the high labials are always highly characteristic.

The specimen figured of natural size on plate IV, is much smaller than those alluded to from Sonora and the Copper Mines, and upon which the foregoing description has been based.

IV. PITUOPHIS, Holbr.

This genus, closely allied to *Heterodon*, is characterized by a prominent snout, the rostral plate elevated and convex, without, however, being recurved. There are two pairs of postfrontal plates instead of one, and occasionally also two verticals; three or four postorbitals; generally two, sometimes only one anteriorly. The scales are carinated along the back, smooth on the sides, and constituting from 29–35 dorsal rows. The preanal or postabdominal scutellæ is entire.

The ground-colors are either whitish or reddish yellow, with a triple series of patches, those of the medial series the largest, and several series of smaller blotches on the sides. Abdomen unicolor or spotted, with an outer row of blotches. Head of same color as the body, maculated with black spots. A narrow band of black across the upper surface between the eyes, and a postocular vitta on each side, extending obliquely from the eye down to the angle of the mouth. A black vertical patch is often seen beneath the eye.

The names of Bull, Pine, and Pilot snake are commonly given to different species of this genus, which are all of great size, including in fact some of the largest serpents of North America. Some of the species utter a hissing or blowing sound.

5. Pituophis McClellanii, B. & G.

Zoology, Pl. V.

Spec. char.—Head subelliptical. Rostral plate very narrow. Anteorbitals 2; postorbitals 4. Dorsal rows 33–35; the 7 outer rows smooth. Tail forming 1·9 or 1-10 of total length. Postocular vitta brown, and rather broad. Suborbital black patch conspicuous; commissure of labials black. Color of body reddish yellow, with a series of 53 blotches from head to origin of tail. Blotches of adjoining series, on either side, confluent across the light spaces between medial blotches. Flanks covered with small blotches, forming 3 or 4 indistinct series. Twelve transverse jet-black bars across the tail. Abdomen yellowish, thickly maculated with black patches.

Syn.—*Pituophis McClellanii*, B. & G. Cat. N. Amer. Rept. I, 1853, 68.—Pilot-snake.

Desc.—Head proportionally large, ovoid, distinct from the body. Snout pointed. Occipital plates small. Vertical broad, subpentagonal, slightly concave on the sides. Superciliaries large. Internal postfrontals rather narrow, elongated; external postfrontals quadrilateral, a little broader forwards. Prefrontals irregularly quadrangular. Rostral very narrow, extending halfway between the prefrontals, convex and raised above the surface of the snout. Nostrils in the middle line between the nasals, the posterior of which is a little the smaller. Loral trapezoidal, proportionally large. Inferior anteorbital very small, resting upon the fourth upper labial. Postorbitals varying in comparative size. Temporal shields small, resembling scales. Upper labials 8; 6th and 7th the larger. Lower labials 12; 6th and 7th largest. Posterior mental shields very small, extending to opposite the junction of the 7th and 8th lower labials. Scales proportionally small, in 33–35 rows, the 7 outer ones perfectly smooth and somewhat larger than the remaining rows.

Ground-color yellowish brown, with three series of dorsal black blotches, 53 in number, from the head to opposite the anus, with 12 on the tail, in the shape of transverse bars. Those of the medial series the larger, and covering 8 or 9 rows of scales. On the anterior part of the body they are subcircular, embracing longitudinally four scales; posteriorly they become shorter by one scale. The light spaces between are a little narrower than the blotches themselves for the twelve anterior blotches, and wider than the blotches for the remaining length of the body. The blotches of the adjoining series alternate with those of the medial series, being opposite to the light intermediate spaces, across which the blotches of either sides are generally united by a transverse narrow band. The flanks are densely covered with small and irregular blotches, forming three indistinct series, confluent in vertical bars towards the origin of the tail. Inferior surface of the head yellowish, unicolor Abdomen dull yellow, with crowded brownish black blotches in series on the extremity of the scutellæ.

Two specimens of this species were caught the 28th of June. The largest is figured, of natural size, on plate V. It is 38¼ inches in length; the tail measuring nearly 5 inches. Abdominal scutellæ 231; subcaudal ones, 52.

V. SCOTOPHIS, B. & G.

The scales in this genus are very slightly carinated on the back, and perfectly smooth on the sides. Preanal scutella bifid. One large ante-orbital plate and two postorbitals. The colors are brown or black, in quadrate blotches on the back and on the sides, separated by lighter intervals; beneath usually coarsely blotched with darker. In one species there are dark stripes on a light ground.

6. SCOTOPHIS LAETUS, B. & G.

ZOOLOGY, Pl. VI.

SPEC. CHAR.—Similar to *S. confinis*, but postfrontals larger. Vertical plate longer than broad. Dorsal rows 29. Abdominal scutellæ 227. Subcaudals 72. Blotches fewer than in *S. confinis*.

SYN.—*Scotophis laetus*, B. & G. Cat. N. Amer. Rept. I, 1853, 78.

DESC.—This species bears a close resemblance to *S. confinis*, and its characters may be best given by comparison with the latter. It differs, therefore, in the greater number of dorsal rows, 29 instead of 25. The whole body and head are much stouter. Exterior eight rows smooth, rest slightly carinated. The vertical is broad before, rather acute behind. A probably monstrous feature is seen in the union of the two postfrontals, except for a short distance before, and in the loral and postnasal coalescing into one trapezoidal plate. Blotches less numerous. A broad vitta across the back part of the postfrontals, passing backwards and downwards through the eye, and terminating acutely on the posterior upper labial. A blotch across the back part of the vertical, and extending through the occipitals on each side to the nape. The spots are larger, longitudinal throughout, with occasional exceptions.

Its affinities to *S. vulpinus* are close. The vertical, however, is narrow, the eyes much larger, dorsal rows 29 instead of 25. The blotches on the back are longitudinal, and fewer in number. For a complete description of this species it will be necessary to procure larger specimens.

The specimen figured on Plate VI is of natural size. The only one caught of this species is 18 inches long. Length of tail $3\frac{1}{4}$ inches.

VI. OPHIBOLUS, B. & G.

The body is rather thick, and the tail short. The scales smooth and lustrous, and disposed in 21 or 23 rows, which scarcely overlap. The preanal scutella is entire. A small anteorbital plate and two postorbitals. Eyes very small.

The ground-colors are black, brown, or red, crossed by lighter intervals, generally bordered by black.

Seven species, besides the two here described, have hitherto been found in North America.

7. Ophibolus Sayi, B. & G.

Zoology Pl. VII.

Spec. char.—Black, each scale above with a large circular or subcircular white or yellow spot in the centre. Sometimes only transverse lines of these spots across the back.

Syn.—*Herpetodryas getulus*, Schl. Ess. Phys. Serp. Part. descr. II, 1837, 198.

Coronella Sayi, Holbr. (non Schl.) N. Amer. Herp. III, 1842, 99. Pl. xxii.

Coluber Sayi, Dekay, New York Fauna, Rept. 1842, 41.

Ophibolus Sayi, B. & G., Cat. N. Amer. Rept. I, 1853, 84.

Desc.—Body, as in most of the other species of the same genus, very tense and rigid, with difficulty capable of being extended after immersion in alcohol. Vertical plate triangular, wider than long; outer edge slightly convex, an angle being faintly indicated at the junction of the superciliaries and occipitals; shorter than the occipitals, which are short, longer than broad. Postfrontals large, broad; prefrontals smaller. Rostral small, not projecting, slightly wedged between prefrontals. Eye very small, orbit about as high as the labial below it; centre of the eye a little anterior to the middle of the commissure, over the junction of the 3d and 4th labials. One anteorbital, vertically quadrate; loral half its height, square. Upper labials 7, increasing to the penultimate. Lower labials 9; 4th and 5th largest.

Scales nearly as high as long, hexagonal, truncated at each end. Dorsal rows 21, exterior rather larger, and diminishing almost imperceptibly to the back, although all the scales in a single oblique row are of very nearly the same shape and size.

The scales on the back and sides are lustrous black, each one with a central elliptical or subcircular spot of ivory-white, which on the sides occupy nearly the whole of the scale, but are smaller towards the back, where they involve one half to one third of the length. Beneath yellowish white, with broad distinct blotches of black, more numerous posteriorly. Skin between the scales brown. The plates on the top and sides of the head have each a yellowish blotch; the labials are yellow, with black at their junction.

Other specimens agree except in having bright yellow instead of white as described; the spots, too, are rather smaller, and manifest a slight tendency to aggregation on adjacent scales, so as to form transverse bands. This is seen more decidedly where the back is crossed by about 70 short dotted yellow lines; the 56th opposite the anus. The scales between have very obsolete spots of lighter, scarcely discernible. The sides are yellow, with black spots corresponding to the dorsal lines; indeed, there may be indistinctly discerned two or three lateral series of alternating blotches.

In larger specimens from the West, this tendency in the spots to aggregation is still more distinct. The back is crossed by these dotted lines of the number and relation indicated, at intervals of four or five scales; the spots on the intervening space being obsolete. These lines bifurcate at about the 9th outer row, the branches connecting with those contiguous, so as to form hexagons; and these extending towards the abdomen again, decussate on about the third outer row, thus enclosing two series of square, dark spots on each side. These lateral markings are, however, not very discernible, owing to the confusion produced by the greater number of yellow spots. On the edge of the abdomen are dark blotches, one opposite each dorsal dark space, the centres of the scutellæ being likewise blotched, but so as rather to alternate with those just mentioned.

The specimen represented on Plate VII was caught the 16th of May, between Cache creek and Red river. Total length 33¼ inches; tail 3¾ inches; abdominal scutellæ 224; subcaudals 49.

8. Ophibolus gentilis, B. & G.

Zoology, Pl. VIII.

Spec. char.—Muzzle more convex and acute than in *O. doliatus*. Body red, encircled by about 25 pairs of broad black rings enclosing a white ring: the white mottled with black on the sides. Black rings broader than in *O. doliatus*. Upper part of head entirely black.

Syn.—*Ophibolus gentilis*, B. & G. Cat. N. Amer. Rept. I, 1853, 90.

Desc.—Ground-color deep-red, encircled by 25 pairs of black rings, the 21st opposite the anus, each pair enclosing a third ring of white, the latter becoming yellowish by immersion in alcohol. The black rings are conspicuously broader above, the three crossing eight scales on the vertebral row anteriorly, and towards the anus about five. Anteriorly the intervals between successive pairs consist of about five scales, posteriorly only of two or three, thus diminishing considerably. The black rings contract as they descend, those of each pair receding slightly from each other, so as to cause the yellow portion to expand about one scale. The black rings are continuous on the abdomen, those of contiguous pairs (not of the same pair) sometimes with their intervening spaces black. The scales in the white rings are always more or less mottled with black, especially along the sides of the body, this mottling being very rarely observable on the red portion. The anterior black ring of the first pair is extended so as to cover the whole head above, except the very tip; the white ring behind it involves the extreme tip of the occipitals.

A large specimen is much duskier in its colors. The black rings extend on the back so that the contiguous rings of adjacent pairs run into each other. There are 28 pairs of rings, the 25th opposite the anus.

Dorsal row of scales, 21; abdominal scutellæ, 198; subcaudal ones, 45. Total length, 20 inches; length of tail, 2¾ inches. Plate VIII represents the largest of two specimens, caught June 14, near Sweetwater creek.

VII. MASTICOPHIS, B. & G.

The prominent feature of this genus consists in a very slender and elongated tail, forming one-third or one-fourth of the length. It bears a close relationship to the black snakes (*Bascanion*), from which it differs chiefly in the structure of the plates on the upper jaw. The scales are smooth and disposed in fifteen or seventeen dorsal rows. The preanal scutella is divided. The vertical plate is long and narrow. There are two anteorbitals and two postorbitals, these resting against the fifth labial.

9. MASTICOPHIS FLAVIGULARIS, B. & G.

SPEC. CHAR.—Light dull yellow, tinged with brown above. Beneath, two longitudinal series of blotches distinct anteriorly. In alcohol, and especially when the epidermis is removed, the whole animal appears of a soiled white.

SYN.—*Psammophis flavigularis*, Hallow. Proc. Acad. Nat. Sc. Philad. VI, 1852, 178.

Masticophis flavigularis, B. & G. Cat. N. Amer. Rept. I, 1853, 99.

DESC.—Size very large. Vertical plate broad before, tapering to the middle, where it is about half as wide as anteriorly, thence it runs nearly parallel. Vertical rather shorter than occipitals. Greatest breadth across superciliaries less than half the length of the portion covered by plates. Occipitals moderate. Centre of eye considerably anterior to the centre of commissure, over the junction of the 4th and 5th labials. Labials 8 above, increasing in size to the 5th, which is elongated vertically, the 7th elongate and largest. The 5th forms part of the inferior and posterior wall of the orbit, as in all the species of the genus, resting above against the lower postorbital, with which the 6th labial is not in contact. Dorsal scales broader than in *Bascanion constrictor*, their sides perfectly straight, slightly truncate, with the corners rounded. Exterior row largest, rest gradually diminishing. Scales on the tail widely truncate.

The general color, both above and below, may be described as a dull straw-yellow, tinged with light olivaceous brown above. This latter tint exists in the form of a shading on the centres and towards the tips of the scales, leaving the bases yellowish. The proportion of brown increases towards the back, and in older specimens sometimes suppresses the yellow. In all instances a darker shade is seen towards the tip of each scale. The skin between the scales is yellowish. The scutellæ anteriorly exhibit each two rather large brownish blotches, one on each side of the median line, constituting two rows on the abdomen, which fade out posteriorly. Sometimes the series are not discernible, the blotches spreading so as to constitute a dark shade to the margins and exterior edges of the scutellæ. The posterior portions of the plates under, and on the sides of the head, are similarly blotched; the same tendency being observable on the posterior edges of the plates on the top of the head, by the deeper shade of the olivaceous brown there prevalent. Anteorbitals yellow. One specimen was procured $57\frac{1}{2}$ inches long. Abdominal scutellæ 191+2. The tip of the tail is missing.

In smaller specimens the blotching beneath is rather more decided. In addition to the colors described, the back is crossed by indistinct bars of darker, eight or nine scales wide and half a scale long. This color is also seen on the skin between the scales under the dark bars, where the bases of the scales themselves are darker instead of light. There is a tendency towards stripes on the side: first one of light brown, on the outer edge of the abdomen; then an interrupted yellow one at the junction of the abdominal scutellæ and outer scales; then brown again through the centres of the rows. This, however, is not very conspicuous. Sometimes the dark shades on the sides are tinged with reddish. The obsolete transverse bars are seen at intervals of one or two scales.

This species may prove to be the *Coluber testaceus* of Say. A specimen from Fort Webster, or the Copper Mines, collected by the U. S. and Mexican Boundary Commission, shows the stripes on the sides much more distinctly, running through all the dorsal rows anteriorly, and crossed by the indistinct bars already referred to. The contrast between the dark chestnut-brown spots on each side, and its deeper centre, with the clear yellow of the edges, is very distinct. Beneath yellow, with the blotches reduced to mere dull spots.

VIII. LEPTOPHIS, Bell.

The body is elongated and very slender, the tail forming more than the third of the entire length. The scales are disposed in 17 dorsal rows and keeled, except the two outer rows, which are smooth. The nostril is situated in the middle of a single plate. The eyes are large. The preanal scutella is bifid or divided. The color uniformly green.

10. Leptophis majalis, B. & G.

Zoology, Pl. IX.

Spec. char.—Reddish green above, yellowish white beneath. Body proportionally stouter and tail shorter than in *L. æstivus*. Snout and whole head, including vertical, longer than in latter species. Dorsal scales in 17 rows.

Syn.—*Leptophis majalis*, B. & G. Cat. N. Amer. Rept. I, 1853, 107.

Desc.—Head more pointed, broader on the temporal region, and more tapering on the snout than in *L. æstivus*. Vertical plate subhexagonal, broader, and postfrontals proportionally larger in comparison with the prefrontals, than in *L. æstivus*. Occipitals maintaining more their width posteriorly, obtuse-angled behind. Nasal more elongated; loral smaller, and longer than high. Two large temporal shields and a few small ones behind. Scales strongly carinated, except the outer row, which is perfectly smooth, and the second row, which is but slightly carinated. The scales of both of these rows are broader than the rest.

Total length, 28½ inches; length of tail, 9¾ inches. Abdominal scutellæ, 163+1; subcaudal, 111. The specimen figured on Plate IX was caught on the 13th of July at the head of Cache creek, near old Witchita village.

LIZARDS.

Six species of saurians, or lizards, belonging to six different genera, were collected during the exploration of Red river. One of them has proved to be new to science; two were recently described for the first time, whilst the three others have been long known to herpetologists.

I. PHRYNOSOMA, Wiegm.

This genus, including the so-called horned toads or horned frogs, more properly horned lizards, is recognizable by a depressed, broad, and subelliptical body, covered above with irregular scales, the majority very small, others quite large, pyramidal, raised above the surface of the skin, and scattered all over the back, sides, and tail. The head is subtriangular and provided with powerful spines or horns, giving to it rather a formidable appearance, although all the species of this genus are perfectly inoffensive. There are external auditive apertures as in most of the lizard tribe.

In a monograph of the genus appended to Stansbury's Exploration of the Valley of the Great Salt Lake of Utah, six species are described as indigenous to North America; another has since been added to the list. The single species collected is the most abundant of the genus.

1. Phrynosoma cornutum, Gray.

Spec. char.—Nostrils situated within the internal margin of the superciliary ridge; occipital and temporal spines longer and more acute than in *Ph. orbiculare;* a double row of pyramidal scales on the sides of the abdomen; scales on the inferior surface of the head small and slightly keeled, of a general uniformity, except one row on each side, somewhat larger, pyramidal, acute, slightly raised, and directed outwards and backwards; a series of very large inframaxillary plates, sharp on their outer edge, the posterior one of which is transformed into a spine. The plates lining the margin of the jaws are not prominent. The scales of the belly are proportionally small, subquadrangular, keeled, and posteriorly very acute; femoral pores undeveloped, or rudimentary in the female.

Syn.—*Phrynosoma cornutum*, Gray, Syn. Rept. in Griff. Anim. Kingd. IX, 1831, 45. Holbr. N. Amer. Herp. II, 1842, 87. Pl. xi.—Girard in Stansbury's Expl. Val. Great Salt Lake, 1852, 360. Pl. viii, fig. 1—6.

Agama cornuta, Harl. Med. and Phys. Res. 1835, 141. Plate, figs. 1 and 2.

Phrynosoma Harlani, Wiegm. Herp. Mex. 1834, 54.—Dum. and Bibr. Erp. gen. IV, 1837, 314.

Obs.—The color of this species has been well described by Dr. Holbrook. We may add that the ground-color above in some individuals is of a variable shade of ferruginous red—a tint sometimes seen on the inferior surface of the body. The belly is either unicolor, or else spotted as in *P. orbiculare*. Numerous specimens of this species were collected during the exploration of Red river; some on the prairie between Camps 2 and 3; others between Camps 6 and 7, and at Camp 7 also; others still on the south fork of Red river, and several other localities.

II. CROTAPHYTUS, Holbr.

Noticed for the first time during Major Long's expedition. The type of this genus was briefly described by Say in the second volume of Long's Narrative, and there called *Agama collaris*, in allusion to the very striking feature of bearing a double black sub-crescentic band on the sides of the neck. The genus *Crotaphytus* was first established by Dr. Holbrook, and is characterized by the presence of small, polygonal plates on the whole surface of the head. The odd occipital plate itself is inconspicuous; the auditive apertures are very broadly open. Teeth are found on the jaws and palate. There is a fold of the skin under the throat; the head is large and sub-triangular; the body covered with minute scales; and the tail very long and tapering. Femoral pores present.

This genus now includes four North American species; three we have lately described under the names of *C. Wislizenii*, from New Mexico, *C. Gambelii*, from California, and *C. dorsalis*, from the desert of Colorado; the fourth is the following:

2. CROTAPHYTUS COLLARIS, Holbr.

SPEC. CHAR.—Tail conical, very long and tapering; head large, sub-triangular, rounded at the snout; two subcrescentic black bars, margined with white on each side of the neck; the largest extends from the origin of the fore-legs to near the dorsal line; the second of these black bars is smaller, and situated between the latter and the head.

SYN.—*Crotaphytus collaris*, Holbr. N. Amer. Herp. II, 1842, 79. Pl. x.

Agama collaris, Say, in Long's Exp. Rocky Mts. II, 1823, 252.—Harl. Med. and Phys. Res. 1835, 142.

OBS.—The specimens on hand exhibit several varieties of coloration worthy of being noticed: thus two specimens from Gypsum Bluffs, on Red river—a rocky locality—present a green ground-color above, with large blue patches and bright yellow spots; underneath light-green, almost uniform, except under the head, which is deeper and provided with blue, irregularly elongated spots; another specimen from the same locality has brown as the predominating tint. Light-brown stripes are seen on the legs and tail; similar spots on the body and head; four rows of red spots on the back; belly light-brown; light reddish-brown under the tail and feet.

Specimens from the head of the south fork of Red river have either a bluish-gray back, with white spots, a bluish-white belly, and the inferior surface of fore-legs reddish, or else the back is yellow and green.

The above memoranda, on the coloration, were taken on the spot by Capt. Marcy. The general distribution of color appears to indicate sexual differences: thus all the specimens before us in which the spots have a tendency to arrange themselves in transverse bands, or even where transverse narrow bands take the place of the spots, have proved to be females. The ground-color, however, varies in both sexes.

III. HOLBROOKIA, Girard.

The genus *Holbrookia* bears a striking resemblance to the one just described; it has the same general form, the same sub-triangular head, covered with small polygonal plates, a fold under the throat, small scales on the back, and femoral pores. The tail is perhaps smaller in proportion to the size of the body. The absence of an external auditive aperture will, however, at once characterize it generically from all its allies. The absence of teeth on the palatine bones is another organic character by which the genus *Holbrookia* can be distinguished from *Crotaphytus*. From *Homalosaurus* it differs only by the absence of an external auditive aperture.

The species upon which the genus was originally based is the one collected by the expedition.

Three other species were found in Texas, and described by us under the names of *H. affinis*, *propinqua*, and *texana*, (see Proceedings of the Academy of Natural Sciences of Philadelphia, August, 1852).

3. Holbrookia maculata, Girard.

Spec. char.—Above light-brown, with two dorsal series of irregularly crescent-shaped black spots convex posteriorly, and provided with an olivaceous margin; flanks with small crowded yellowish or reddish spots; two, occasionally three, deep-bluish black spots on the sides of the abdomen; beneath unicolor, either of a soiled white or yellow tint; sometimes irregular bluish vittæ under the head.

Syn.—*Holbrookia maculata*, Girard, Proc. Amer. Assoc. Adv. of Sc. IV, (1850), 1851, 201; and in Stansbury's Expl. Valley of Great Salt Lake, 1852, 342. Plate vi, fig. 1—3.

Obs.—A full description, as well as a figure of this species, may be found in the Report of Captain Howard Stansbury on the Valley of the Great Salt Lake of Utah; rendering it unnecessary to reproduce either here.

Numerous specimens were collected on the Canadian river and surrounding localities.

IV. SCELOPORUS, Wiegm.

The genus has the general appearance of *Holbrookia*, but is provided with large auditive apertures, large imbricated and carinated scales on the back in most instances; and smooth scales on the belly. The subguttural fold of the former, however, is not to be seen here and on the surface of the head; the plates, though small, are larger, especially the occipital. There are no teeth on the palate.

Most of the species of this genus are Mexican: one is common in the United States, and known as the brown or fence lizard. A second species was discovered by Captain Stansbury in the valley of the Great Salt Lake. Another species inhabits the western States, and a fourth is peculiar to Oregon.

4. Sceloporus consobrinus, B. & G.

Zoology, Pl. X, Figs. 5–12.

Spec. char.—Ground color above brownish, with a series of small black spots, eight or ten in number, on each side of the dorsal line. A yellowish stripe outside of the spots, and a black band beneath the stripe. A greenish area between the black band above and the elongated blue patch on the sides of the abdomen. Beneath, greenish blue.

Desc.—This species bears a close relationship to *S. graciosus*, from the valley of the Great Salt Lake of Utah, in the description of which the remarkably large size of the dorsal scales was mentioned as constituting one of its most distinguishing features, when compared to *S. scalaris*. In the present species the dorsal scales are proportionally still larger than in *S. graciosus*. Its body and head are also more slender and narrower. The tail is more tapering and elongated, and constitutes almost three-fifths of the total length. In coloration the differences between the two species are very striking.

The head is subelliptical, depressed, declive towards the snout, which is rather pointed. The superciliary region is but slightly raised above the plane of the vertex. The rostral plate is subtriangular, very low, and elongated transversely. The nostrils are almost circular, situated in the middle of a small plate, separated from the rostral by two small intervening ones. There are ten or twelve internasal and very small plates, and nine somewhat larger frontals, the middle one the largest. There are two verticals (or frontals), the anterior one the largest. The occipital is large and pentagonal, surrounded by four or six smaller plates, two anterior of medium size, contiguous to the postvertical, two lateral, larger and triangular, exteriorly to which two smaller ones may be observed. There are three or four subhexagonal, transversely elongated, plates on the superciliary region, surrounded internally by one row and externally by two rows of minute plates. The superciliary edge is formed by five sharp and imbricated thin plates; it is continued in the shape of a ridge to the nostril by means of two sharply keeled plates. The suborbitals are two in number, the posterior one much the longest. There is a small loral.

The plates which line the jaws are subquadrangular, very narrow and elongated, four above and five below, the latter considerably larger. Above the series of plates of the upper jaw, and between the suborbitals, two series of small and irregular plates may be observed. Four or five inframaxillary plates constitute a series on each side of the lower surface of the head, joined anteriorly by the subpentagonal symphysal plate. Between the inframaxillary series and the series lining the lower jaw exists a series of four or five elongated and small plates.

The auditive aperture, which is proportionally large, is oval, and almost vertical in its longest diameter. At its anterior margin may be seen two or three scales, larger and more pointed than those on the temporal region. Behind the auditive aperture, and situated obliquely on the neck, is a slight fold of the skin.

The neck is somewhat contracted, the body slender and depressed, with the back, however, slightly arched, and the belly flat. The tail, as already mentioned, is quite long and slender, depressed at its base, and hence conical towards the tip.

The fore-legs, when stretched backwards along the sides, extend nearly to the groin, while the hind-legs, when brought forwards, reach almost to the ear.

The scales are imbricated and keeled on the back and sides. Their general shape is that of a lozenge, terminated posteriorly by an acute spine. There are ten longitudinal rows along the back, with five on each side, which are somewhat oblique, and smaller. Underneath, the scales are smooth, posteriorly tricuspid on the belly, whilst under the head and throat they have but two posterior spines. The scales below as well as above the fore-legs are keeled. Those on the upper part of the hind-legs are also keeled, whilst on the thighs they are smooth. The fingers and toes are surrounded with carinated scales to their very tips.

The femoral pores, thirteen or fourteen in number on each side, are conspicuous and situated in the middle of one single small plate.

The black spots in the series along the back are comparatively small, and separated from each other by a space greater than their diameters. The yellow stripe extends from the origin of the neck to beyond the anus, the black from above the shoulder to the groin. The blue patch is elongated and narrow, terminated posteriorly by a black stripe which runs for a little distance along the thigh. There is an elongated black spot on the shoulder.

One specimen was collected on the 6th of June.

Plate X, fig. 5, represents the species in profile and of the natural size.

Fig. 6. The side of the head enlarged, to show more distinctly the structure of its plates.

Fig. 7. Head from above, enlarged in the same proportion as fig. 6.

Fig. 8. Head from below, enlarged.

Fig. 9. The right arm and fingers, seen from below.

Fig. 10. The right leg and toes, seen from below.

Fig. 11. Dorsal scales, enlarged four times.

Fig. 12. Scales from the belly, slightly enlarged.

V. CNEMIDOPHORUS, Wagl.

This genus is characterized by a bifid tongue; a double transverse fold of skin under the throat; teeth on the palate; maxillary teeth compressed, the posterior one tricuspid; femoral pores; broad plates under the thighs; fingers not carinated underneath; and a subcylindrical, very long and tapering tail. The body above is covered with minutely crowded scales; whilst on the belly there are eight longitudinal rows of subquadrangular, transversely elongated plates, or scutellæ. On the tail the scales are quite large and very conspicuous, strongly carinated and constituting circular rows or whorls.

The explorations of the last few years in Texas and New Mexico have brought to light several other species of the genus *Cnemidophorus*, all provided with eight longitudinal rows of abdominal scutellæ. These are *C. gracilis*, from the desert of the Colorado; *C. perplexus*, from the upper valley of the Rio Grande; *C. gularis*, *C. Grahamii*, and *C. marmoratus*, from different localities in Texas.

The discovery of *C. gularis* in Arkansas is an interesting fact in regard to its geographical distribution.

5. Cnemidophorus gularis, B. & G.

Zoology, Pl. X, fig. 1–4.

Spec. char.—Ground color brownish, with six longitudinal stripes, green or yellow; beneath yellowish white, unicolor. Scales on the subguttural fold quite large and conspicuous in proportion to those in other species.

Syn.—*Cnemidophorus gularis*, B. & G. Proc. Acad. Nat. Sci. Philad., vi, 1852, 128.

Desc.—This species is very closely allied to *C. sexlineatus*, having, like the latter, six longitudinal stripes, three on each side of the body, running from head to some distance along the tail. It has, also, the same general form; but on a close comparison it will soon be observed that the body is proportionally shorter, the limbs more developed, whilst the scales on the back appear to be actually larger. The head is proportionally smaller and narrower. But the most striking organic character consists in the presence of somewhat large and conspicuous scales on the margin of the subguttural fold of the skin. The following indications of color are derived from the notes of the Expedition: The upper surface of the head is reddish brown; three longitudinal yellow or greenish stripes extending from the head to the origin of the tail; the middle stripe on each side may be followed on the tail to a considerable distance. The dorsal space between the two uppermost stripes on each side is brown, or reddish brown, like the head above. The space between the uppermost and middle stripes is of a deep black, and extends from the upper angle of the orbit down to a certain distance along the tail. The space between the middle and lower stripes, and between the latter and the abdominal scutellæ, is green, or greenish brown. The legs are brownish red, and the belly white or bluish white. The tail underneath is yellowish red; above, brownish, or reddish brown.

Specimens were collected on the 5th and 6th of June.

Plate X, fig. 1, represents *Cnemidophorus gularis* of natural size.

Fig. 2. Head seen from above, to exhibit the plates.

Fig. 3 shows the scales on the subguttural fold and the hand from beneath, as well as the submaxillary plates.

Fig. 4 represents the femoral pores, the preanal plates, and also the plates at the inferior surface of the hind-legs, and the lower surface of the feet.

VI. LYGOSOMA, Gray.

This genus includes small scincoid lizards, the nostrils of which open in one single plate, the nasal. The supranasals are wanting. The palate is without teeth, and provided with a triangular notch situated far back. The scales, broader than long, are all smooth.

All the species of *Lygosoma* belong to the Old World, except the one here mentioned.

6. Lygosoma lateralis, Dum. and B.

Spec. char.—Upper part of head and body chestnut-brown; a black lateral band extending from the snout across the eye to a considerable distance along the tail. Flanks grayish-brown, with longitudinal indistinct, darker, interrupted vittæ. Abdomen yellowish, and tail beneath bluish; circumference of scales mottled with gray. Tail longer than the body. Limbs very small.

Syn.—*Scincus lateralis*, Say, in Long's Exp. Rock. Mts. II, 1823, 324.—Harl. Jour. Acad. Nat. Sc., V, 1827, 221, and VI, 1829, 12.—Holbr. N. Amer. Herp., first ed, I, 1836, 71. Pl. viii.

Scincus unicolor, Harl. Journ. Acad. Nat. Sc. Philad., V. i, 1825, 156.

Tiliqua lateralis, Gray, Syn. Rept., in Griff. Anim. Kingd., Cuv. IX, 1831, 70.

Lygosoma lateralis, Dum. and B., Erp. gen. V, 1839, 719. Holb. N. Amer. Herp, second ed., II, 1842, 133. Pl. xix.

This small and graceful species appears to be spread over a large portion of the United States. It is always met with running on the surface of the ground in forests, among dead leaves, never ascending either trees or shrubs like many other lizards.

The body is sub-quadrangular, the head continuous with it, and, like it, flattened above. The tail is sub-circular, tapering into a point. The plates of the head correspond with the descriptions which we have before us, except that the frontonasals are not contiguous, but separated by a small odd plate directly in advance of the vertical (sometimes called frontal). But this peculiarity of structure is not indicative of any specific difference.

The auditive apertures are large, circular, and their margin simple. The fore-legs, when extended forward, reach the eye. The hind-legs are a little longer and stouter than the fore-legs. The scales are perfectly smooth, uniform above and below, and disposed in thirty longitudinal rows around the body. The two middle preanal scutellæ considerably larger.

One specimen was procured near the mouth of Cache creek, on the 16th of May.

BATRACHIANS.

Of this order of reptiles only two species were procured—a toad and a frog.

1. Bufo cognatus, Say.

Zoology, Pl. XI.

Spec. char.—Greenish brown above, with a lighter yellowish dorsal line. Patches of blackish-brown scattered over the sides and legs. Beneath unicolor of a dingy yellow. Head short, groove on its upper surface, not extending to the anterior rim of the eye.

Syn.—*Bufo cognatus*, Say, in Long's Exp. to Rock. Mts. II, 1823, 190.

Obs.—It is not without hesitation that we have referred the present species to *Bufo cognatus;* the description of Say as cited is exceedingly brief, applying almost equally well to several allied species. The colors of our specimen vary considerably from the *B. cognatus* as described by Say; but the characters of the groove of the crown agree better. The mark of "head with a groove which hardly extends anteriorly to the line of the anterior canthus of the eye," although not strictly in accordance with our species, may, with some allowance, be made to answer to it. It is much to be regretted that the original specimen of Say was destroyed in the conflagration of the Philadelphia Museum, and thus all hopes of identification are lost. If, however, further explorations in Arkansas should yield many additional specimens, all differing as much as the present from Say's description, it will become necessary to assign a new name to it, especially if the true *B. cognatus* be at the same time detected.

Description.—The head is very short, the snout obtuse and truncate, with the nostrils subterminal. Upper surface of head grooved; groove subelliptical and short, not extending anteriorly to the anterior rim of the eye (fig. 2). The superciliary ridges thicken from before backwards, extending to the tympanum in passing obliquely behind the eyes, and in contact also with the parotid glands, which are subovoidal and of medium size. Tympanum rather small, subelliptical; its longest diameter almost vertical. The fore and hind legs are well proportioned to the size of the body. The under surface of the hand is provided with small crowded tubercles, a more conspicuous and a larger one at the articulations of the fingers; the fingers themselves are depressed or flattened. A larger disc-like knob is observed on the middle and at the base of the hand (fig. 3).

The toes (fig. 4) are but slightly webbed, and, like the fingers, depressed. The fourth is conspicuously the longest, and the third a little longer than the fifth. The under surface of the feet (fig. 5) is covered with smaller tubercles than those of the hands. A large spade-like process exists at the base of the first or inner toe, exteriorly to which, and at the base still of the metatarsus, is a small knob-like tubercle. The body is thickly covered with papillæ, with some large ones more conspicuous along the sides of the back; on the flanks they are smaller, similar to those of the intervening spaces on the back; on the abdomen the papillae are smaller still; upper part of hands and feet minutely granulated. The snout alone is smooth.

The dark patches scattered over the upper part of the animal are margined with a light yellowish line. Sinuating yellowish lines may be observed on the sides of the belly, or flanks and legs. A rather large spot is seen beneath the eye, and another in advance and beneath the tympanum near the angle of the mouth.

One specimen procured near the Water-hole between Camps 6 and 7.

Plate XI, fig. 1, represents *Bufo cognatus* of natural size.

Fig. 2. The head from above.

Fig. 3. Left hand seen from below.

Fig. 4. Right foot from above.

Fig. 5. Right foot from below.

2. Rana pipiens, Latr.—Bullfrog.

Spec. char.—Toes webbed to their extremity, fourth toe one fourth longer than the third and fifth. An elongated tubercle at the base of the first toe; sub-articular tubercles of fingers and toes but slightly developed. Vomerine teeth on two rounded and separated elevations situated between the internal nostrils. Diameter of tympanum (in the specimen before us) greater than the diameter of the eye.

Syn.—*Rana pipiens*, Latr. Hist. Nat. Rept. II, 1802, 153. Harl. Amer. Jour. Sc. X, 62. Med. & Phys. Res., 1835, 101; and Jour. Acad. Nat. Sc. Philad. V, 1827, 335. Holbr. N. Amer. Herp. IV, 1842, 77. Pl. xviii.

Rana mugiens, Merr. Tent. Syst. Amph. 1820, 175. Dum. & B. Erp. gen. VIII, 1841, 370.

The bullfrog is quite a common animal in the United States, though its northern, western, and southern limits are not yet accurately known.

A large specimen was found in a cold spring near the head of the south fork of Cache creek, in the Witchita mountains. The upper parts of body and limbs are covered with warty eminences, more crowded on the body. These warts are perfectly smooth, like the skin itself. The ground-color is greenish brown above, with crowded deep brown or blackish spots. Beneath, dull yellow, with clouded bluish patches. The lower surface of the feet has the same marmorated appearance as the back. The jaws and snout are greenish brown, and perfectly smooth.

The specimen before us is remarkable for the size of its tympanum, which is much larger than the eye.

J.H. Richard. Ackerman Lith. 379 Broadway N.Y. W.H.Dougal.

CROTALUS CONFLUENTUS, Say

J.H.Richard. Ackerman Lith. 379 Broadway N.Y. W.H.Dougal.

EUTAENIA PROXIMA, B. & G.

J. H. Richard — Ackerman Lith. 379 Broadway N.Y. — W. H. D'ougal

EUTAENIA MARCIANA, B. & G.

J. H. Richard. W. H. Dougal.

HETERODON NASICUS, B. & G.

J. H. Richard. W. H. Dougal.

PITUOPHIS MC'CLELLANII, B. & G.

J. H. Richard. W. H. Dougal.

SCOTOPHIS LAETUS, B. & G.

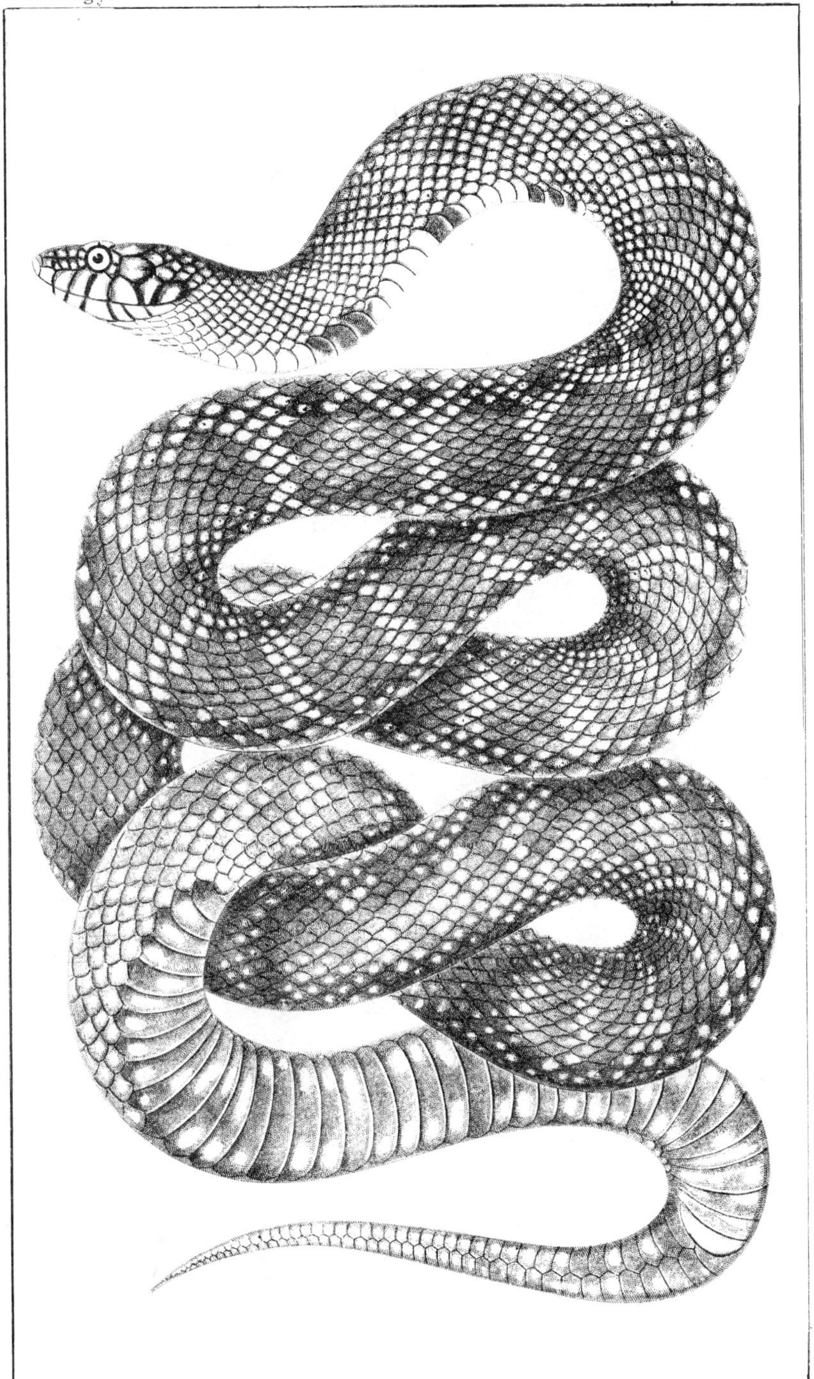

J.H.Richard. W.H.Dougal.

OPHIBOLUS SAYI, B. & G.

J.H.Richard. W.H.Dougal.

OPHIBOLUS GENTILIS, B. & G.

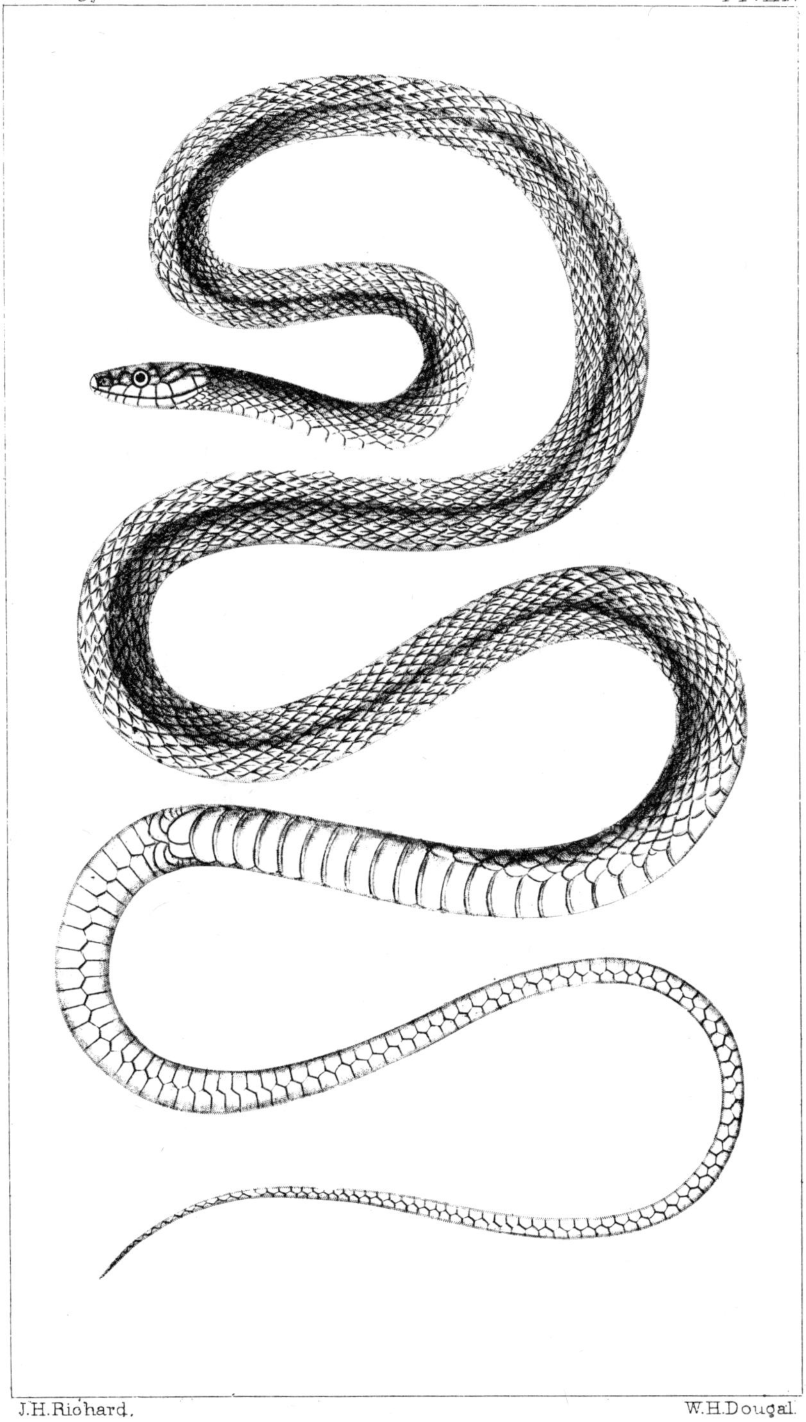

J.H.Richard. W.H.Dougal.

LEPTOPHIS MAJALIS, B. & G.

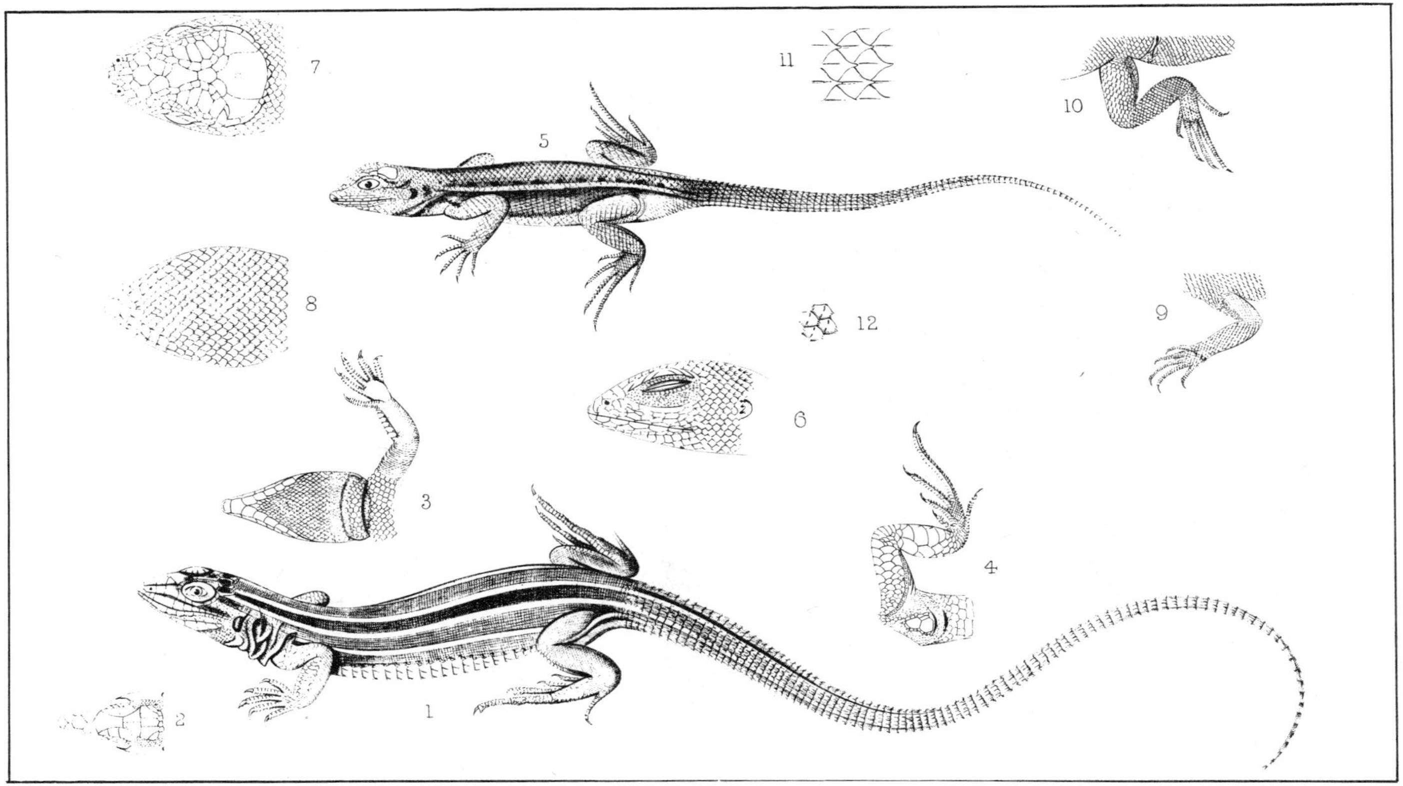

J.H.Richard. Ackerman Lith 379 Broadway N.Y. W.H.Dougal.

Figs. 1-4. CNEMIDOPHORUS GULARIS, B. & G. Figs. 5-12. SCELOPORUS CONSOBRINUS, B. & G.

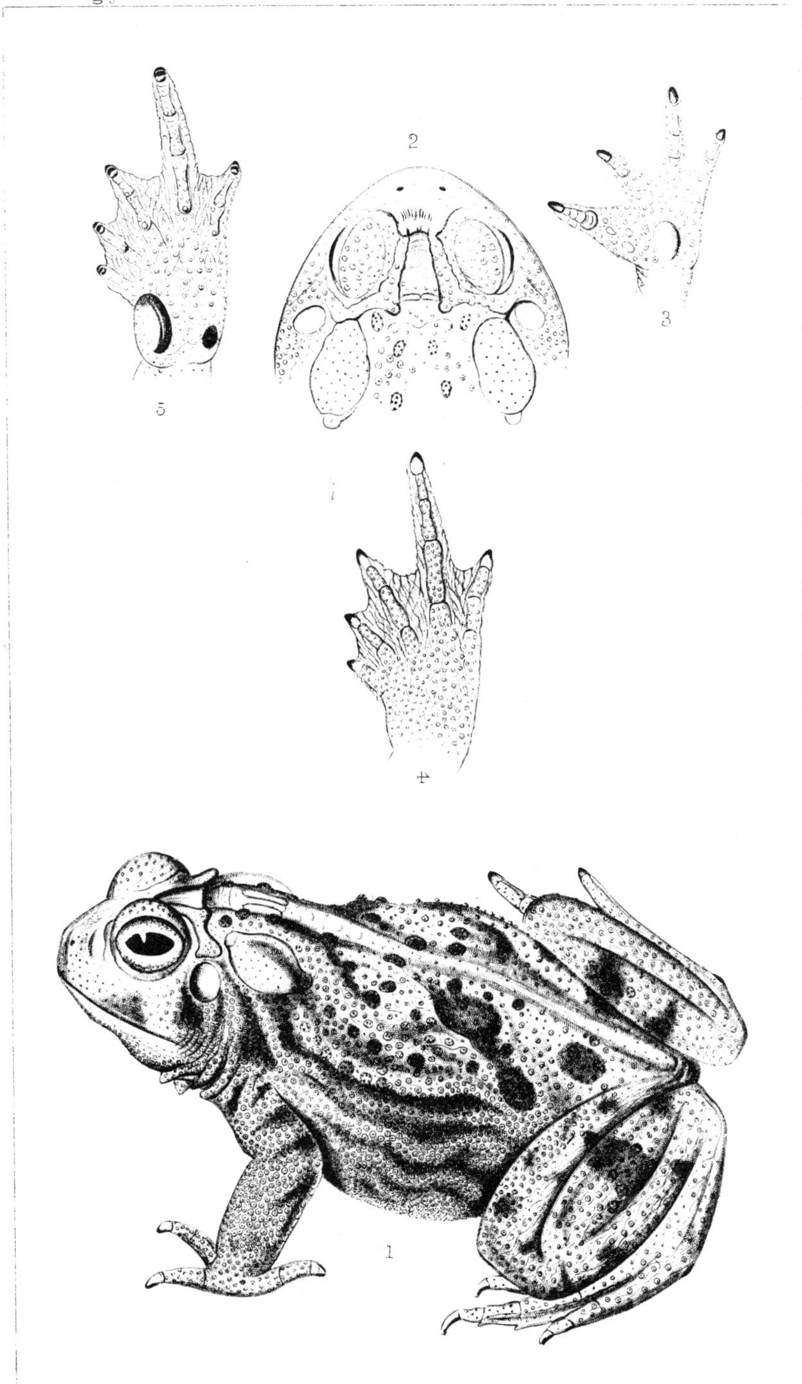

J.H.Richard. Ackerman Lith. 379 Broadway N.Y. W.H.Dougal.

BUFO COGNATUS, Say.

REPORT ON REPTILES COLLECTED ON THE SURVEY

S[pencer] F[ullerton] Baird

No. 4.

REPORT ON REPTILES COLLECTED ON THE SURVEY.

BY S. F. BAIRD.

SCELOPORUS GRACIOSUS, B. & G.

Sceloporus graciosus, B. & G. Pr. A. N. Sc. VI, April, 1852, 69.—IB. Stansbury's Expl. Great Salt Lake, 1852, 346; pl. v, fig. 1.

Sceloporus gracilis, B & G. Pr. A. N. Sc. VI, Oct. 1852, 175.—GIRARD, Herp. U. S. Ex. Ex. 1858, 386; pl. xx figs. 1—9.

2832. Upper Pit river. Dr. Newberry.

SCELOPORUS OCCIDENTALIS, B. & G.

Sceloporus occidentalis, B. & G. Pr. A. N. Sc. VI, Oct. 1852, 175.—GIRARD, Herp. U. S. Ex. Ex. 1858, 383; pl. xix.

Sceloporus frontalis, B. & G. Pr. A. N. Sc. VI, Oct. 1852, 175.—GIRARD, Herp. U. S. Ex. Ex. 1858, 384; pl. xix, figs. 1—7.

2866. Upper Willamette valley. Dr. Newberry. 2838. Benicia, California. Do.

TAPAYA DOUGLASII, Girard.

Agama douglasii, BELL, Trans. Linn. Soc. Lond. XVI, 1833, 105; pl. x.

Phrynosoma douglasii, HOLBROOK, N. Am. Herp. II, 1842, 101; pl. xiv.

Tapaya douglasii, GIRARD, Herp. U. S. Ex. Ex. 1858, 398; pl. xxi, figs. 1—5.

219. Upper Klamath valley. 220. Klamath Lake. Dr. Newberry.

ELGARIA SCINCICAUDA, B. & G.

Tropidolepis scincicauda, SKILTON, Am. Jour. Sc. VII, 1849, 202; plate —, figs. 1—3.

Elgaria scincicauda, B. & G. Pr. An. S. VI, Ap 1852, 69.—IB. Stansbury's expl. 1852, 348; pl. iv.—GIRARD, Herp. U. S. Ex. Ex. 1845, 210; pl. xxiii, figs. 1—9.

3100. Bodega, California. Dr. Newberry.

PLESTIODON SKILTONIANUS, B. & G.

Plestiodon skiltonianus, B. & G. Pr. A. N. Sc. VI, April, 1852, 69.—IB. Stansbury's Rep. 1852, 349; Reptiles, pl. iv, figs. 4—6.

Eumeces quadrilineatus, HALLOWELL, Pr. A. N. Sc. VII, June, 1854, 94.

3168. California. Dr. Newberry. 3148. Pit river, California. Do.

CROTALUS LUCIFER, B. & G.—Oregon Rattlesnake.

PLATE XI.

Crotalus lucifer, B. & G. Pr. A. N. Sc. VI, 1852, 177.—IB. Catal. N. Am. Serp. 1853, 6.—GIRARD, Herp. U. S. Ex. Ex. 1858, 187, pl. xv. figs. 1—6.

SPEC. CHAR.—Head short, rather broad and deep, angles much rounded; nose less pointed than in *C. confluentus*. Labials above 15–16; below 15–16. Dorsal rows 25. A dorsal series of brown blotches not margined by whitish; two small, irregular lateral series on each side. Posteriorly 15 or 20 half rings, becoming blackish in old specimens. Light stripe from orbit below superciliary to angle of jaw above upper labials Light stripe before the eye expanded upon the whole of the upper labials and front and sides of face below the crown and in front of the orbits. Single light transverse line on superciliaries often obsolete Rostral not edged with lighter.—(Kennicott.)

No. 379. Lost river, Oregon Territory. Dr. Newberry.

EUTAENIA INFERNALIS, B. & G.

? *Coluber infernalis*, BLAINVILLE, Nouv. Ann. du Mus. III, 1834, 59, pl. xxvi, fig. 3.

Eutaenia infernalis, B. & G. Catal. N. Am. Serp. 1853, 26 —GIRARD, Herp. U. S. Ex. Ex. 1858, 148, pl. xiv, figs. 11—16.

SPEC. CHAR —Most slender of all the species of the section. Head and eye large. Above black; a series of about 110 triangular reddish yellow spots, confluent with the indistinct lateral stripe, itself confluent with the greenish white sides and abdomen. Lateral stripe on second and third lateral rows. Dorsal rows 19.

No. 893. Fort Reading, California. Dr. Hammond.

EUTAENIA CONCINNA, B. & G.

Tropidonotus concinnus, HALLOWELL, Pr. A N. Sc. VI, 1852, 182.

Eutaenia concinna, B. & G. Catal. N. Am. Serp. 1853, 146.

SPEC. CHAR.—Dorsal rows of scales 21, all carinated. Black, with a dorsal light stripe, and the usual lateral stripes replaced by a series of distinct salmon-colored spots.

No. 937. Portland, Oregon. Dr. J. S. Newberry.

EUTAENIA ELEGANS, B. & G.

Eutaenia elegans, B. & G. Catal. N. Am. Serp. 1853, 35.

Tropidonotus trivittatus, HALLOW. Pr. A. N. Sc. VI, Jan. 1853, 237.

SPEC. CHAR.—Body moderately stout, compact, and cylindrical; stouter than *E. pickeringii*, and the tail proportionally shorter. Head short, labials largely developed, eight above, sixth largest. Dorsal rows 19–21. Scales smaller than in *E. pickeringii*. In color strongly resembles *E. faireyii*, but with the lateral stripes differently situated. Uniform black above. Uniform light greenish below, not darker posteriorly. A broad light yellow dorsal stripe distinct to the end of the tail; the lateral stripe, on the second and third rows, broad and well defined of a greenish white color.—(Kennicott.)

No. 879. Canoe Creek, Oregon Territory.—831. Willamette valley, Oregon Territory.—884. Pit River valley, California. Dr. J. S. Newberry.—No. 878. Fort Reading, California. Dr. Hammond.

EUTAENIA COUCHII, Kennicott.

SPEC. CHAR.—Body moderately stout. Tail less than one-fourth the total length and very small. Head exceedingly elongated and narrow, more so than in any other of the genus. Snout very long, narrow, and pointed. Crown plane above; eye large; 8 upper labials, 6 largest, 7 scarcely smaller. Dorsal rows 21; outer row large, very faintly carinated; the second smaller but broader than the third; the rest rather large, elongated, and strongly carinated. Dorsal stripe narrow and very indistinct. An indistinct dull whitish lateral stripe on second and third rows. Dark olive brown or black above the lateral stripe. Abdomen dark ashy olive or black.—(Kennicott.)

No. 866. Bank of Pit river, California. Dr. J. S. Newberry.

PITYOPHIS CATENIFER, B. & G.

Coluber catenifer, BLAINVILLE, Nouv. Ann. Mus. III, 1834; pl. xxvi, fig. 2.
Pituophis catenifer, B. & G. Catal. Am. Serp. 1853, 69.—GIRARD, Herp. Ex. Ex. 1858, 135; pl. viii, figs. 1—7.

SPEC. CHAR.—Head elongated, conical anteriorly, narrow; snout obtuse. Rostral very broad, vertical plate pentagonal elongated, narrow, broader anteriorly. The ante-orbitals 2, post-orbitals 3. Tail forming $\frac{1}{7}$ to $\frac{1}{6}$ the total length. Dorsal rows of scales 29—31; 7 to 9 of the external rows smooth. Ground color yellowish, deepest on the back. A series of about 60—70 sub-quadrate blotches from head to anus, 3 or 4 smaller series on each side. Two or three of the lateral series perfect to the tail, one continued on it. The lateral blotches not assuming the form of a series of vertical bars posteriorly. Fifteen or twenty broad, short transverse bands on the tall which are separated from the sub-caudal scuttellae by the width of two or three scales.—(Kennicott.)

No. 1423. Willamette valley, Oregon. 1810. Banks of Pit river, California. Dr. J. S. Newberry.

OPHIBOLUS BOYLII, B. & G.

Ophibolus boylii, B. & G. Catal. N. Am. Serp. 1853, 82.
Coronella balteata, HALLOW. Pr A. N. Sc. VI, Jan. 1853, 236.

SP. CH.—Black, with upwards of 30 broad ivory white transverse bands widening on the sides. Dorsal rows of scales 23.

1703. Fort Reading, California. Dr. Newberry.

MASTICOPHIS TAENIATUS, B. & G.

Leptophis taeniata, HALLOWELL, Pr. A. N. Sc. VI, 1852, 181.
Masticophis taeniatus, B. & G. Catal. Serp. 1853, 103.

SP. CH.—A broad brown dorsal stripe margined by a darker line. The four outer rows of scales on each side yellow, with a dark line through the centre of each. A dark line along the edge of the abdomen, making six dark lines on each side. Beneath yellowish. Dorsal rows 15. Tail about one-third the total length.

1983. Canoe creek, California. Dr. Newberry.

BASCANION VETUSTUS, B. & G.

Bascanion vetustus, B. & G Catal. N. Am. Serp. 1853, 97.—GIRARD, Herp. U. S. Ex. Ex. 1858, 127; pl. viii, figs. 12—19.

SP. CH.—Stouter than *B. flaviventris*. Breadth of head across superciliaries equal to half the interval between tip of rostral and posterior end of occipitals. Lower postorbital in notch between the fourth and fifth labials. Olive-brown above, greenish white beneath.

No. 1765. Pit river valley, California. 1785. Klamath Lake, Oregon. Dr. Newberry.

DIADOPHIS PULCHELLUS, B. & G.

Diadophis pulchellus, B. & G. Catal. N. Am. Serp. 1853, 115.

SPEC. CHAR.—Body above uniform dark greenish brown: beneath orange with small black spots scattered irregularly all over from head to tail; these spots sometimes absent. Color of abdomen extending upon lateral dorsal scales. Occipital ring broad. Dorsal scales in 15 rows.

No. 2071. California. Dr. Newberry.

WENONA PLUMBEA, B. & G.

Wenona plumbea, B. & G. Pr. A. N. Sc. VI, 1852, 176.—IB. Catal. N. A. Serp. 1853, 139.—GIRARD, Herp. U. S. Ex. Ex, 1858, 112; pl. vii, figs. 1—7.

SPEC. CHAR.—Uniform bluish lead color above; uniform yellowish white beneath. Three pairs of frontal plates. Middle pair united with the loral, and thus extending to the labials. Labials not entering into the orbit. Dorsal scales in 45 rows.

Upper Willamette valley, Oregon. Dr. Newberry.

BUFO COLUMBIENSIS, B. & G.

Bufo columbiensis, B. & G. Pr. A. N. Sc. VI, 1853, 378.—GIRARD, Herp. U. S. Ex. Ex. 1858, 77; pl. v, figs. 4—9.

SPEC. CHAR.—Upper surface of head plane, skin adhering to the skull and granulated. Parotids and tympanum small. A membranous tarsal fold. Toes palmated. Color light greenish with black blotches and reddish spots. A dorsal white vitta. An oblique dark patch beneath the eye. Beneath soiled yellow, spotted.—(Girard.)

2578. Upper Pit river.—2580. Upper Klamath. Dr. Newberry.

SCAPHIOPUS HAMMONDII, Baird.

PLATE XXVIII, FIG. 2.

SPEC. CHAR.—Tongue very large, orbicular, without notch behind. Spade highly developed. Color above dark olive brown, with very indistinct blotches of darker. Summits of dorsal pustulation whitish. Beneath whitish, the chin black. Head and body two inches long. Hind leg 2½.

This species is easily distinguished by its nearly uniform and very dark color, without the light lines of *S. holbrookii*, figured for comparison on figure 1 of the same plate. The tongue appears to be much broader.

3695. Fort Reading, California. Dr. J. F. Hammond, U. S. A.

HYLA REGILLA, B. & G.

PLATE XXVIII, FIGS. 3.

Hyla regilla, B. & G. Pr. A. N. Sc VI, 1852, 174.—IB. 1853, 301.—GIRARD, Herp. U. S. Ex. Ex. 1858, 60; pl. iii, figs. 13—18.

Hyla scapularis, HALLOWELL, Pr. A. N. Sc. V, 1852, 183.

SPEC. CHAR.—Olive or ashy green, coarsely marked on each side with blotches in two longitudinal series, bordered on either side with scattered smaller and more rounded ones. A triangular blotch between the eye. A dark narrow line from snout to eye; a broad postocular vitta to the arm, beneath which is a bar of grayish white about half the width. Dorsal blotches sometimes wanting.

3229. Yreka, California, and 3231, Klamath lake. Dr. Newberry.

RANA BOYLII, Baird.

Rana boylii, BAIRD, Pr. A. N. Sc. VII, April, 1854, 62.

SPEC. CHAR.—A broad depressed ridge of skin on each side of back. Skin finely tubercular above. Head broader than long. Tympanum scarcely evident, pustulated. Tibia more than half the length of body; hind foot less than half this length; webbed entirely to the horny tips; outer toe decidedly longer than the third. An elongated tubercle at base of inner toe, with another opposite to it. Above dull reddish olivaceous, with indistinct blotches on the back, and fasciae on the legs. Beneath yellowish, mottled anteriorly. Two inches long.

Sacramento river. Dr. Newberry.

AMBLYSTOMA CALIFORNIENSE, Gray.

PLATE XXX, FIGS. 1, 2, and 3.

Amblystoma californiense, GRAY, Pr. Zool. Soc. Lond. Jan. 1853, 11; pl. vii.

SPEC. CHAR.—Black. Sides of lips, lower part of neck, body and tail, and limbs, with large white spots. Palatine teeth in an elongated angular transverse line, bent forwards in the middle and extending to the outer edge of the hinder part of the internal nostril.—(Gray.)

Petaluma, California. E. Samuels.

ANAIDES LUGUBRIS, Baird.

PLATE XXX, FIG. 4.

Salamandra lugubris, HALLOWELL, Pr. A. N Sc. IV, 1848, 26.
Anaides lugubris, BAIRD, Iconographie Encyclopedia, II, 1849, 256.—GIRARD, Herp. U. S. Ex. Ex. 1858, 8; pl. i, figs. 26—33.

Columbia river. Dr. Newberry.

SIREDON GRACILIS, Baird.

PLATE XLIV, FIG. 2.

SPEC. CHAR.—Body slender; head short; branchial arches almost vertical, their pectination coarse and distant. General color reddish brown, marmorated and blotched, (almost vermiculated,) obscurely with blackish; quite distinctly so on the belly. Length 6½ inches. Head and body about 3½.

This species differs from *S. lichenoides*[1] in the variegation of reddish brown and dark brown in nearly equal proportions. The body is more slender; the branchial arches more perpendicular to the axis of the body. The feet are quite well developed. The form is much more slender than in *S. mexicanus*, the limbs stouter, the color quite different.

4080. Cascade mountains, near latitude 40°. Dr. Newberry.

[1] A full grown specimen of *S. lichenoides*, from the Rocky mountains, is figured on the same plate for comparison.

b

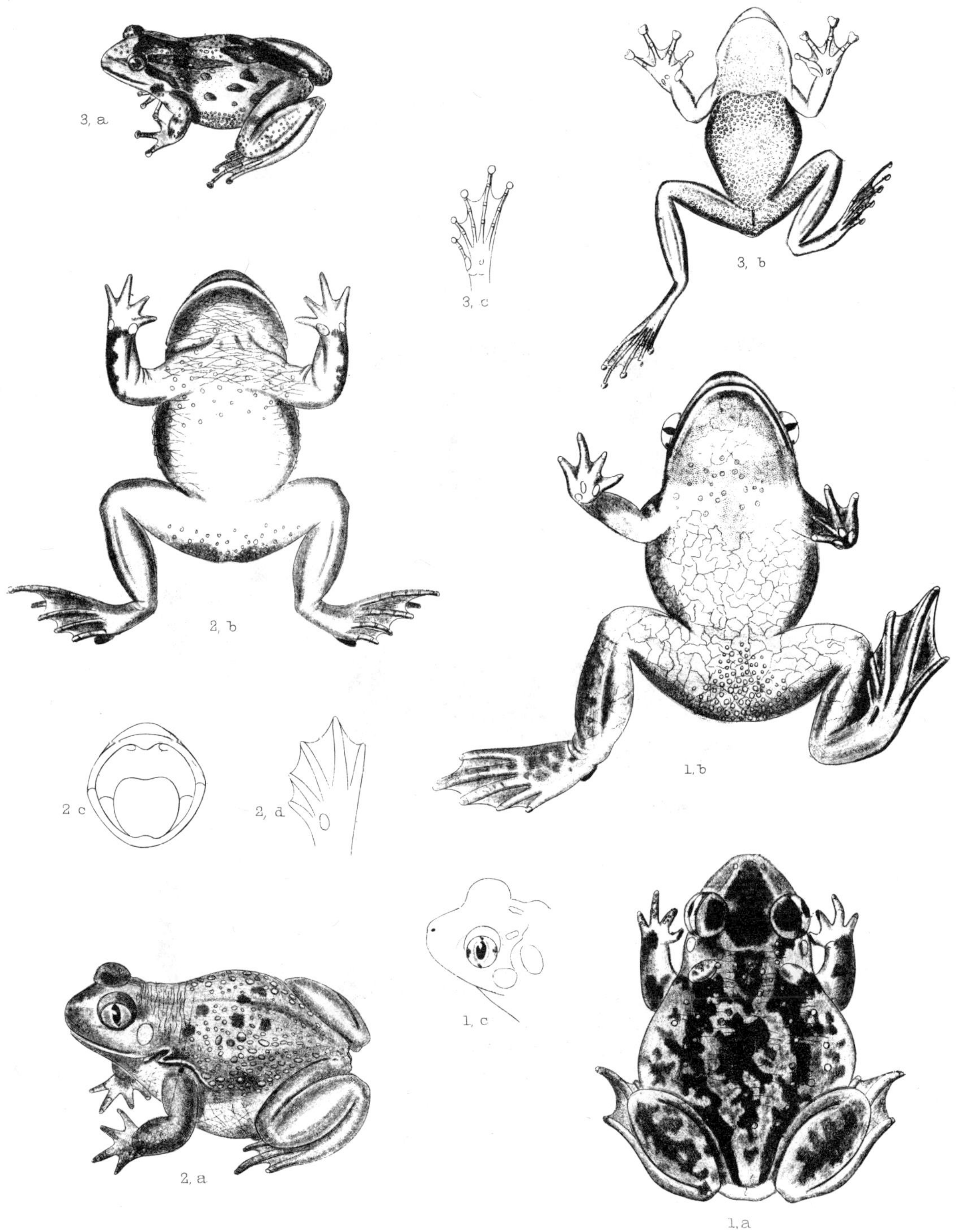

J. H. Richard del. C. Metzeroth sc.

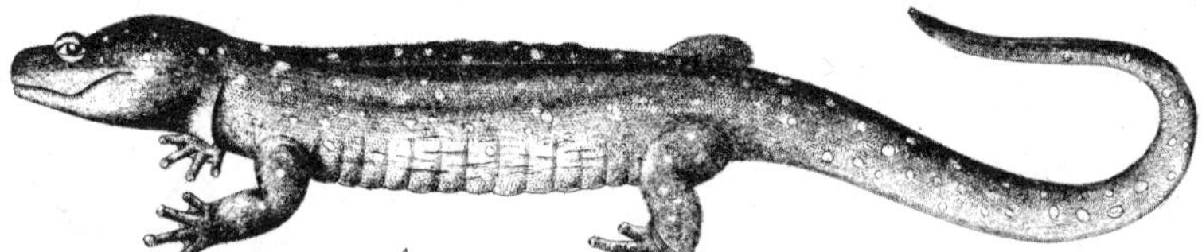

4

3

2

1

J.H. Richard del.

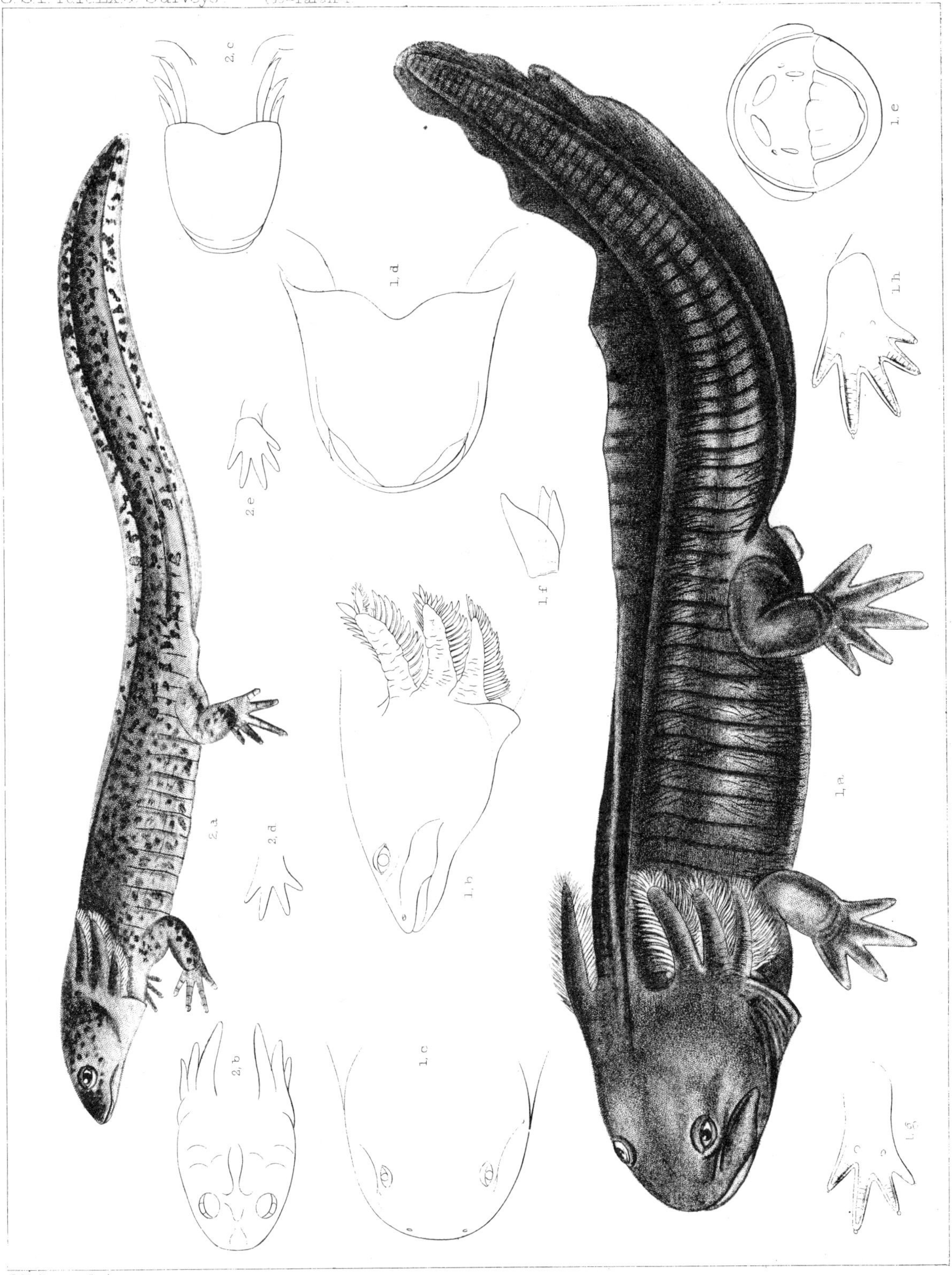

J. H. Richard del.

REPORT ON REPTILES
COLLECTED ON THE SURVEY

S[pencer] F[ullerton] Baird

No. 3.

REPORT ON REPTILES COLLECTED ON THE SURVEY.

BY S. F. BAIRD.

SCELOPORUS GRACIOSUS, B. & G.

Sceloporus graciosus, B. & G. Pr. A. N. Sc. VI, April, 1852, 69.—Ib. Stansbury, Great Salt Lake, 1852, 346; pl. v, fig. I.
Sceloporus gracilis, B. &. G. Pr. A. N. Sc. Oct. 1852, 175.—Girard, Herp. U. S. Ex. Ex. 1858.

2842. Valley of Salt Lake City.

?SCELOPORUS OCCIDENTALIS, B. &. G.

Sceloporus occidentalis, B. & G. Pr. A. N. Sc. VI, Oct. 1852, 175.—Girard, Herp. U. S. Ex. Ex. 1858, 383; pl. xix, figs. 8—14.

2837. Salt Lake valley. The specimen is not in very good condition, and may possibly belong to the next species.

?SCELOPORUS LONGIPES, Baird.

Sceloporus longipes, Baird, Pr. A. N. Sc. Dec. 1858.

2863. Salt Lake valley. 2863. Salt Lake.

CROTAPHYTUS COLLARIS, Holbrook.

Plate XXIV, Fig. 1.

Agama collaris, Say, Long's Exped. II, 1823, 252.
Crotaphytus collaris, Holbrook, N. Am. Herp. II, 1842, 79; pl. x.

2729, 2697, 2721. Upper Arkansas to Rocky mountains. These specimens exhibit the usual variety of coloring of the species.

CROTAPHYTUS WISLIZENII, B. &. G.

Crotaphytus wislizenii, B. & G. Pr. A. N. Sc. VI, April, 1852, 67.
Crotaphytus gambelii, B & G. Pr. A. N. Sc. VI, Aug. 1852, 126.
Crotaphytus fasciatus, Hallow. Pr. A. N. Sc. VI, Dec. 1852, 207.

2698. Salt Lake Basin. 2695. Head of Humboldt. This species was found to be quite abundant in Utah.

CALLISAURUS VENTRALIS, Baird.

Homolosaurus ventralis, Hallowell, Pr. A. N. Sc. VI, Oct., 1852, 179.
Callisaurus ventralis, Baird, Herp. Mex. Bound. Surv. 1859.

2803. Salt Lake Basin. This specimen is more northern in its locality than any collected by the other expeditions.

HOLBROOKIA MACULATA, Girard.

Holbrookia maculata, GIRARD, Pr. Am. Assoc. IV, for 1850, 1851, 201.—IB. Stansbury's Report, 1852, 342; pl. vi, figs. 1—3.

2657. Salt Lake Basin. This was found to be exceedingly abundant along the Arkansas.

TAPAYA BREVIROSTRIS, Girard.

Tapaya brevirostris, GIRARD, Herp. U. S. Ex. Ex. 1858, 397.

207. Rocky mountains, near 38°.

TAPAYA DOUGLASSII, Girard.

Agama douglassii, BELL, Trans. Linn. Soc. XVI, for 1828, 1833, 105.
Phrynosoma douglassii, WAGLER, Naturl. Syst. Amph. 1830, 146.—HOLBROOK, N. Am. Herp. II, 1842, 101; pl. xiv
Tapaya douglassii, GIRARD, Herp. U. S Ex. Ex. 1858, 398; pl. xxi, figs. 1—5.

229. Utah Basin.

DOLIOSAURUS PLATYRHINOS, Girard.

Phrynosoma platyrhinos, GIRARD, Stansbury's Ex. 1852, 361; pl. vii, figs. 1—5.
Doliosaurus platyrhinos, GIRARD, Herp. U. S. Ex. Ex. 1858, 407.

190. Salt Lake valley.

CNEMIDOPHORUS TESSELATUS, Baird.

Ameiva tesselata, SAY, Long's Exped. II, 1823, 50.

3026. Salt Lake Basin. The specimen collected appears more nearly referable to this long lost species of Say than to any other, although closely related to *C. tigris*.

PLESTIODON SKILTONIANUS, B. &. G.

Plestiodon skiltonianus, B. & G. Pr. A. N. Sc. VI, April, 1852, 69.—IB. Stansbury's Report, 1852, 349; pl. iv, figs. 4—6.
Eumeces quadrilineatus, HALLOWELL, Pr. A. N. Sc. VII, June, 1854, 94.

3166. Fort Humboldt, California. This locality is more than usually eastern.

PLESTIODON GUTTULATUS, Hallowell.

Lamprosaurus guttulatus, HALLOW. Pr. A. N. Sc. VI, Dec. 1852, 206.
Plesticdon guttulatus, HALLOW. Pr. A. N. Sc. VII, Dec. 1857, 215.

3115. Upper Arkansas?

PLESTIODON SEPTENTRIONALIS, Baird.

PLATE XXIV, FIG. 2.

Plestiodon septentrionalis, BAIRD, Pr. A. N. Sc. Dec. 1858.

One post-nasal plate which does not separate the inter-nasals and post-frontals. Color above olive, with four equidistant and equal dark stripes on adjacent half rows of scales. Two narrow white lines on each side, traversing the centres of single rows, and margined above and below by black. Upper lateral light stripes separated by six rows of scales. Beneath light greenish.

No. 1356. Minnesota and Nebraska.

EUTAENIA ORDINOIDES, B. & G.

Tropidonotus ordinoides, B. & G., Pr. A. N. Sc. VI, 1852, 176.
Eutaenia ordinoides, B. & G., Catal. N. Am. Serp 1853, 33.—GIRARD, Herp. U. S. Ex. Ex. 1858, 153; pl. xiv. figs. 1, 4.

SPEC. CHAR.—The largest of the genus. Body very stout. First row of dorsal scales largest, but slightly carinated; second much smaller, only slightly larger than the third. Caudal scales strongly carinated. A dorsal and two lateral stripes. Lateral stripe on the second and third rows. Two series of black spots on each side, with the interval reddish brown; these, in large specimens, are usually confluent, forming two distinct black stripes, separated by a well defined reddish one. First row of scales below the lateral stripe, greenish, slightly darker than the abdomen. Abdomen uniform greenish; dorsal rows, 19.—(KINNICOTT.)

No. 889. Between Salt Lake and California. Lt. Beckwith.

EUTAENIA VAGRANS. B. & G.

PLATE XVII.

Eutaenia vagrans, B. & G. Catal. N. Am. Serp. 1853, 35.
GIRARD, Herp. U. S. Ex. Ex. 1858, 154; pl. xiv, figs. 5—10.

SPEC. CHAR.—Head large and high, eight labials above; sixth and seventh very large, higher than wide, the sixth extending above the level of the lower edge of the eye; the seventh much larger than the fifth. Light olive brown above, with two series of black spots on each side, the upper of which encroaches upon the dorsal stripe, constricting it at regular intervals, while the lower encroaches upon the indistinct lateral stripe. Lateral stripe on second and third rows; dorsal rows, 21.—(KENNICOTT.)

No. 919. Camp of May 17, 921, 922, 923. Utah. Lt. Beckwith.

NERODIA ERYTHROGASTER, B. & G.

PLATE XVIII.

Coluber erythrogaster, SHAW, Gen. Zool. III, 1804, 458.
Tropidonotus erythrogaster, HOLBROOK, N. Am. Herp III, 1842, 33; pl. viii.
Nerodia erythrogaster, B & G. Catal. N. Am. Serp. 1853, 41.

SPEC. CHAR.—Head small, elongated, narrowing forwards; occipital region flattened. Vertical plate pentagonal, very large, as are also the occipitals. Three post-orbitals. An elevated loral. Dorsal rows of scales, 23 in number, all very strongly carinated. Uniform dark bluish black above, lighter on the sides; external band of dull blue extending on the abdominal scutellæ. Body beneath uniform dull coppery yellow, sometimes with the anterior edge of each scutellæ bluish. No distinct spots on the abdomen as in *N. sipedon*. Indications of three series of dorsal blotches, as in *N. Woodhousii*, in young specimens.

No. 1320. Cedar creek, Arkansas. Lt. Beckwith.

HETERODON NASICUS, B. & G.

Heterodon nasicus, B. & G. Stansbury's Report Salt Lake, 1852, 352.—IB. Catal. N. Am. Serp. 1853, 61.

SPEC. CHAR.—Vertical broader than long. Rostral excessively broad and high. Azygos plate surrounded behind and on the sides by many small plates (12–15.) A second loral. Labials short and excessively high. Dorsal rows of scales 23, exterior alone smooth. A dorsal series of about 50 blotches, with four or five others on each side. Body beneath black. A narrow white line across the middle of the superciliaries; a second behind the rostral. A broad dark patch from the eye to the angle of the mouth, crossing the last two labials.

No. 1263. Rocky Mountains. Lt. Beckwith.

PITYOPHIS BELLONA, B. & G.

Churchillia bellona, B. & G., Stansbury's Report, 1852, 350.
Pituophis bellona, B. & G., Catal. N. Am. Serp., 1853, 66.
Pituophis affinis, HALLOWELL, Pr. A. N. Sc. VI, 1852, 181.

SPEC. CHAR.—Head broad behind, tapering to the snout. Snout rather pointed but less so than in *P. sayi*, elevated. Crown flattened between the eyes and posteriorly. Vertical plate broad anteriorly, elongated and narrow posteriorly. Auti-orbitals 2, sometimes 1; post-orbitals 3 or 4. Frequently a supplemental plate before the vertical. Dorsal rows of scales 29 to 35, the seven outer rows smooth. Head spotted with black; transverse frontal bar from one orbit to the other, well marked; the oblique post-orbital stripe rather narrow. Color of the body whitish yellow; a dorsal series of 45—65 subquadrate blotches from head to anus, transversely elongated posteriorly; 3 or 4 smaller series on each side. The lateral blotches longitudinally elongated near the head; vertically elongated posteriorly, where they form a series of jet black vertical bars. Ten or fifteen transverse black bands on the tail. Abdomen yellow with an external series of black spots on each side.—(KENNICOTT.)

Nos. 1547—1522. Lt. Beckwith.

MASTICOPHIS TAENIATUS, B. & G.

PLATE XXIII.

Leptophis taeniata, HALLOWELL, Pr. A. N. Sc. VI, 1852, 181.
Masticophis taeniatus, B. & G., Catal. Serp., 1853, 103.

A broad brown dorsal stripe margined by a darker line. The four outer rows of scales on each side yellow, with a dark line through the centre of each. A dark line along the edge of the abdomen, making six dark lines on each side. Beneath yellowish. Dorsal rows 15, tail about one-third the total length.

1984, 1980. Utah. Lt. Beckwith.

BUFO WOODHOUSII, Girard.

Bufo dorsalis, HALLOW, Pr. A. N. Sc., VI, 1852, 181.—IB. Sitgreaves Rep., 1853, 142; pl. xix.
Bufo woodhousii, GIRARD, Pr. A. N. Sc., VII, May, 1854, 86.

SP. CH.—Head short and thick; upper central surface but little depressed, not to say grooved; the suborbital ridge being slightly elevated. The occipito-temporal ridge is thicker, and hence a little more conspicuous. Snout rounded, nostrils terminal. Mouth wide, upper jaw emarginated. Tympanum and parotids of moderate size. Limbs rather short and stout. First finger much longer than the second. A large metacarpal disk. Toes semipalmated. Two metatarsal tubercles, a very large and a very small one. No membranous fold at the inner lower edge of the tarsus. Papillæ of medium size upon the back. Inferior surface, with rather small and crowded granular warts. Above dark brown, with numerous lines of yellow. A dorsal yellowish vitta running the whole length of the body. Transverse blotches of black upon the thighs and fore arms. Beneath ochraceus.—(GIRARD.)

2641. Upper Arkansas. Captain Beckwith.

AMBLYSTOMA MAVORTIUM, Baird.

Ambyostoma mavortium, BAIRD, J. A. N. Sc., 2d series, I, October, 1849, 28.

4062. Rocky mountains. Captain Beckwith.

SIREDON LICHENOIDES, Baird.

Siredon lichenoides, BAIRD, Pr. A. N. Sc., pl. vi, April, 1852, 68.—IB. Stansbury's Report, 1852, 336; pl. i.

4076. Rocky mountains. Captain Beckwith.

H. Richard del.

J. H. Richard del.

H. Richard del.

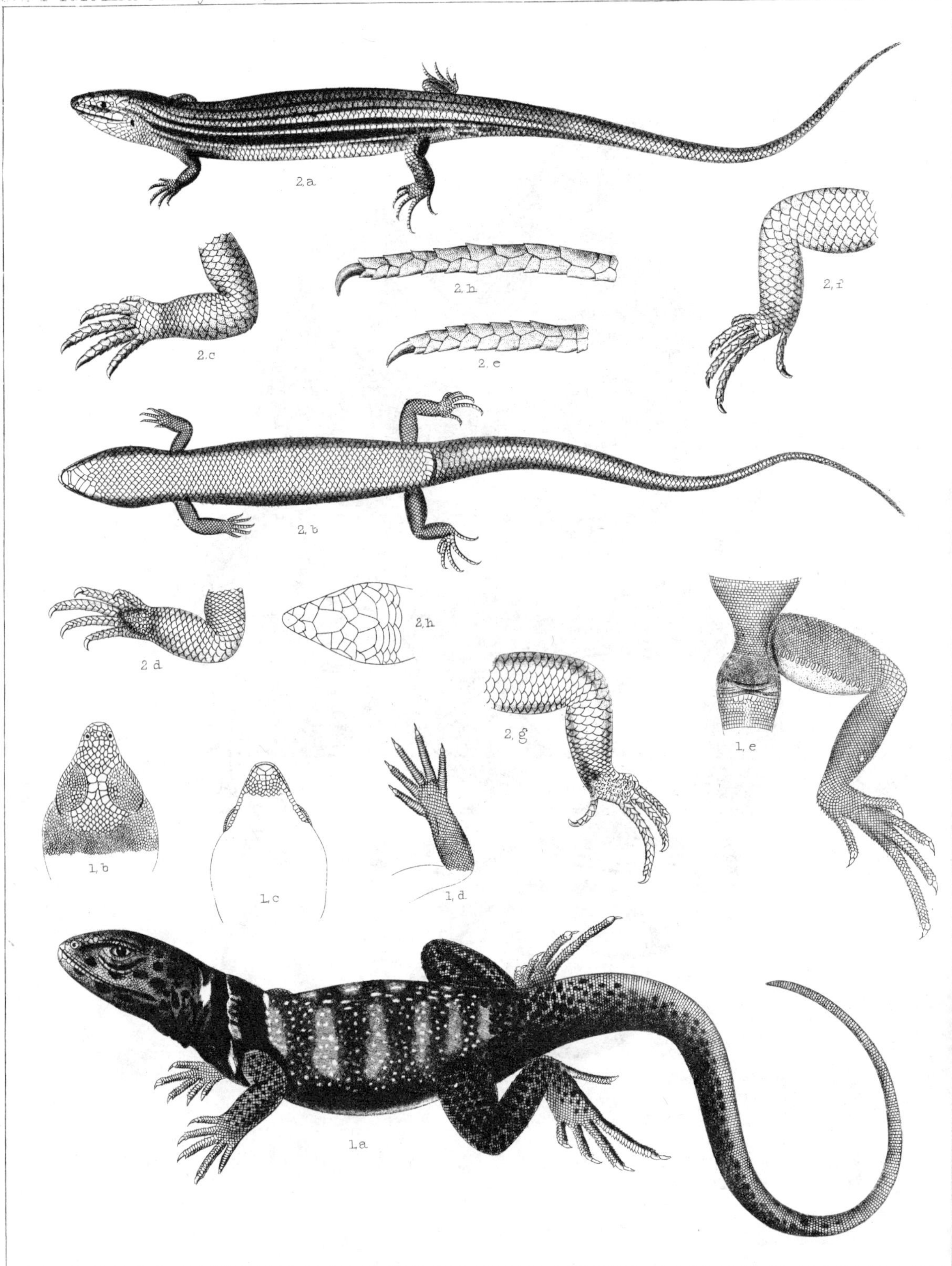

J. H. Richard del.

EXPLORATIONS AND SURVEYS FOR A RAILROAD ROUTE FROM THE MISSISSIPPI RIVER TO THE PACIFIC OCEAN.
WAR DEPARTMENT.

REPTILES:

BY SPENCER F. BAIRD,

ASSISTANT SECRETARY OF THE SMITHSONIAN INSTITUTION.

WASHINGTON, D. C.
1859.

WAR DEPARTMENT,
Office of Explorations and Surveys, February 14, 1859.

The General Natural History Reports having been extended so much beyond the limits originally contemplated, the War Department has considered it advisable to omit the publication of the Report on Reptiles. The plates to accompany this report having been prepared and printed, they, with a brief explanation of the figures composing them, are herewith given.

A. A. HUMPHREYS,
Captain Topographical Engineers, in charge.

EXPLANATION OF THE PLATES.

As the general report on the reptiles of western North America, observed by the different exploring parties, has been excluded from the series for want of room, all that can be given here is an explanation of the plates prepared for this report. These represent the details of external form in different species of North American serpents. All the outlines from the same specimen have a common number, to which is added a system of lettering common to all, and indicating the several details, as follows:

u, the upper surface of the head.
l, the lower surface of the head.
s, the side of the head, (usually the left.)
f, the head from before.
a, the anal region, showing the peculiar scutellation.
sc, the scales on the side of the body, showing several oblique rows, with the upper portion of the abdominal scales. Each figure illustrates usually the median dorsal row and those between it and the abdominal plates on the left side. The entire number of dorsal rows of scales will therefore be twice the number represented, less one. The figure is generally taken at about the posterior end of the anterior third of the length of body, including head and tail.

In some cases a different series of letters has been adopted; these are explained in their place.

The figures have, as far as possible, been taken from the type specimens of the species, especially those described in the catalogue of serpents in the museum of the Smithsonian Institution, (1853,) to which the page column refers.

Plate.	Fig.	Name.	Page.	Locality.	Details given.
XXIV	1	*Crotalus durissus*, L. p.	1	Carlisle, Pennsylvania	All.
	2	do. *adamanteus*, Beauv.	3	Pensacola, Florida	All.
	3	do. *atrox*, B. & G.	5	Indianola, Texas	u. l. s. f.
	4	do. *confluentus*, Say.	8	Red River	All.
	5	do. *molossus*, B. & G	10	Sonora	u. l. s. f. sc.
	6	do. *oregonus*, Holb.	145	Oregon?	u. sc.
	7	*Crotalophorus miliarius*, Holb.	11	Charleston, South Carolina	All.
	8	do. *consors*, B. & G.	12	Indianola, Texas	u. sc.
XXV	9	*Crotalophorus tergeminus*, Holb.	14	Racine, Wisconsin	All.
	10	do. *edwardsii*, B. & G.	15	Rio Grande, Texas	All.
	11	do. *kirtlandii*, Holb	16	Ohio	All.
	11 bis.	do. do. (young)	16	Ohio	s.
	12	*Ancistrodon contortrix*, B. & G.	17	Near San Antonio, Texas	All.
	13	*Toxicophis piscivorus*, B. & G.	19	Red River, Louisiana	All.
	14	do. *pugnax*, B. & G.	20	Indianola	All.
	15	*Elaps fulvius*, Cuv.	21	Charleston, South Carolina	All.
	16	do. *tener*, B. & G.	22	San Felipe, Texas	All.
	17	do. *tristis*, B. & G.	23	Kemper county, Mississippi	All.
	18	*Dipsas septentrionalis*, Kennicott[1]		Brownsville, Texas	All.
XXVI	19	*Eutaenia saurita*, B. & G.	24	Carlisle, Pennsylvania	All.
	20	do. *faireyi*, B. & G.	25	Red River, Louisiana	All.
	21	do. *proxima*?, B. & G.	25	Indianola, Texas	All.
	22	do. *ornata*, B. & G.[2]	28	San Antonio to El Paso	All.
	23	do. *sirtalis*, B. & G.	30	Westport, New York	All.
	24	do. *ordinata*, B. & G.	32	Riceboro', Georgia	All.
	25	do. *radix*, B. & G.	34	Racine, Wisconsin	All.
	26	do. *marciana*, B. & G.	36	Red river	u. l. s. f. sc.
	1	*Nerodia transversa*, B. & G.	148	Red river	All.
	2	*Eutaenia dorsalis*, B. & G.	31	Texas	All.
	3	do. *ordinoides*, B. & G.	33	California	All.[3]
XXVII	27	*Nerodia sipedon*, B. & G.	38	Carlisle, Pennsylvania	All.
	28	do. *erythrogaster*, B. & G.	40	Red river, Louisiana	u. l. s. f. sc.
	29	do. *taxispilota*, B. & G.	43	Liberty county, Georgia	All.
	30	do. *holbrookii*, B. & G.	43	Red river, Louisiana	All.
	31	do. *niger*, B. & G.	147	Massachusetts	All.
	32	*Regina leberis*, B. & G.	45	Carlisle, Pennsylvania	All.
	33	do. *rigida*, B. & G.	46	Pennsylvania	All.
	34	do. *grahamii*, B. & G.	47	Texas	All.
	35	do. *clarkii*, B. & G.	48	Indianola	u. l. s. f. sc.
	36	do. *kirtlandii*, Kenn.[4]		Illinois	All.
	37	*Ninia diademata*, B. & G.	49	Orizaba, Mexico	s. u.
XXVIII	38	*Heterodon platyrhinos*, Latr.	51	Carlisle, Pennsylvania	All.
	39	do. *cognatus*, B. & G.	54	Indianola	All.
	40	do. *niger*, Troost	55	Carlisle, Pennsylvania	All, nasal pl.
	41	do. *atmodes*, B. & G.	57	Charleston, South Carolina	All, n. p.
	42	do. *simus*, Holb.	59	do. do.	All.
	42 bis.	do. *simus*, (second spec.)	59	do. do.	u.
	43	do. *nasicus*, B. & G.	61	Eagle Pass, Texas	All, n. p.

[1] U. S. Boundary Report, II, 1859; Reptiles, p. 16.

[2] U. S. Boundary Report, II, 1859; Reptiles, 16: *Eutaenia parietalis*, Catal. Serpents, 28.

[3] a, head from above; b, do. from side; c, do. from below; d, anal region; e, side scales.

[4] KENNICOTT, Pr. A. N. Sc. VIII, April, 1856, 95.

Plate.	Figure.	Name.	Page.	Locality.	Details given.
XXIX	44	*Pityophis melanoleucus*, Holb	65	Carolina	u. l. s. f. sc.
	45	....do..*sayi* B & G	151	Fort Snelling	All.
	46	....do..*bellona*, B. & G	66	Rio Grande	All.
	47	....do..*mc clellanii*, B. & G	68	Red river	All.
	48	....do..*annectens*, B. & G	72	San Diego, California	All.
	49	*Scotophis alleghaniensis*, B. & G	73	Carlisle, Pennsylvania	u. l. s. f. sc.
	50	....do..*lindheimeri*, B. & G	74	New Braunfels, Texas	All.
	51	....do..*vulpinus*, B. & G	75	Racine, Wisconsin.	All.
XXX	52	....do..*confinis*, B. & G	76	Anderson, South Carolina	u. s.
	53	....do..*laetus*, B. & G	77	Red river, Arkansas	All.
	54	....do..*guttatus*, B. & G	78	Kemper county, Mississippi	All.
	55	....do..*4-vittatus*, B. & G	80	Florida	u. l. s. f. sc.
	56	....do..*emoryi*, B .& G	157	Howard Springs, Texas	All.
	57	*Ophibolus boylii*, B. & G	82	Eldorado county, California	u. l. s. f. sc.
	58	....do..*splendidus*, B. & G	83	Sonora	All.
	59	....do..*sayi*, B. & G	84	Red river, Arkansas	u. l. s. f. sc.
	60	....do..*rhombomaculatus*, B. & G	86	Georgia	All.
	61	....do..*eximius*, B. & G	87	Warren, Massachusetts	All.
	62	....do..*clericus*, B. & G	88	Clark county, Virginia	All.
	63	....do..*doliatus*, B. & G	89	Kemper county, Virginia	All.
	64	....do..*gentilis*, B. & G	90	Red river, Arkansas	All.
XXXI	65	....do..*getulus*, B. & G	85	Charleston, South Carolina	All.
	66	*Georgia obsoleta*, B. & G	158	Brownsville, Texas	All.
	67	*Bascanion constrictor*, B. & G	93	Carlisle	All.
	68	....do..*fremontii*, B. & G	95	California	All.
	69	....do..*foxii*, B. & G	96	Michigan ; both sides of head drawn to show difference in labials	All.
	70	....do..*flaviventris*, B. & G	96	Texas?	All.
	71	*Masticophis flagelliformis*, B. & G., (old)	98	Pensacola, Florida	All.
XXXII	72	....do...*flagelliformis*, (young)	149	Georgia	u. l. s. f.
	73	....do...*flavigularis*, B. & G., (young)	99	Indianola, Texas	u. l. s. f.
	74	....do...*mormon*, B. & G	101	Salt Lake	All.
	75	....do...*ornatus*, B. & G.	102	San Antonio to El Paso	All.
	76	....do...*taeniatus*, B. & G	103	California	All.
	77	....do...*schottii*, B. & G	160	Eagle Pass	All.
	78	*Salvadora grahamiae*, B. & G	104	Sonora	u. l. s. f. sc.
	79	*Leptophis aestivus*, Bell	106	Anderson, South Carolina	u. l. s. f. sc.
	80	....do..*majalis*, B. & G	107	Indianola	u. s.
	81	*Chlorosoma vernalis*, B. & G	108	Westport, New York, 2 diam.	s. a.
	1	? *Diadophis docilis*, B. & G	114	Tucson, 2 diam.*	All.
	2	*Diadophis*		Santa Magdalena, 2 diam *	All.
	3	*Wenona*		Oregon*	All.
XXXIII	82	*Diadophis punctatus*, B. & G	112	Carlisle, Pennsylvania	u. s.
	83	...do...*amabilis*, B. & G	113	San Jose, California	All.
	84	...do...*docilis*, B. & G	114	San Pedro, Comanche Spring	All.
	85	...do...*pulchellus*, B. & G	115	Eldorado county, California	All.
	86	...do...*regalis*, B. &. G	116	Sonora, Mexico	All.
	87	*Taeniophis imperialis*, B. & G [1]		Brownsville, Texas	All.
	88	*Sonora semi-annulata*, B. & G	117	Sonora	s. u.
	89	*Rhinostoma coccinea*, Holb	118	Riceboro', Georgia	All.
	90	*Rhinocheilus lecontii*, B. & G	120	California	u. l. s. f. sc.

* References as in plate XXVI, figure 1.—3.

[1] *Taeniophis imperialis*, B. & G., Mexican Boundary Report, II, 1859 ; Reptiles, 23.

Plate.	Fig.	Name.	Page.	Locality.	Details given.
XXXIII	91	*Haldea striatula*, B. & G	122	Richmond, Virginia, 2 diam	u. l. s. f. sc.
	92	*Farancia abacura*, B. & G	123	Red river, Louisiana	u. l. s. f. sc.
	93	*Abastor erythrogrammus*, Gray	125	Southern States	u. l. s. f. sc.
	94	*Virginia valeriæ*, B. & G	127	Maryland, 2 diam	u. l. s. f. sc.
	95	*Celuta amoena*, B. & G.	129	Carlisle, 2 diam	u. l. s. f. sc.
	96	*Tantilla coronata*, B. & G.	131	Kemper county, Mississippi, 2 diam	u. s.
	97	*Osceola elapsoidea*, B. & G	133	Charleston, South Carolina, 2 diam	u. s.
	98	*Storeria dekayi*, B. & G.	135	Framingham, Massachusetts, 2 diam	u. s.
	99	--do-- *occipito-maculata*, B. & G.	137	Madrid, New York, 2 diam	u. s.
	100	*Rena dulcis*, B. & G.	142	San Pedro, Can. Sp., Tex., 3 diam	u. l. s.
XXXIV	1	*Georgia couperi*, B. & G	92	Georgia	a. b. c. d. e [1]
	2	*Nerodia rhombifer*, B. & G	147	Arkansas	As in fig. 1.
	3	--do-- *woodhousii*, B. & G	42	Texas	As in fig. 1.
	4	--do-- *fasciata*, B. & G	39	South ?	As in fig. 1 [2]
	5	*Eutaenia radix*, B. & G	34	Wisconsin ?	As in fig. 1.
	6	? *Microps lineatus*, Hall[3]		Fort Chadbourne, Texas	As in fig. 1.
	7	———		?	
XXXV	1	*Crotalus tigris*, Kennicott[4]			As in fig. 1.
	2	*Crotalus*		Colorado Bottom	As in fig. 1.
	3	*Crotalus*		Sierra Verde, California	As in fig. 1.
	4	*Crotalus cerastes*, Hallow.[5]			As in fig. 1.
	5	--do--		Sierra Verde, California	As in fig. 1.
	6	*Lamprosoma occipitale*, Hall[6]		Colorado Desert	As in fig. 1.
	7	*Lamprosoma occipitale*, Hall[7]		California	As in fig. 1.
	8	*Toluca lineata*, Kenn[8]		Valley of Mexico	As in fig. 1.
XXXVI	1	*Crotalus lucifer*, B. & G	6	Oregon	As in fig. 1.
	2	*Eutaenia leptocephala*, B. & G	29	Oregon	As in fig. 1.
	3	--do-- *pickeringii*, B. & G	27	Oregon	As in fig. 1.
	4	*Pityophis catenifer*, B. & G	69	California	As in fig. 1.
	5	--do-- *wilkesii*, B. & G	71	Oregon	As in fig. 1.
	6	*Bascanion vetustus*, B. & G	97	Oregon	As in fig. 1.
	7	*Contia mitis*, B. &	110	Oregon, 2 diam	As in fig. 1.
	8	*Lodia tenuis*, B. & G	116	Oregon, 2 diam	As in fig. 1.

[1] All, except head from front.
[2] Add f. dorsal scales ; g, side view of ditto, showing a peculiar serration of the carination.
[3] HALLOWELL, Pr. A. N. Sc. VIII, 1856, 240.
[4] Mex. Bound. Rep. II, 1839 ; Reptiles, 14.
[5] HALLOWELL, Pr. A. N. Sc. VII, June, 1854, 95.
[6] U. S. Mex. Bound. II, 1859 ; Reptiles, 21.
[7] HALLOWELL, Pr. A. N. Sc. *ut supra*.
[8] KENNICOTT, *ut supra*, 23.

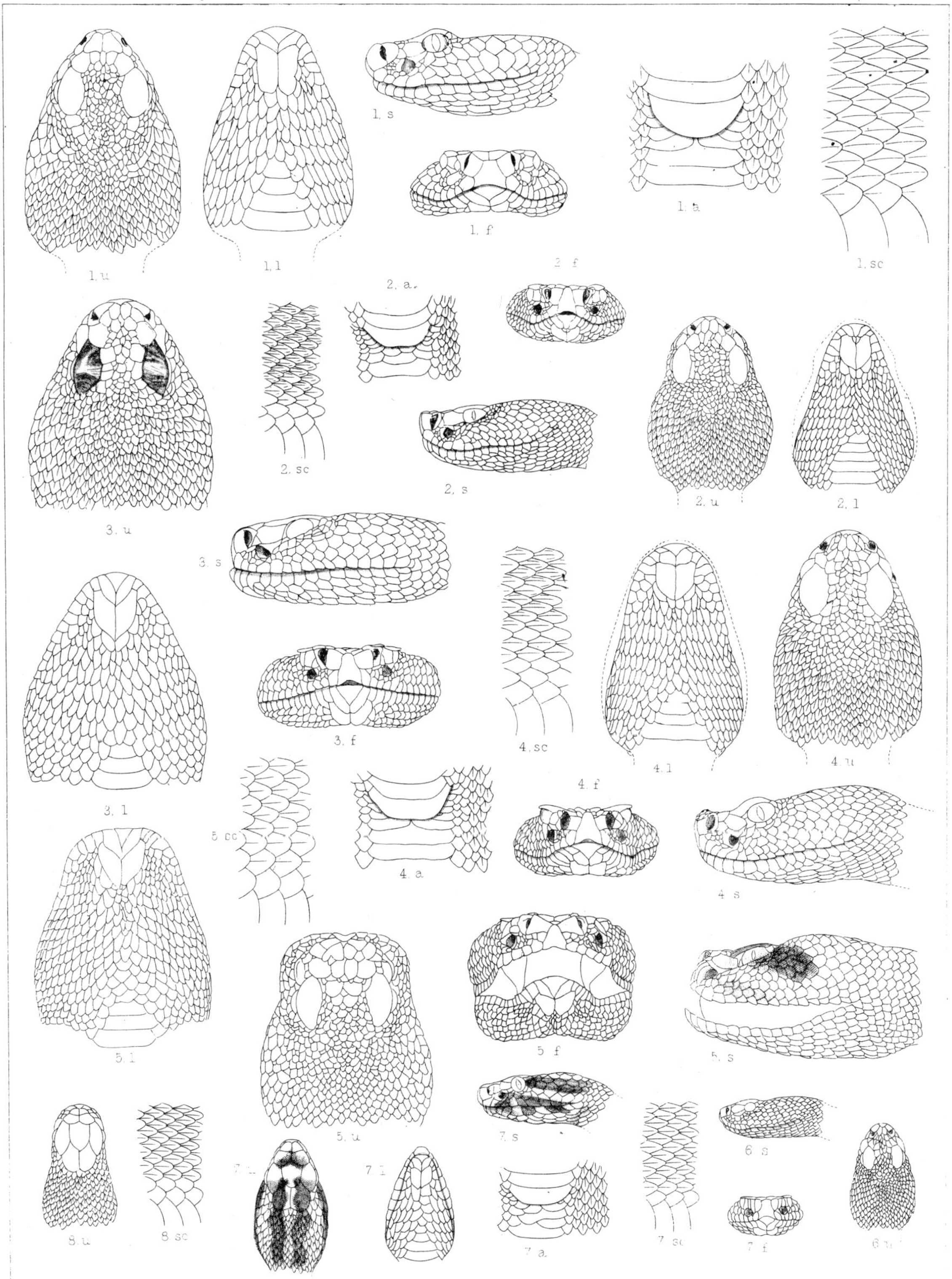
1. u
1. l
1. s
1. f
1. a
1. sc
2. a.
2. f
2. sc
2. s
2. u
2. l
3. u
3. s
3. f
3. l
4. sc
4. l
4. u
4. f
4. a
4. s
5. sc
5. l
5. u
5. f
5. s
6. s
6. u
7. s
7. l
7. a
7. sc
7. f
8. u
8. sc

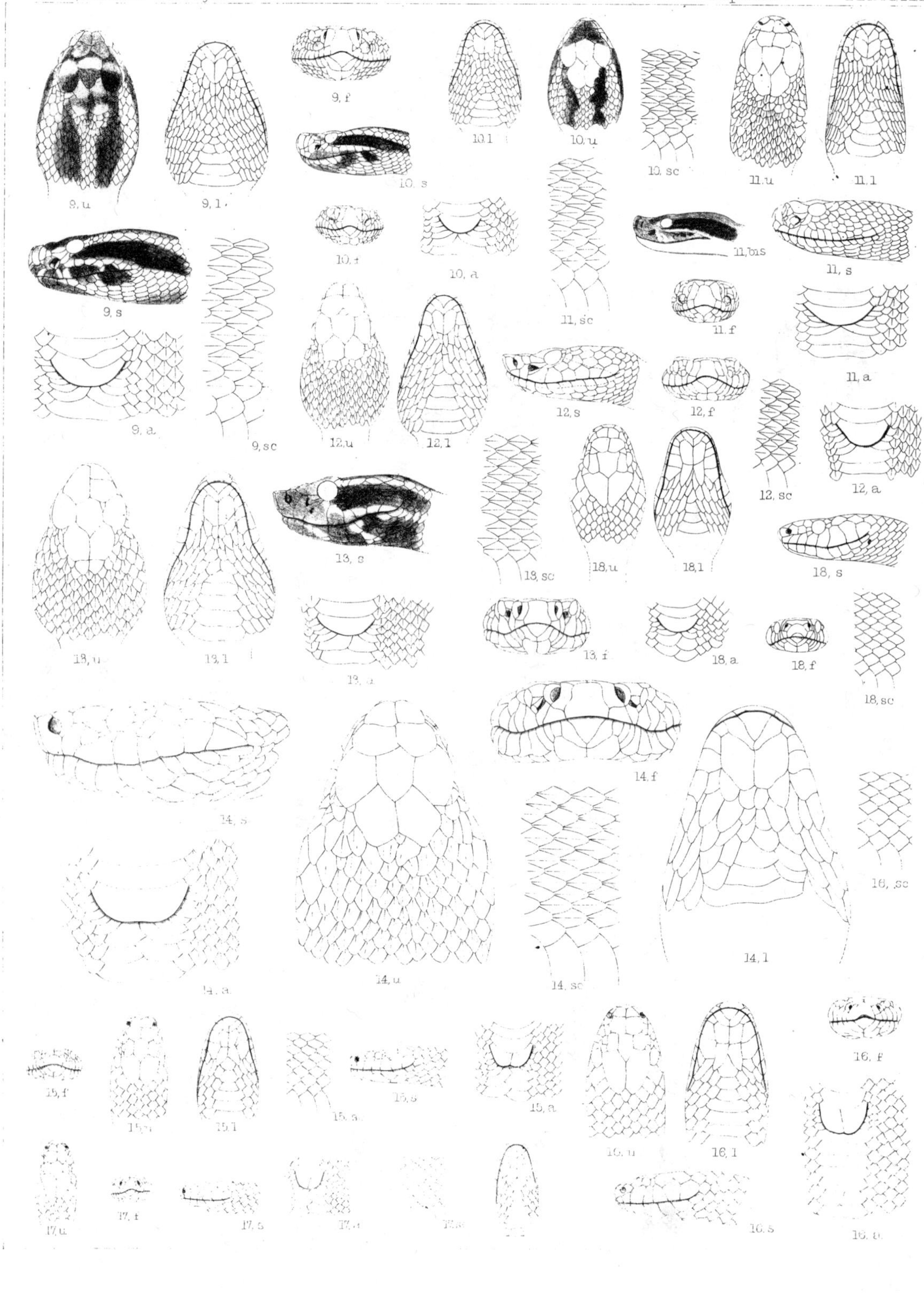
9, u
9, l
9, f
10, s
10, l
10, u
10, sc
11, u
11, l
9, s
10, f
10, a
11, sc
11, bis
11, s
11, f
11, a
9, a
9, sc
12, u
12, l
12, s
12, f
12, sc
12, a
13, u
13, l
13, s
13, sc
18, u
18, l
18, s
13, a
13, f
18, a
18, f
18, sc
14, s
14, f
14, u
14, l
16, sc
14, a
14, sc
15, f
15, l
15, a
16, u
16, l
16, f
16, s
16, a
17, u
17, f
17, s

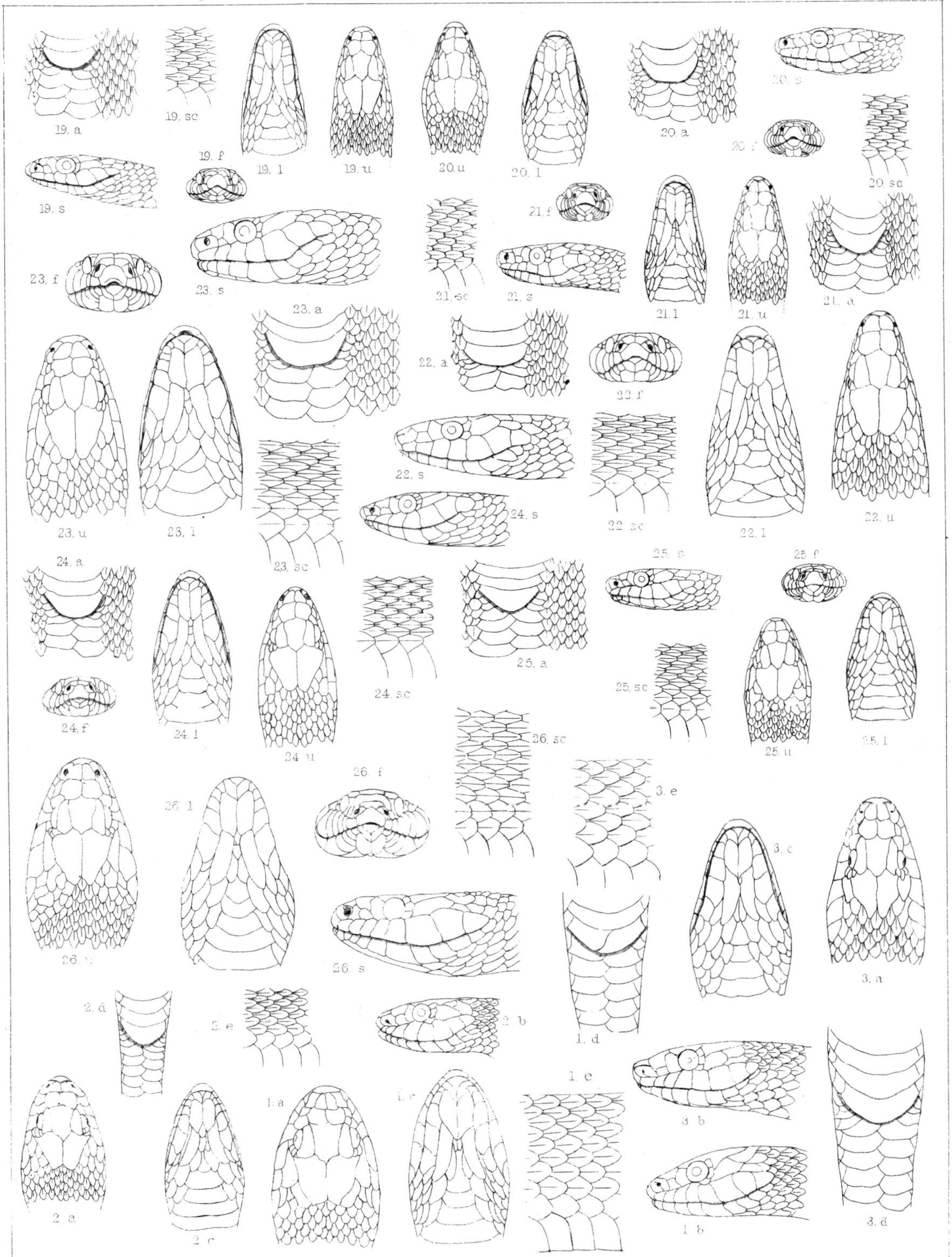
19. a
19. sc
19. f
19. l
19. u
20. u
20. l
20. a
20. s
20. sc
19. s
23. f
23. s
21. sc
21. s
21. f
21. l
21. u
21. a
23. a
22. a
22. f
23. u
23. l
23. sc
22. s
24. s
22. sc
22. l
22. u
24. a
24. f
24. l
24. u
24. sc
25. a
25. s
25. f
25. sc
25. u
25. l
26. sc
26. f
26. l
3. e
3. c
3. a
26. u
26. s
2. d
2. e
2. b
1. d
1. e
3. b
2. a
2. c
1. a
1. c
1. b
3. d

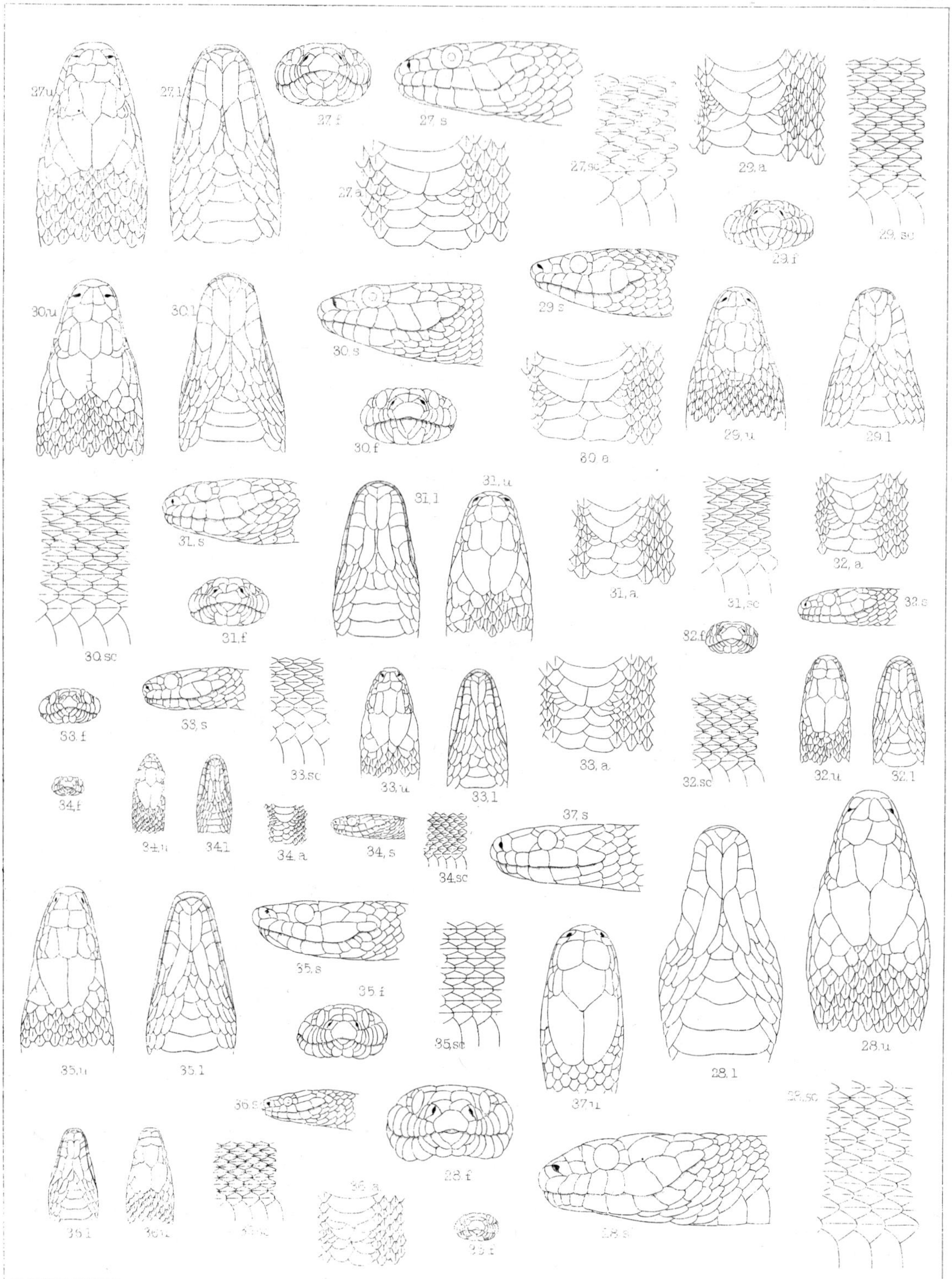
27.u
27.l
27.f
27.s
27.a
27.sc
28.a
29.sc
29.f
29.s
30.u
30.l
30.s
30.f
30.a
29.u
29.l
30.sc
31.s
31.f
31.l
31.u
31.a
31.sc
32.a
32.s
32.f
33.f
33.s
33.sc
33.u
33.l
33.a
32.sc
32.u
32.l
34.f
34.u
34.l
34.a
34.s
34.sc
37.s
35.s
35.f
35.sc
35.u
35.l
37.u
28.l
28.u
28.sc
36.s
28.f
36.l
36.u
36.a
28.s

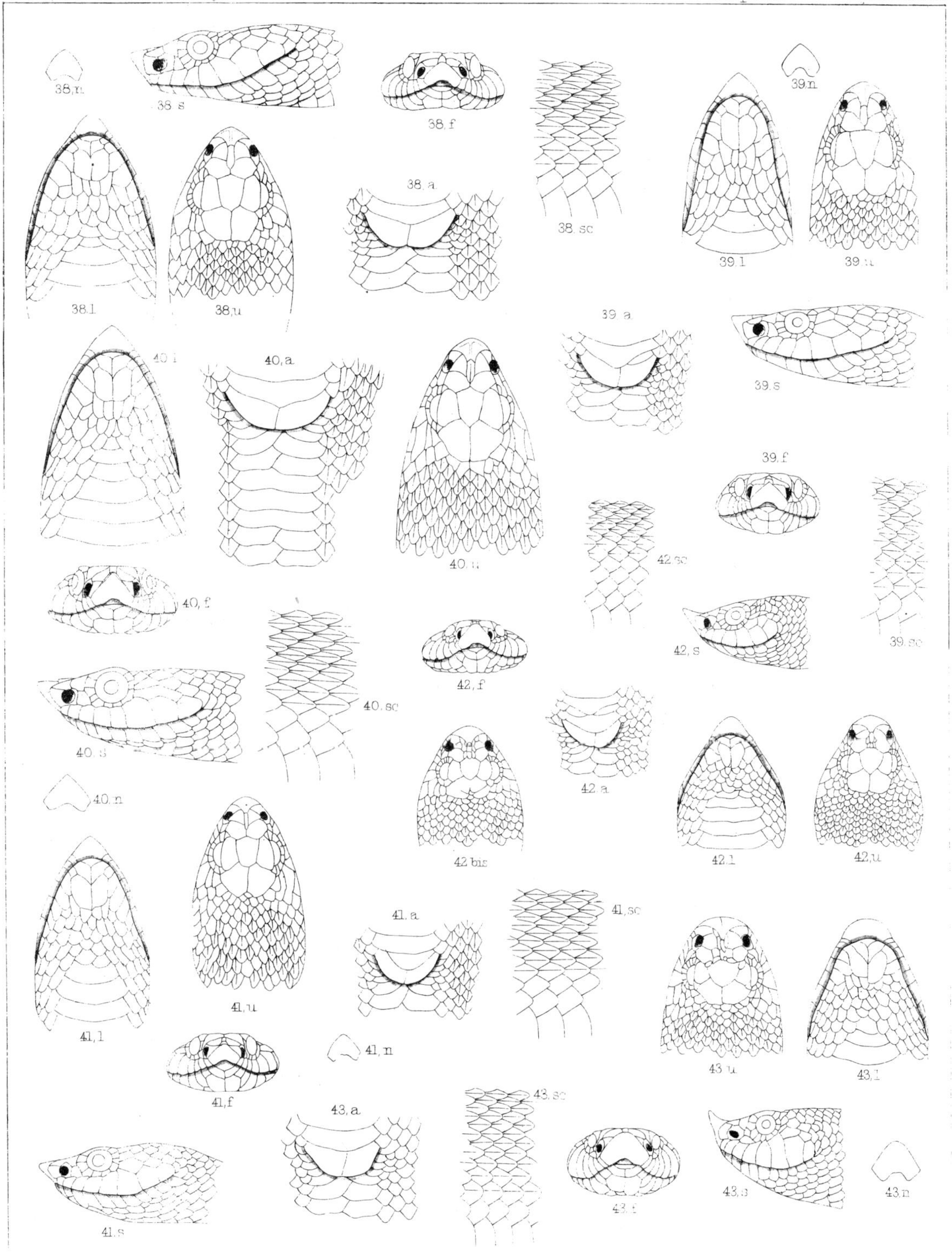
38, n
38. s
38. f
38. sc
39. n
38. l
38, u
38. a
39. l
39. u
40. l
40, a
39. a
39. s
40. u
39. f
42. sc
40, f
42, s
39. sc
42, f
40. sc
40. s
42. a
40. n
42 bis
42. l
42, u
41. a
41, sc
41, l
41, u
41, n
43. u
43, l
41, f
43. sc
43, a
43. s
43. n
43. f
41. s

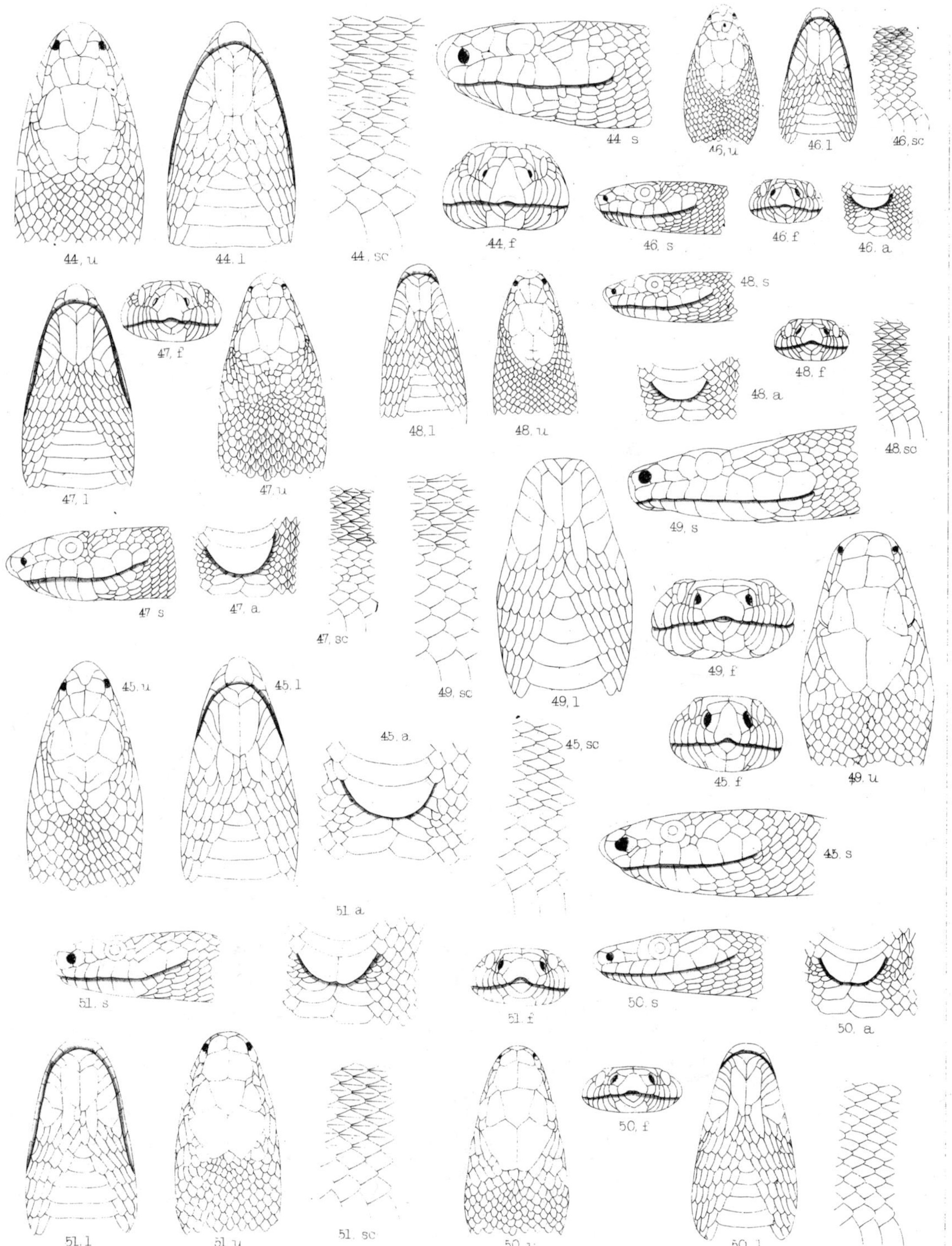
44, u
44. l
44, sc
44. s
44, f
46, u
46. l
46, sc
46. s
46. f
46. a
47, f
47. l
47. u
47. s
47. a
47. sc
48. s
48. f
48. a
48. l
48. u
48. sc
49. s
49. l
49. sc
49. f
49. u
45. u
45. l
45. a
45. sc
45. f
45. s
51. a
51. s
51. f
51. l
51. u
51. sc
50. s
50. a
50. f
50. u
50. l
50. sc

52, u
52, s
55, s
53, u
53, l
53, s
53, f
53, sc
53, a
54, f
54, u
54, l
56, a
55, sc
55, u
55, f
55, l
54, sc
54, a
54, s
57, s
56, u
56, l
56, f
56, s
56, sc
57, f
58, sc
57, u
57, l
57, sc
59, u
59, l
58, u
58, l
58, a
58, s
58, f
64, s
59, sc
59, f
59, s
60, a
64, sc
64, a
64, u
64, l
60, f
64, f
61, f
61, a
61, u
61, l
61, sc
60, u
60, l
60, s
61, s
60, sc
62, f
62, sc
62, l
62, u
62, a
62, s
68, u
63, f
63, s
63, a
63, l
63, sc

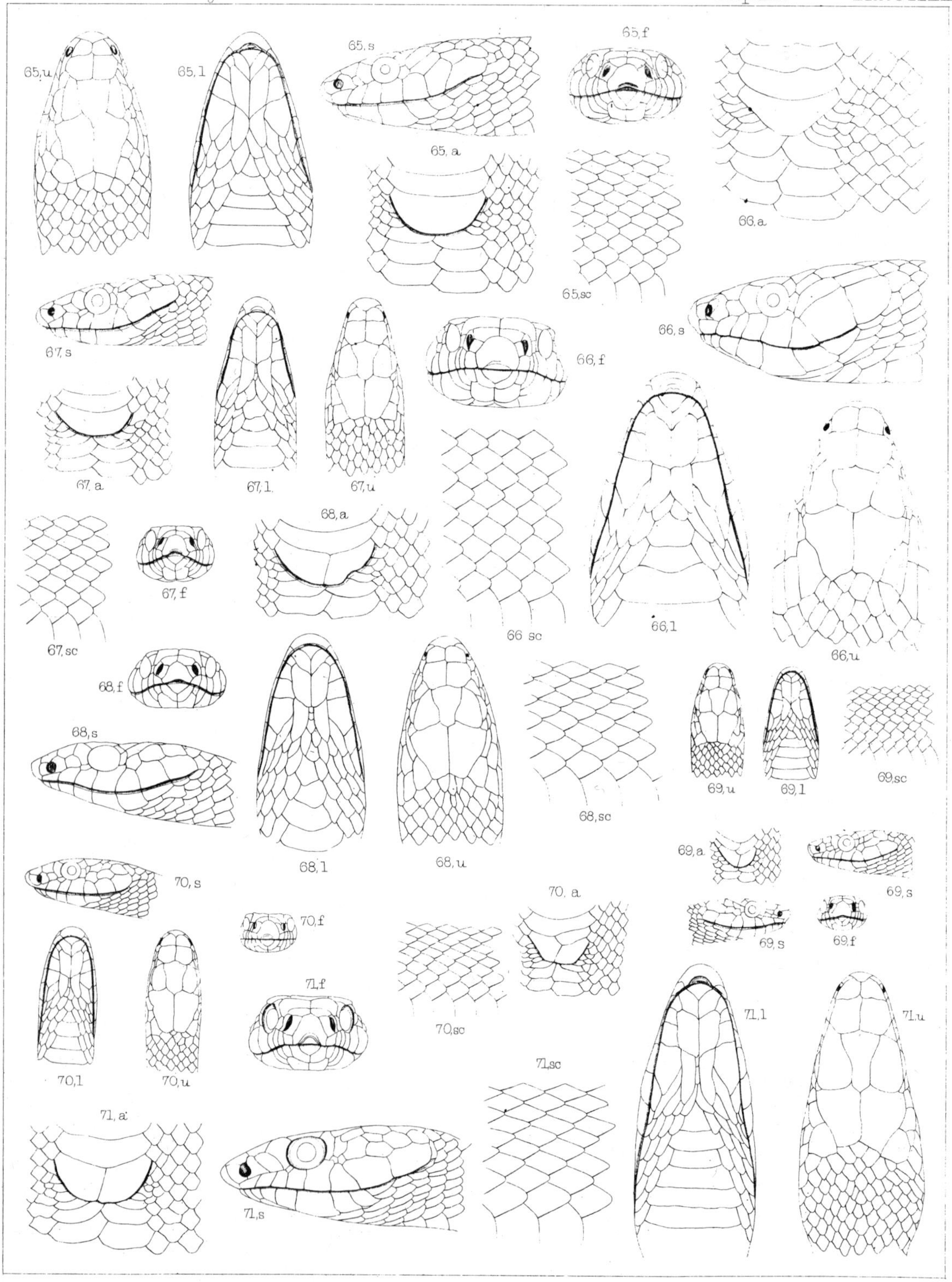
65,u
65,l
65,s
65,f
65,a
65,sc
66,a
67,s
66,s
66,f
67,a
67,l
67,u
68,a
67,f
67,sc
66 sc
66,l
66,u
68,f
68,s
68,sc
69,u
69,l
69,sc
68,l
68,u
69,a
69,s
70,s
70,a
70,f
69,s
69,f
71,f
70,sc
71,l
71,u
70,l
70,u
71,sc
71,a
71,s

72, u
72, l
72, s
72, f
73, f
73, l
73, u
73, sc
73, s
76, a
75, f
76, s
75, s
75, sc
75, a
75, l
75, u
76, f
76, l
76, u
76, sc
74, s
79, sc
79, l
79, s
74, f
74, sc
74, a
74, l
74, u
79, a
79, u
77, f
78, s
77, s
78, sc
78, u
78, l
78, f
77, a
77, sc
77, u
77, l
81, s
2, a
2, c
2, d
2, b
2, e
1, e
1, d
1, b
80, u
3, b
81, u
80, s
3, e
3, d
3, a
3, c
1, c
1, a

82, s
83, f
83, a
84, s
82, u
83, s
83, u
83, l
83, sc
84, u
84, a
84, sc
86, a
85, sc
85, s
85, l
85, u
85, f
86, u
86, l
85, a
87, f
84, l
86, f
86, s
89, s
87, s
87, f
89, a
89, f
86, sc
89, sc
90, f
89, l
89, u
87, l
87, u
87, a
87, sc
90, s
91, f
88, s
90, l
90, u
90, sc
91, s
91, l
91, u
91, sc
88, u
93, s
92, sc
93, sc
93, u
93, l
92, s
92, u
92, l
92, f
93, f
94, sc
94, f
94, l
94, u
95, f
95, sc
95, u
95, l
95, s
96, u
94, s
97, a
96, s
99, s
98, s
98, u
99, u
100, u
100, s
100, l

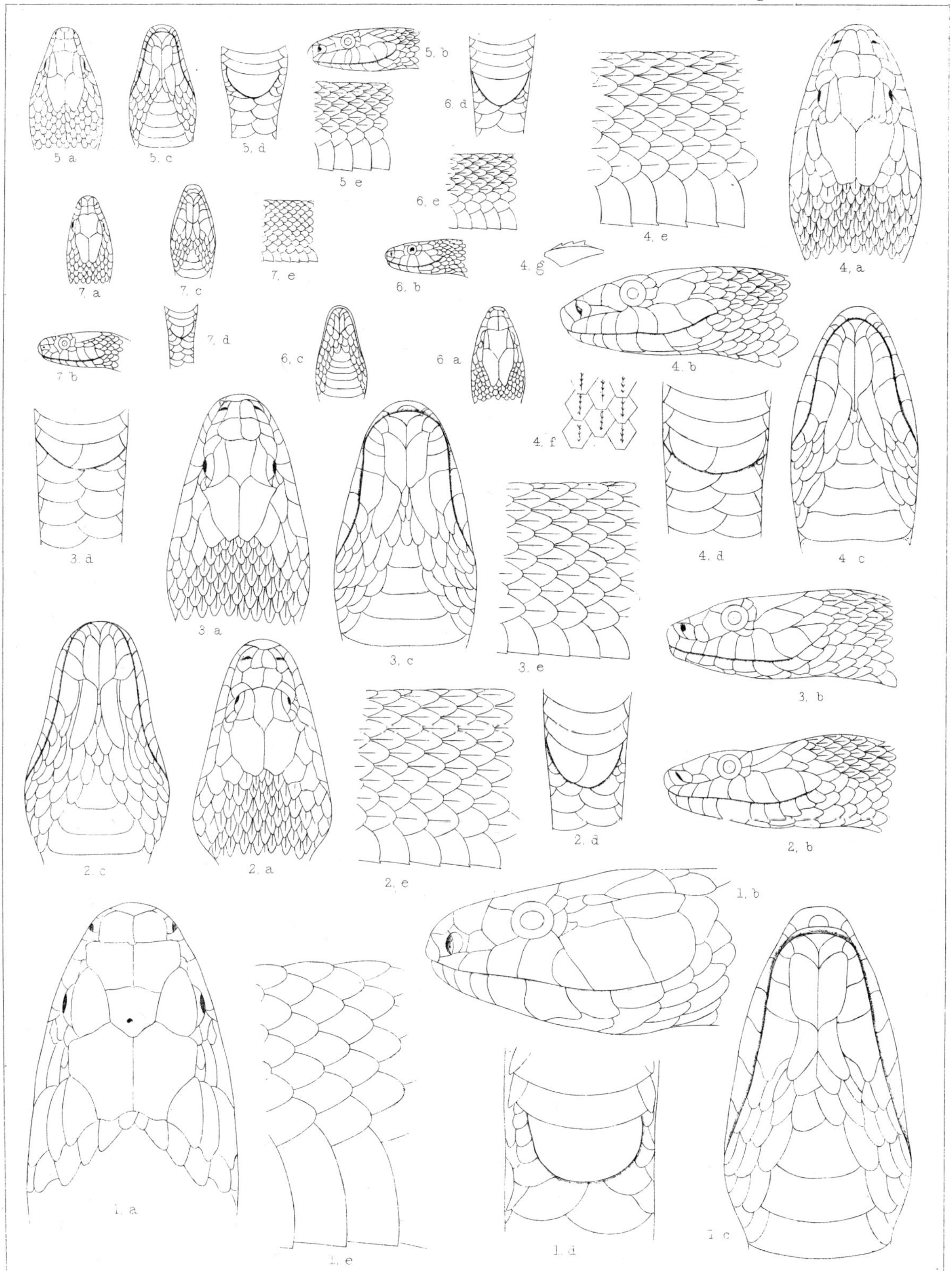
5 a
5. c
5. d
5. b
5 e
6. d
6. e
4. e
4. a
7. a
7. c
7. e
6. b
4. g
7. b
7. d
6. c
6 a
4. b
4. f
3. d
3. a
3. c
3. e
4. d
4 c
3. b
2. c
2. a
2. e
2. d
2. b
1. b
1. a
1. e
1. d
1. c

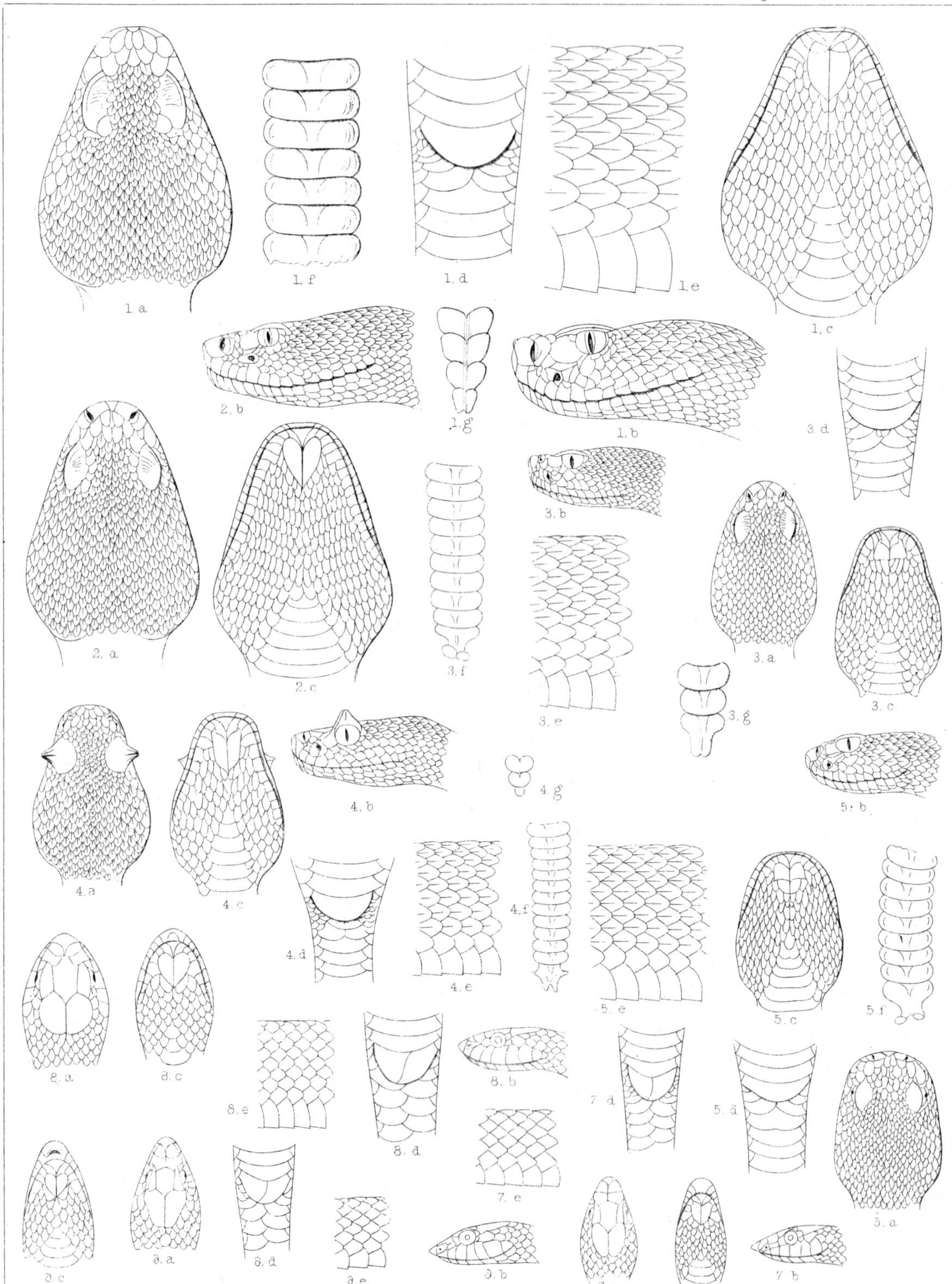
1.a
1.f
1.d
1.e
1.c
2.b
1.g
1.b
3.d
2.a
2.c
3.f
3.b
3.a
3.c
3.e
3.g
4.a
4.c
4.b
4.g
5.b
4.d
4.e
4.f
5.e
5.c
5.f
8.a
8.c
8.e
8.d
8.b
7.d
7.e
5.d
5.a
9.c
9.a
9.d
9.e
9.b
7.a
7.c
7.b

1. a
1, b
1, c
1, d
1, e
3, a
3, b
3, e
3, c
3, d
3, f
2, c
2, d
2, b
2, a
2, f
2, e
5, e
5, f
5, a
5, b
5, d
4, a
4, b
4, d
4, c
4, f
4, e
5, c
6, g
6, c
6, f
6, d
6, b
6, a
6, e
8, a
8, b
7, d
7, c
7, b
7, a

REPORT UPON THE REPTILES OF THE ROUTE

S[pencer] F[ullerton] Baird

No. 4.

REPORT UPON THE REPTILES OF THE ROUTE.

BY S. F. BAIRD.

SCELOPORUS UNDULATUS, Wiegm.

Stellio undulatus, LATREILLE, Hist. Rep. II, 1802, 40.
Tropidolepis undulatus, CUVIER, R. A.—HOLBROOK, N. Am. Herp. II, 1842, 73; pl. ix.
Sceloporus undulatus, WIEGMANN, Isis, 1828, 369.

2874. Near mouth of Poteau river, Arkansas. B. Möllhausen.

SCELOPORUS SPINOSUS, Wiegm.

Sceloporus spinosus, WIEGM. Isis, 1828, 369.
Tropidolepis spinosus, GRAY, Syn. Rept. Griff. An. King. IX, 1831, 43.

2938. San Antonio, Texas. Dr. Kennerly.

SCELOPORUS CONSOBRINUS, B. & G.

Sceloporus consobrinus, B. & G. Marcy's Report, 1853, 237.

2910. Canadian Fork. B. Möllhausen.

SCELOPORUS THAYERI, B. & G.

Sceloporus thayeri, B. & G. Pr. A. N. S. VI, August, 1852, 127.

2907. Coal creek, Arkansas. B. Möllhausen.

CROTAPHYTUS COLLARIS, Holbrook.

Agama collaris, SAY, Long's Exped. II, 1823, 252.
Crotaphytus collaris, HOLBROOK, N. Am. Herp. II, 1842, 79; pl. x.

2701 Fort Smith, Arkansas.—2707. Near the Gold mountains, Arkansas.—4106. Coal creek of Arkansas.—2715. Near Canadian. B. Möllhausen.—2690. Pecos to Rio Grande. Dr. Kennerly.

CROTOPHYTUS WISLIZENII, B. & G.

Crotophytus wislizenii, B. & G. Pr. A. N. Sc. VI, April, 1852, 69.

New Mexico. Dr. Kennerly.

UTA STANSBURIANA, B. & G.

Uta stansburiana, B. & G., Pr. A. N. Sc. VI, April, 1852, 69.—IB. Stansbury's Great Salt Lake, 1852, 345; Reptiles, plate v.

2682. Albuquerque. Kennerly and Möllhausen.

[1] The diagnoses of most of the serpents have been prepared by Mr. Robert Kennicott, and of the toads by Dr. Charles Girard.

HOLBROOKIA MACULATA, Girard.

Holbrookia maculata, GIRARD, Pr. Am. Assoc. for 1850, 1851, 201.—IB. Stansbury's Report, 1852, 342, pl. vi, fig. 1, 3.

2810. Near Canadian. B. Möllhausen.

HOLBROOKIA TEXANA, B. & G.

Cophosaurus texanus, TROSCHEL, Wiegmann's Archiv for 1850, 1852, 389; pl. vi.
Holbrookia texana, B. & G., Pr. A. N. Sc. VI, August, 1852, 125.

2797. Pecos to Rio Grande. Dr. Kennerly.

TAPAYA HERNANDEZII, Girard.

Tapaya hernandezii, GIRARD, Herp U. S. Ex. Ex., 1858, 395.

198. Santa Fé, New Mexico. Kennerly and Möllhausen.

TAPAYA ORNATISSIMA, Girard.

Phrynosoma orbiculare, HALLOWELL, Sitgreaves' Exped. 1853, 125; pl. viii.
Tapaya ornatissima, GIRARD, Herp. U. S. Ex. Ex., 1858, 396.

199.—203. Along Canadian river to Albuquerque. Möllhausen and Kennerly.

PHRYNOSOMA CORNUTUM, Gray.

Agama cornuta, HARLAN, J. A. N. Sc. IV, II, 1825, 299.
Phrynosoma cornutum, GRAY, Syn. Rept. Griff. Cuv. IX, 1831, 45.—HOLBROOK, N. Am. Herp. II, 1842, 87; pl. xi.—GIRARD, Herp. U. S. Ex. Ex. 1858, 403; pl. xxi, fig. 6, 9.

138. Pecos to Rio Grande. Dr. Kennerly.—140. Near Canadian. B. Möllhausen.

DOLIOSAURUS MODESTUS, Girard.

Phrynosoma modestum, GIRARD, Stansbury's Report, 1852, 361; pl. vi, fig. 4--5.
Doliosaurus modestus, GIRARD Herp. U. S. Ex. Ex., 1858, 409.

172. Near Canadian river. B. Mllöhausen.—173. Near Albuquerque. B. Möllhausen.

CNEMIDOPHORUS SEX-LINEATUS, Dum. Bib.

Lacerta sex-lineata, LINN. Syst. Nat. 1766, 364.
America sex-lineata, HOLBROOK, N. Am. Herp. I, 1838, 63; pl. vi.
Cnemidophorus sex-lineatus, DUM. BIB. Erp. Gen. V, 1839, 131.

3005. Fort Smith, Arkansas. B. Möllhausen.

CNEMIDOPHORUS GULARIS, B. & G.

Cnemidophorus gularis, B. & G. Pr. A. N. Sc. VI, August, 1852, 128.
Cnemidophorus guttatus, HALLOWELL, Pr. A. N. Sc. VII, October, 1854, 192.

2990. San Antonio to Fort Inge, Texas. Dr. Kennerly.—3016. Near Canadian. B. Möllhausen.

HELODERMA HORRIDUM, Wiegmann.

Heloderma horridum, WIEGMANN, Isis, 1829, 627.—IB. Herp Mex. 1834; plate.
BAIRD, Reptiles, U. S. and Mex. Boundary; plate.

Mohave river. Kennerly & Möllhausen.

PLESTIODON FASCIATUS.

Lacerta fasciata, LINN. Syst. Nat. I, 1758, 290.
Scincus fasciatus, HOLBROOK, N. Am Herp. II, 1842, 127; pl. xviii.
Lacerta quinquelineata, LINN. Syst. Nat. 1766, 366.
Scincus quinquelineatus, HOLBROOK, N Am. Herp. II, 1842, 121; pl. xvii.
Plestiodon erythrocephalus, HOLBROOK, N. Am. Herp. 11, 1842, 117; pl. xvi.

3176. Fort Smith. Dr. Shumard.

PLESTIODON OBSOLETUS, B. & G.

Plestiodon obsoletum, B & G. Pr. A. N. Sc. VI, August, 1852, 129.

3113. Coal creek, Arkansas. B. Möllhausen.

This specimen is very young and is the only one in the collection which exhibits the typical coloration of the species.

LYGOSOMA LATERALE, Dum. Bib.

Scincus lateralis, SAY, Long's Exped. II, 1823, 324.—HOLBROOK, N. Am. Herp. II, 1st ed. I, 1836, 71.
Lygosoma lateralis, DUM BIB Erp. Gen. V, 1839, 719.—HOLBROOK, N Am. Herp. II, 1842, 133; pl. xix.

3132. Fort Inge, Texas.—3135. San Antonio. Dr. Kennerly.

CROTALUS DURISSUS, Linn.—Rattlesnake.

Crotalus durissus, LINN. Syst. Nat I, 1766, 372.—HOLBROOK, N. Am. Herp. III, 1842, 9; pl. i.—B. & G Catal. N. Am. Serpents, 1853, 1.

SPEC. CHAR.—Head small, narrow; superciliaries, frontals, and other large plates smooth and uncorrugated. Sub-orbital chain continuous; two rows between this and the labials. Labials 12–14 above, 13–16 below. Dorsal rows of scales 23–25, all carinated; carinations of outer row obsolete. No defined light lines on head or body, and no lighter edgings to the plates of the head: tail black. Above yellow or yellowish brown, with a double dorsal series of brown or blackish confluent rhomboids, with another series opposite on each side. A distinct narrow reddish dorsal stripe, three or four scales in width from head to tail. A broad dark stripe from the lower and posterior angle of the orbit across the angle of the mouth; this stripe sometimes obsolete.—(Kennicott)

No. 277. Fort Smith, Arkansas. Dr. G. G. Shumard.

CROTALUS ATROX, B. & G.

Crotalus atrox, B & G. Catal. N. Am. Serp. 1853, 5.

SPEC. CHAR.—Nose broad, obtuse. Plates of head elongated and much imbricated; two rather small anterior frontals in contact; two larger imbricated plates between these and each superciliary; space enclosed occupied by scales of smaller size but larger than those between the posterior half of the superciliaries. Superciliaries usually bordered by a row of larger scales, of which the anterior is much largest. All the larger plates of the crown very much corrugated, and the anterior plates generally with their edges turned up; this, together with their corrugated surfaces, presenting a very rough appearance. Three rows of scales between the sub-orbital chain and labials. Labials 15–16 above, 15–17 below. Dorsal rows 25–27; central row strongly carinated, with additional small ridges converging to the apex of each scale. Tail with black rings. Color dull yellowish brown, with a dorsal series of rhomboidal blotches, margined by lighter lines much the same as in *C. adamanteus*, but the lateral markings represented in *C adamanteus* all obsolete. Abdomen without spots. A single transverse light line on the superciliary, sometimes obsolete Light stripe from posterior angle of orbit below the superciliary to the upper labials in front of the angle of the mouth. No light lines in front of the nostril, or light edgings on rostral —(Kennicott)

No. 4225. Rocky Dell creek, Indian territory. B. Möllhausen.

CROTALUS CONFLUENTUS, Say.

Crotalus confluentus, Say, Long's Exped. II, 1823, 48.—B. & G. Catal. N. Am. Serp. 1853, 8
Crotalus lecontii, Hallow. Pr. A. N. Sc. VI, 1851, 8.

Spec Char.—Plates on crown irregular, angulated, imbricated, and frequently tuberculated, smoother and less elongated than in *C. atrox*. Three, rarely four, rows of scales between the sub-orbital series (which extends to the centre of the orbit) and the labials. Labials 14-18 above, 14-18 below. Dorsal rows of scales 25-29. Dorsal blotches brown, margined by narrow whitish lines, sub-quadrate, emarginate before and behind; posteriorly intervals greater and spots transversely quadrate, ultimately becoming twenty-five to thirty half rings. Colors not deeper posteriorly in old specimens. Two transverse lines on superciliary enclosing about one-third, sometimes obsolete. Light stripe from orbit below superciliary to angle of jaw passes angle of mouth on the second or third row of scales above the labials. Rostral margined with lighter.—(Kennicott.)

No. 291. Canadian river. B. Möllhausen.

CROTALOPHORUS MILIARIUS, Holbrook.

Crotalus miliarius, Linn. Syst. Nat. I, 1766, 372.
Crotalophorus miliarius, Holbrook, N. Am Herp. III, 1842, 25; pl xv.—B. & G. N. Am. Serp. 1853, 11.

Spec. Char.—Twenty-one to twenty-three dorsal rows of scales all carinated, the two lateral rows but slightly. Greyish ash with black blotches. A vertebral reddish line. Vertical plate obtuse posteriorly, widening but little anteriorly, slightly concave on the sides. A narrow white line from the lowest point of the orbit passes obliquely backward to the angle of the mouth.—(Kennicott.)

No. 489. Fort Smith, Arkansas. B. Möllhausen.

TOXICOPHIS PISCIVORUS, B. & G.

Trigonocephalus piscivorus, Holbrook, N. Am. Herp. 1st ed. II, 1838, 63: pl. xiii.—Ib. 2d ed. III, 1842, 33; pl. vii.
Toxicophis piscivorus, B. & G. Catal. N. Am. Serp. 1853, 19.

Spec. Char.—No loral. Inferior wall of orbit constituted by third labial. Twenty-five dorsal rows. Dark chestnut brown, with indistinct vertical dark bars. Line from superciliary along the edge of the head, through the middle of the second supra-labial row. A second line from the lowest point of the orbit parallel to the first.

No. 823. Fort Smith, Arkansas. B. Möllhausen.

EUTAENIA PROXIMA, B. & G.

Coluber proximus, Say, Long's Expedition, I, 1823, 187.
Eutainia proxima, B. & G. Catal N Am Serp. 1853, 25.

Spec. Char.—Body stouter than *E. saurita* or *E. fairei*. Brown or blackish above. Three longitudinal stripes; the dorsal ochraceous yellow or brown; lateral greenish white or yellow on the third and fourth lateral rows. Dorsal stripe one and over two half scales wide. Lateral row of scales usually the color of the abdomen. Total length about 3½ times that of the tail. Dorsal rows 19.—(Kennicott.)

No. 762. San Antonio, Texas.—742. Between Pecos and Rio Grande. Dr. Kennerly.

EUTAENIA DORSALIS, B. & G.

Eutainia dorsalis. B. & G. Catal. N Am. Serp 1853, 31.

Dimensions of *E. sirtalis*. Color olivaceous green. Dorsal stripe broad, whitish yellow, margined with black. A row of spots above the lateral stripe. Lateral stripe on second and third rows. Dorsal rows 19.

No. 978. Fort Conrad, New Mexico. Dr. Kennerly.

EUTAENIA VAGRANS, B. & G.

Eutainia vagrans, B. & G. Catal N. Am. Serp. 1853, 35.—GIRARD, Herp U. S Ex. Ex 1858, 154; pl. xiv, figs. 5—10.

SPEC. CHAR —Head large and high. Eight labials above, sixth and seventh very large, higher than wide; the sixth extending above the level of the lower edge of the eye; the seventh much larger than the fifth. Light olive brown above, with two series of black spots on each side, the upper of which encroaches upon the dorsal stripe, constricting it at regular intervals, while the lower encroaches upon the indistinct lateral stripe. Lateral stripe on second and third rows. Dorsal rows 21.—(Kennicott.)

No. 926. Near the Gold mountains. Möllhausen.

EUTAENIA MARCIANA, B. & G.

Eutainia marciana, B. & G. Catal. N. Am. Serp. 1853, 37.

SPEC. CHAR.—Head triangular, short, very broad posteriorly; nose pointed. Upper labials eight, sixth largest, seventh nearly as large. Body rather stout, sub-cylindrical; tail short. Prominent color light brown; a vertebral paler line and one lateral on each side, more or less indistinct. Three series of square black spots on each side, of about 56–60 in each series, from occiput to anus. Sides of head black, with a crescentic patch of yellowish posterior to the labial plates. Three and sometimes four black vittæ radiating from the eye across the jaws. A double white spot with a black margin on the suture of occipital plates. Dorsal rows 21.

No. 856. Pecos to Rio Grande, Texas.—1424. San Antonio, Texas. Dr. Kennerly.

NERODIA WOODHOUSII, B. & G.

Nerodia woodhousii, B. & G. Catal. N. Am. Serp. 1853, 42.

SPEC CHAR.—Form of *N. erythrogaster*, but the head broader behind and more flattened above. Twenty-five dorsal rows of scales, all carinated. Three rather large post-orbitals; ante-orbital narrow; loral large. Three series of quadrangular dark blotches on a brownish clay-colored ground. The middle series separated by narrow white lines, the latter by intervals of the ground color, wider than themselves; the three series perfect to the head. A double yellow occipital spot; a yellow spot between the superciliary and vertical plates; an indistinct black line from the posterior rim of the eye to the angle of the mouth. Abdomen unspotted.—(Kennicott.)

No. 1318. Antelope creek, Arkansas.—1325. Near Canadian river, Arkansas. H. B. Möllhausen.

NERODIA ERYTHROGASTER, B. & G.

Coluber erythrogaster, SHAW, Gen. Zool. III, 1804, 458.
Tropidonotus erythrogaster, HOLBROOK, N. Am. Herp. 2d ed. III, 1842, 33; pl. viii.
Nerodia erythrogaster, B. & G. Catal. N. Am. Serp. 1853, 40.

SP. CH.—Head elongated, narrowing forwards; occipital region flattened; convex on the snout. Vertical plate pentagonal, very large, as are also the occipitals. Three post-orbitals. An elevated loral. Dorsal rows of scales 23 in number, all very strongly carinated. Uniform dark bluish black above, lighter on the sides; a lateral or external band of dull blue extending on the abdominal scutellæ. Body beneath uniform dull coppery yellow, sometimes with the anterior edge of each scutellæ bluish. No distinct spots on the abdomen, as in *N. sipedon*. Indications of three series of dorsal blotches, as in *N. woodhousii*, in young specimens.

No. 1324. Fort Smith, Arkansas. Dr. Shumard.

HETERODON NASICUS, B. & G.

Heterodon nasicus, B. & G. Stansbury's Expl. Salt Lake, 1852, 352.—IB. Catal. N. Am. Serp. 1853, 61.

SPEC. CHAR.—Vertical broader than long. Rostral excessively broad and high. Azygos plate surrounded behind and on the sides by many small plates, (12–15.) A second loral. Labials short and excessively high. Dorsal rows of scales 23, exterior alone smooth. A dorsal series of about 50 blotches, with four or five others on each side. Body beneath, black. A narrow white line across the middle of the superciliaries; a second behind the rostral. A broad, dark patch from the eye to the angle of the mouth, crossing the last two labials.

No. 1280. Near Canadian, Arkansas. Möllhausen.

PITYOPHIS BELLONA, B. & G.

Churchillia bellona, B. & G., Stansbury's report, 1852, 350.
Pituophis bellona, B. & G., Catal. N. A. Serpents, 1853, 66.
Pituophis affinis, HALLOWELL, Pr. A. N. Sc. VI, 1852, 181.

SPEC. CHAR.—Head broad behind, tapering to the snout. Snout rather pointed, but less so than in *P. sayi*, elevated. Crown flattened between the eyes and posteriorly. Vertical plate broad, anteriorly narrow, and elongated posteriorly. Ante-orbitals 2, sometimes 1; post-orbitals 3 or 4, Frequently a supplemental plate before the vertical. Dorsal rows of scales 29 to 35, the 7 outer rows smooth. Head spotted with black; transverse frontal bar from one orbit to the other well marked; the oblique post orbital stripe rather narrow. Color of the body whitish yellow; a dorsal series of 45–65 subquadrate blotches from head to anus, transversely elongated posteriorly; 3 or 4 smaller series on each side. The lateral blotches longitudinally elongated near the head; vertically elongated posteriorly when they form a series of jet black vertical bars. Ten or fifteen transverse black bands on the tail. Abdomen yellow with an external series of black spots on each side.—(Kennicott.)

No. 1528. San Antonio to Austin, Texas. Möllhausen.

ARIZONA ELEGANS, Kennicott.

Arizona elegans, KENNICOTT, in Mex. Bound. Report, II, 1859; Baird Rep. Reptiles, 18, plate xiii.

SP. CH.—Body rather more slender than in *Pityophis sayi* and *P. bellona*, and head narrower; otherwise bearing a general resemblance in form to these and other species of *Pityophis*, especially in the protruding and recurved rostral. The tail forms nearly one-sixth of the total length Head depressed anteriorly, arched, and much more elevated posteriorly, where it is not much wider than high, nor much wider than the neck. Vertical plate sub-pentagonal, broad in front, tapering, and very acute posteriorly. Occipitals large, perfect, longer than the vértical. Supercilliaries small. Pre-frontals of greater longitudinal extent than post frontals, separated for more than two-thirds of their length by the apex of the nostril, the anterior angle extending down in front of the pre-nasal to below the level of the nostril. Post-nasal more than twice as large as pre-nasal. Loral very narrow, as long as both nasals together, and longer than the post-frontals. One ante-orbital, sometimes a second very small one below; two post orbitals of about equal size. Two narrow and much elongated temporal shields just behind the post-orbitals, entirely filling the space between the seventh upper labial and the occipital; behind this, small scale-like temporal shields. Eight upper labials, seventh twice as large as any other. Fourteen lower labials, seventh largest. Dorsal scales in 29 to 31 rows all perfectly smooth; central rows not much smaller; outer row largest, but not as high as long.

Body whitish yellow above, with a dorsal series of transversely quadrate light olive brown blotches and two smaller lateral series on each side. Abdomen uniform clear whitish. The dorsal series of blotches are indistinctly edged white blackish; they cover three or four scales longitudinally and twelve or thirteen rows transversely, and are separated by regular intervals of one and a half scales of the ground color. The lateral blotches become more or less indistinct in age from a dark suffusion over the ground color. Each dorsal scale occupying a dark blotch is edged with lighter. In the young, the head above is light brown, with a blackish bar across the post frontals and through the eyes to the angles of the mouth; there is also a dark blotch below the eye and some smaller ones on the crown posteriorly. In older specimens these markings become obsolete, leaving the head uniform light glossy olive brown.—(Kennicott.)

1705. Between Pecos river and Rio Grande. Dr. Kennerly.

SCOTOPHIS ALLEGHANIENSIS, B. & G.

Coluber alleghaniensis, HOLBROOK, N. Am. Herp. I, 1836, 111; pl. xx.—IB. 2d ed. III, 1842, 85; pl. xix.
Scotophis alleghaniensis, B. & G. Catal. Serp. 1853, 73.

SPEC CHAR.—Snout broad, rounded, much depressed. Vertical plate longer than broad. Superciliaries proportionally small. Anterior frontals large. Occipitals rather small. Loral, nasals and anterior upper labial less elevated than in *S. obsoletus* and *S. lindheimeri*. 25–27 dorsal rows, central rows elongated, moderately carinated. Entirely pitch black above. Abdomen slate colored posteriorly; yellow with quadrangular black blotches anteriorly. Sometimes lighter in the young, with dark dorsal blotches on a light ground.—(Kennicott.)

No. 2257. Fort Smith, Arkansas. H. B. Möllhausen.

SCOTOPHIS EMORYI, B. & G.

Scotophis emoryii, B. & G Catal Serpents, 1853, 157.

SPEC. CHAR.—Head narrow, elongated, widening suddenly behind the eyes. Snout elongated, narrow, elevated, obtuse; outline anterior to the eye subquadrangular. Crown flattened, eyes very large. Vertical plate much longer than wide, narrow posteriorly, much wider in front; superciliaries narrow. Loral large, trapezoidal, acutely angled behind. Dorsal rows 25–27, carinations very faint, barely perceptible on the central rows. Above ash grey with a dorsal series of transverse brown blotches, on each side of which are two others of smaller size; indistinct traces of a third. A frontal brown vitta passing back through the eye, and crossing the angle of the mouth on to the side of the neck. Two blotches on the upper labials which are not margined with black.—(Kennicott.)

No. 1716. Pecos to Rio Grande, Texas. Dr. C. B. Kennerly.

OPHIBOLUS EVANSII, Kennicott.

SPEC. CHAR.—Light olivaceous brown or grey, with a dorsal series of about sixty subquadrangular dark chestnut brown blotches, emarginate anteriorly and posteriorly, and two smaller lateral series on each side. Dorsal scales in 25 rows.—(Kennicott.)

No. 1702. Canadian river, Arkansas. H. B. Möllhausen.

OPHIBOLUS SPLENDIDUS, B. & G.

Ophibolus splendidus, B. & G. Catal. N. A. Serpents, 1853, 83.

SPEC. CHAR.—Black above; the sides black with a yellowish white spot on each scale. The body crossed by transverse bands consisting of yellowish white spots, one in each scale. Throat marked with black. Dorsal rows 23.—(Kennicott.)

No. 1709. Pecos to Rio Grande, Texas. Dr. C. B. R. Kennerly.

MASTICOPHIS TESTACEUS, B. & G.

Coluber testaceus, SAY, Long's Exped. II, 1823, 84.—HOLBROOK, N. Am. Herp. III, 1842, 63; pl. xiii.
Masticophis testaceus, B. & G. Catal. N. Am. Serp. 1853, 151.
Psammophis flavigularis, HALLOWELL, Pr. A. N. Sc. VI, 1852, 178.
Masticophis flavigularis, B. & G. Catal. N. Am. Serp. 1853, 99.

SPEC. CHAR.—In alcohol, light dull yellow, tinged with brown above. Beneath, two longitudinal series of blotches, most distinct anteriorly. When the epidermis is removed, the whole animal appears of a soiled white. In life tinged with rose color; some specimens entirely brick red. Tail one-fifth of the total length. Dorsal rows 17.—(Kennicott.)

No. 1994. Pecos to Rio Grande. Dr. Kennerly.

LEPTOPHIS MAJALIS, B. & G.

Leptophis majalis, B. & G Catal. N. A. Serp. 1853, 107.

SPEC. CHAR.—Reddish green above, yellowish white beneath. Body proportionally stouter and tail shorter than in *L. æstivus*. Snout and whole head, including vertical, longer than in latter species. Dorsal scales in 17 rows.

No. 1427. Fort Smith, Arkansas. Dr. Shumard.

DIADOPHIS DOCILIS, B. & G.

Diadophis docilis, B. & G. Catal. N. A. Serp. 1853, 114.

SPEC. CHAR.—Body above uniform ash grey; yellowish white beneath, sparsely spotted with black. A proportionately broad yellowish white occipital ring. Dorsal scales in 17 rows.—(Kennicott.)

No. 2078. Pecos to Rio Grande, Texas. Dr. Kennerly.

BUFO AMERICANUS, Leconte.

Plate XXV, Fig. 2.

Bufo americanus, Leconte, Holbrook's N. Am. Herpetology, V, 1842, 17; pl. iv.—Girard, in Reptiles U S. B. S. 1859, 25; pl. xxxix, fig. 1–4.

Spec. Char.—Head above grooved, a ridge from the snout to the occiput, hence at almost right angle to the tympanum. The ridge from either side is slightly diverging from before backwards. Skin upon the head thick, and adhering to the skull. Parotids sub-reniform. Tympanum well developed. A subgular vocal sac in the male sex. Limbs large and stout. First finger larger than the second. A large carpal corneous disk and a small one. No membranous fold at the inner lower edge of the tarsus. Toes semi-palmated. Two metatarsal tubercles, the internal large and spade-shaped, the external small and sub-conical. Large papilla on the back. Greenish or yellowish brown above, with scattered black patches or spots. A vertebral fuscous vitta or streak. Two black patches under the eye, occasionally spotted about the nostrils and over the jaws. Beneath dirty yellowish.—(Girard.)

2558, 2559. Fort Smith, Arkansas. Dr. Shumard.

BUFO NEBULIFER, Girard.

Bufo granulosus, B. & G. Pr. A. N. Sc. VI, 1852, 173. (Not of Spix.)
Bufo nebulifer, Girard, Pr. A. N. Sc. VII, May, 1854, 87.

Spec. Char.—Upper surface of head deeply excavated or concave; concavity bordered with conspicuous ridges. Skin very thin and firmly adhering to the skull. Parotid small; eyes and tympanum rather large. Tongue moderate, broadest posteriorly. Upper jaw emarginated. Inner finger much longer than the second, which is longer than the fourth. Two well developed carpal disks. No membranous fold along the tarsus. Toes semi-palmated; two metatarsal tubercles. Palm of hands and sole of feet tuberculous. Yellowish brown with a dorsal broad streak of the same tint. An interocular black bar on either side of the dorsal streak extending to the whole length of the body. Sides maculated; upper part of limbs barred. Beneath unicolor in the adult, spotted in the young.—(Girard.)

2592. San Pedro, Texas.—2601. San Antonio. Dr. Kennerly.

BUFO WOODHOUSII, Girard.

Plate XXV, Fig. 1.

Bufo dorsalis, Hallowell, Pr. A. N. Sc. VI, 1852, 181.—Ib. Sitgreaves' Report, 1853, 142; pl. xix. (Not of Spix.)
Bufo woodhousii, Girard, Pr. A. N. Sc. VII, May, 1854, 86.

2632. Canadian river; and 2646. Antelope Hills. H. B. Möllhausen.

BUFO COGNATUS, Say.

Plate XXVI.

Bufo cognatus, Say, Long's Exped. II, 1823, 190.—Holbrook, N. Am. Herp. V, 1842, 21; pl. v.—B. & G. Marcy's Report, 1853, 242; pl. xi.

Red river, Arkansas. *Bufo lentiginosus* is represented on Plate XXVII for comparison.

ACRIS CREPITANS, Baird.

Hylodes gryllus, Dekay, N. Y. Zool. III, 1842, 70; pl. xxii, fig. 61. (Not of Leconte)
Acris crepitans, Baird, Pr. A. N. Sc. VII, 1854, 59.

Spec. Char.—Brownish above. The median region of head and body above bright green; a dark triangle between the eyes. Three oblique blotches on the sides, nearly equidistant; the first behind the eye, the last on the flanks and running up on the back; all usually margined with lighter. A narrow white line from the eye to the arm. Beneath yellowish white. Inferior face of thigh plain. Tibia a little more than half the length of the body; foot rather smaller. Head rather obtuse, scarcely longer than broad. Web of hind foot extending to the penultimate articulation of the fourth toe.

Fort Smith, Arkansas.—3270. Gypsum creek. Möllhausen.—3262. San Pedro, Texas. Dr. Kennerly.

RANA CATESBIANA, Shaw.—Bull Frog.

Rana catesbiana, SHAW, Gen. Zool. III, Amphibia, 1802, 106; pl. xxxiii.

Rana boans, LACEP. Hist. Nat. Quad. Ovip. 1798.

Rana pipiens, HARLAN, Sill. Am Jour. X, 1825, 62.—HOLBROOK, N. Am. Herpet. IV, 1842, 77; pl. xviii. (Not of Linnæus.)

Fort Smith, Arkansas. Dr. Shumard.—3336. Shawnee village. H. B. Möllhausen.—3319. San Antonio. Dr. Kennerly.

RANA CLAMITANS, Daud.

Rana clamitans, DAUDIN, in Sonnini and Latreille, Hist. Rept. II, 1802, 157.—HOLBROOK, N. Am. Herp. IV 1842, 85; pl. xx.

Rana clamata, DAUDIN, Hist. Nat. Rept. VIII, 1803, 104.—DUM. BIB. Erp. Gen. VIII, 1841, 373.

Rana fontinalis, LECONTE, Ann. N. Y. Lyc. I, 1825, 281.—HOLBROOK, N. Am. Herp. IV, 1842, 87; pl. xxi.

Ranaria melanota, RAFINESQUE, Annals of Nature, 1820, No. 25.

Rana flavoviridis, HARLAN, Sill. Am. Jour. X, 1825, 58.

Rana horiconensis, HOLBROOK, N. Am. Herp. 1st ed. III, 1838, 91; pl. xviii.

Rana nigricans, AGASSIZ, Lake Superior, 1850, 379; pl. vi, figs. 4, 5.

Fort Smith, Arkansas. Lieutenant Whipple.

RANA HALECINA, Kalm.

Rana halecina, KALM.—DAUDIN, Hist. Nat. Rept. VIII, 1803, 122.—HOLBROOK, N. Am. Herp. IV, 1842, 91; pl. xci.

Rana pipiens, GMELIN, Syst. Nat. 1788, 1052.

Rana utricularis, HARLAN, in Silliman's Journal, X, 1825, 60.

3323. Fort Smith, Arkansas. Dr. Shumard.

RANA BERLANDIERI, Baird.

Rana berlandieri, BAIRD, Rep. Mex. Bound. Survey, II, 1859, Reptiles, 27; pl. xxxvi.

SPEC. CHAR.—Size large. Body stout, robust. Eye distant not quite one and a half times its diameter from tip of snout, and contained two and two-third times in the length of jaw from rictus. Tympanum two-thirds the diameter of the eye. A vocal vesicle on each side of the head. A glandular fold on each side the jaw, and another broad and depressed on each side of the body. Between these is one pair of ridges along the coccyx; several pairs more interrupted anterior to it. Skin corrugated and irregular, quite pustular in some specimens. Feet webbed from the bulb of the toes; web excavated on the inner edges; last joint of longest toe free. Femur about half the length of body, shorter than the tibia.

Color above greenish olive, with distant sub-circular blotches of darker, scarcely areolated in the preserved specimens. Beneath yellowish white, with brown mottlings on the throat. An indistinct whitish line on the side of the head, especially in the young; the lateral ridge bronzed.

San Antonio to Fort Inge. Dr. Kennerly.

NECTURUS LATERALIS, Baird.

Triton lateralis, SAY, Long's Exped. R. Mts. I, 1823, 5.

Menobranchus lateralis, HARLAN, Ann. N. Y. Lyc. I, 1825, 221.—HOLBROOK, N. Am. Herp. III, 1842, 119; pl. xxx.

4058. Fort Smith, Arkansas. Dr. Shumard.

J. H. Richard del.

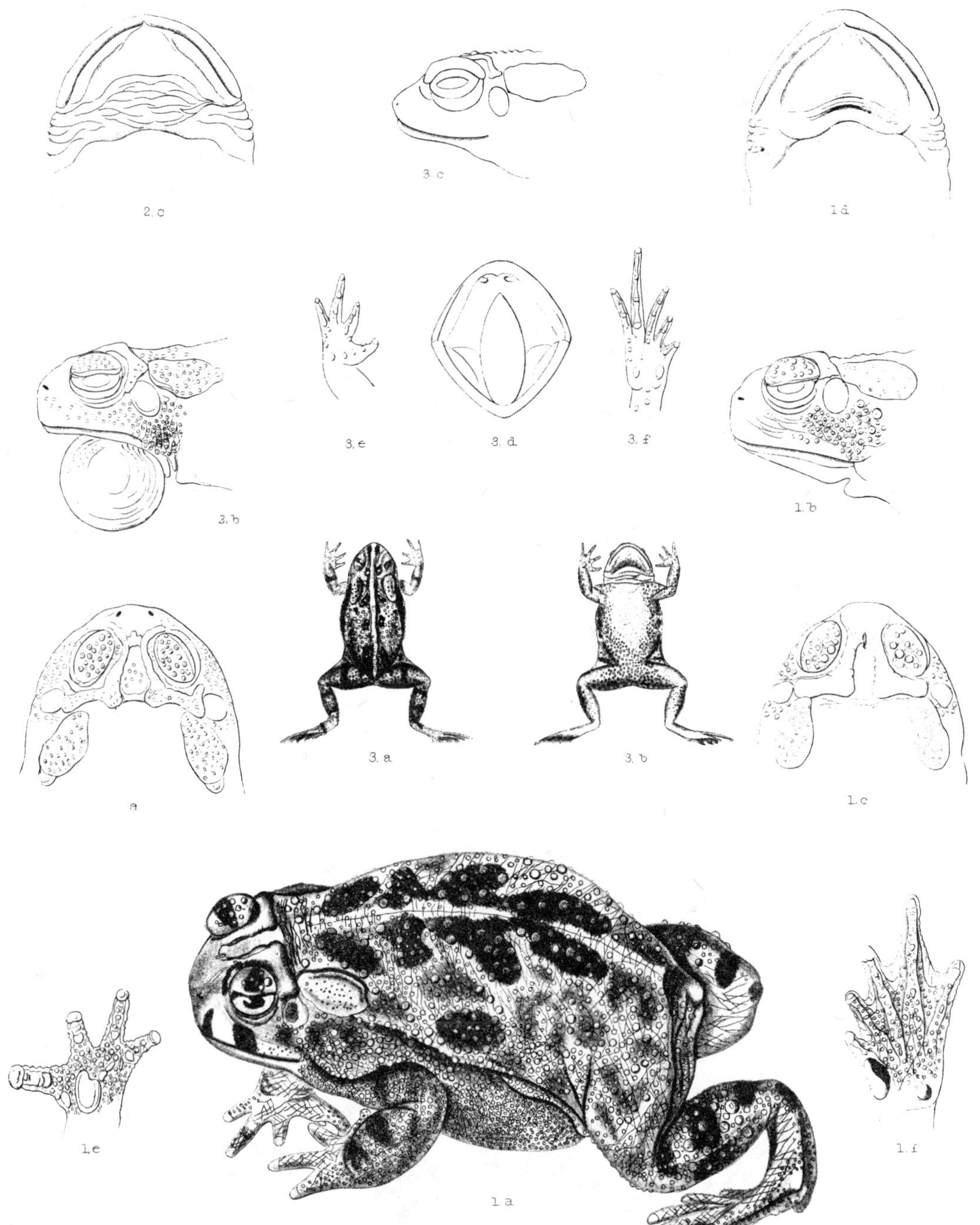

J.H. Richard del. C. Metzeroth sc.

J. H. Richard del.

REPTILES OF THE BOUNDARY

Spencer F[ullerton] Baird

UNITED STATES AND MEXICAN

BOUNDARY SURVEY,

UNDER THE ORDER OF

LIEUT. COL. W. H. EMORY,

MAJOR FIRST CAVALRY, AND UNITED STATES COMMISSIONER.

REPTILES

OF THE BOUNDARY,

BY

SPENCER F. BAIRD,

ASSISTANT SECRETARY OF THE SMITHSONIAN INSTITUTION.

WITH NOTES BY THE NATURALISTS OF THE SURVEY.

CHELONIA.[1]

1. ASPIDONECTES EMORYI, Agassiz.—Soft-shelled Turtle.

Aspidonectes Emoryi, AGASSIZ, Contributions to Nat. Hist. U. S., I, 1857, 407.

Lower Rio Grande of Texas. J. H. Clark.

2. GYPOCHELYS LACERTINA, Agassiz.—Alligator Turtle.

Gypochelys lacertina, AGASS. Cont. I, 1857, 414.
Chelydra lacertina, SCHWEIGGER, Prodromus.
Chelonura temminckii, HOLBROOK.

Lowlands of Texas.

3. OZOTHECA TRISTYCHA, Agassiz.

Ozotheca tristycha, AGASSIZ, Cont. I, 1857, 425.

Medina river, Texas. Dr. Kennerly.

4. THYROSTERNUM SONORIENSE, Agassiz.

Kinosternon sonoriense, LECONTE, Pr. A. N. Sc. VII. 1853, 184.
Thyrosternum sonoriense, AGASSIZ, Contributions, I, 1857, 428; pl. v, figs. 8–11.

Tucson and Guadalupe cañon. Dr. Kennerly.

5. PLATYTHYRA FLAVESCENS, Agassiz.

Platythyra flavescens, AGASSIZ, Contributions, I, 1857, 430; pl. v, fig. 12–15.

Texas, and valley of Gilla.

6. PTYCHEMYS MOBILENSIS, Agassiz.

Emys mobilensis, HOLBROOK, N. Am. Herp. I, 1842, 71; pl. ix.
Ptychemys mobilensis, AGASSIZ, Contributions, I, 1857, 433; pl. iii, figs. 14—16

Guadalupe mountains and Pecos river, Texas.

7. TRACHEMYS ELEGANS, Agassiz.

Emys elegans, MAXIM. Reise Nord. Am. I, 1839, 213.
Trachemys elegans, AGASSIZ, Cont. I, 1857, 435.

Brownsville, Texas.

[1] NOTE.—The descriptions of genera and species in the present article have been either omitted entirely, or in most cases greatly curtailed, owing to the fact that the report, as prepared, greatly exceeded the limits of space assigned.

A number of the species have been determined, or their characters better defined, by Dr. Girard and Mr. Kennicott. The contributions of those gentlemen have their name attached.—(S. F. BAIRD.)

8. CHRYSEMYS OREGONENSIS, Agassiz.

Emys oregonensis, HARLAN, Amer. Jour. Sc. XXXI, 382.
Chrysemys oregonensis, AGASSIZ, Cont. I, 1857, 440.

Guadalupe mountains, Texas.

9. XEROBATES BERLANDIERI, Agassiz.—Texas Gopher.

Xerobates berlandieri, AGASSIZ, Cont. I, 1867, 447.

Lower Rio Grande.

SAURIA.

10. ALLIGATOR LUCIUS, Cuv.

3184. Brownsville, Texas. Capt. Van Vliet.

11. SCELOPORUS CLARKII, B. & G.

Sceloporus clarkii, B. & G. Pr. A. N. Sc. VI, Aug. 1852, 127.
Sceloporus magister, Hallow. Pr. A. N. Sc. VII, 1854, 93.

2940. Sonora. J. H. Clark.—2897. Alamo de Parras, Mex. Lt. Couch.—2964. Janos. Dr. Kennerly.—2965. Sta. Madelina. A. Schott.—2954. Los Nogales, Sonora. Dr. Kennerly.

12. SCELOPORUS SPINOSUS, Wieg.

Plate XXIX, Figs. 4—6.

Sceloporus spinosus, Wiegmann, Isis, 1828, 369—Ib. Herp. Mex. 1834, 50; pl. vii.

2968. El Paso. J. H. Clark.—2943. Indianola, Tex. J. H. Clark.—2945. Laredo to Camargo. A. Schott.

13. SCELOPORUS CONSOBRINUS, B. & G.

Sceloporus consobrinus, B. & G. Marcy's Report, 1853, 237.

4109. R. San Pedro, Tex. J. H. Clark.—2899. Janos to San Luis Springs. Dr. Kennerly.

14. SCELOPORUS TORQUATUS, Wiegm.

Agama torquata, Peale & Green, Pr. A. N. Sc. II, 131.
Sceloporus torquatus, Wiegmann, Isis, 1828, 369.

2959. Laredo to Camargo. A. Schott.

15. SCELOPORUS POINSETTII, B. & G.

Plate XXIX, Figs. 1—3.

Sceloporus poinsettii, B. & G. Pr. A. N. Sc. VI, Aug. 1852, 126.

2952. Sonora. J. H. Clark.—2955. Limpia mountains, Tex. Dr. Kennerly.—2948. San Pedro. J. H. Clark.

16. SCELOPORUS ORNATUS, Baird.

Sceloporus ornatus, Baird, Pr. A. N. Sc. Dec. 1858.

Sp. Ch.—Dorsal scales in about 64 oblique series, with but slight carination, mucronation, and denticulation. Femoral pores, 12. A well marked black cervical collar, complete above, and margined with yellowish. Color dark green above, nearly black towards the median line. Back with small yellowish spots.

2845. Patos, Coahuila. Lt. Couch.

17. SCELOPORUS THAYERI, B. & G.

Sceloporus thayeri, B. & G., Pr. A. N. Sc. VI, Aug. 1852, 127.

2896. San Pedro, Texas, J. H. Clark.—2887. Indianola. J. H. Clark.

18. SCELOPORUS SCALARIS, Wiegmann.

Sceloporus scalaris, Wiegm. Herp. Mex. 1834, 52: tab. viii, f. 2.

2284. Los Nogales, Sonora. Dr. Kennerly.

19. SCELOPORUS MARMORATUS, Hallow.

Sceloporus marmoratus, Hallow. Pr. A. N. Sc. VI, Oct. 1852, 178.
Sceloporus delicatissimus, Hallow. Pr. A. N. Sc. VI, Oct. 1852, 178.

2916. Medina, Texas. J. H. Clark.—4116. Redmond's Ranch, Texas. J. H. Clark.

20. SCELOPORUS COUCHII, Baird.

Sceloporus couchii, Baird, Pr. A. N. Sc. Dec. 1858.

Sp. Ch.—General appearance of *S. marmoratus*, Hall. Cephalic plates smooth. Scales very small. About 80 oblique dorsal rows from head to tail. Femoral pores, 25. Color above dark green, with two lateral light stripes, separated by 18 rows of scales. Back with irregular spots. Sides with a white band from groin. An obsolete dark indigo patch on each side the belly, widely separated below. Sides of jaw banded transversely with blue and whitish. A sub-circular indigo patch in front of shoulder, surrounded by light blotches.

2904. Pesquieria Grande, Mexico. Lt. Couch.—2739. Santa Caterina, New Leon. Lt. Couch.

21. EUPHRYNE OBESA, Baird.

Plate XXVII.

Euphryne obesus, Baird, Pr. A. N. Sc. Dec. 1858.

Sp. Ch.—Width of head nearly equal to distance from nose to ear. Tail shorter than the body. General color of the young, olive green, with five broad transverse bars above from head to base of tail, and about as many on the tail; these rings yellow, dotted with red. Beneath, pea green, dotted with black. With increasing age the bands become obsolete and disappear, the general color becoming reddish olive.

2774 Fort Yuma. Major Thomas.—4172. Do. A. Schott.

22. CROTAPHYTUS COLLARIS, Holbrook.

Agama collaris, Say, Long's Exped. II, 1823, 252.
Crotaphytus collaris, Holb. N. Am. Herp. II, 1842, 72: pl. x.

2694. Sonora. Dr. Kennerly.—2710. San Antonio to El Paso. J. H. Clark.

23. CROTAPHYTUS RETICULATUS, Baird.

Crotaphytus reticulatus, Baird, Pr. A. N. Sc. Dec. 1858.

Sp. Ch.—Infra-orbital chain in a series of six or eight nearly equal plates. Scales on the gular fold much smaller than those between the fore legs. Above ashy gray, with a hexagonal reticulation of lighter; the interstices here and there dark brown. Chin and throat reticulated. Neither black collar nor light spots. More closely related to *Crotaphytus collaris* than to *Crotaphytus* (*Gambelia*) *wislizenii.*

2692. Ringgold Barracks, Texas. J. H. Clark.—2731. Laredo, Texas. A. Schott.

24. CROTAPHYTUS WISLIZENII, B. & G.

PLATE XXXI.

Crotaphytus wislizenii, B. & G. Pr. A. N. Sc. VI, April, 1852, 69.
Crotaphytus gambelii, B. & G. Pr. A. N. Sc. VI, Aug. 1852, 126.
Crotaphytus fasciatus, HALLOWELL, Pr. A. N. Sc. VI, Dec. 1852, 207.

2704. San Antonio to El Paso. J. H. Clark.—2685. Sonora. J. H. Clark.

25. UTA STANSBURIANA, B. & G.

Uta stansburiana, B. &. G. Pr. A. N. Sc. VI, April, 1852, 69.

2745. San Antonio to El Paso. J. H. Clark.—2740. Gila desert. A. Schott.

26. UTA ORNATA, B. & G.

Uta ornata, B. & G., Pr. A. N. Sc. VII, Aug. 1852, 126.

2700. San Pedro, Texas. J. H. Clark.—2764. Eagle Pass, Texas. A. Schott.—2737. Sonora. J. H. Clark.

27. UTA ORNATA VAR. LINEARIS, Baird.

Similar in markings to the described character of *U. ornata*, but with four linear interrupted black stripes instead of transverse bands.

2759. Los Nogales, Sonora. Dr. Kennerly.

28. UTA SYMMETRICA, Baird.

Uta symmetrica, BAIRD, Pr. A. N. Sc. Dec. 1858.

SP. CH.—Larger dorsal scales in four regular series, two on either side of the median smaller ones. Head short, depressed, one and a half times as wide as deep. Tail one and a half times the head and body. General color light brown above, the belly white. Sides with broad transverse bands of blackish. Size of *U. ornata.*

2760. Gila river. A. Schott.

29. UTA SCHOTTII, Baird.

Uta schottii, BAIRD, Pr. A. N. Sc. Dec. 1858.

SP. CH.—Dorsal scales and size as in the last. Head pointed, narrow, nearly or quite as high as wide. General color nearly black, scarcely lighter beneath. Back with small blue spots. Tail banded laterally with the same.

3761. Santa Madelina. A. Schott.

30. UTA GRACIOSA, Baird.

Urosaurus graciosus, HALLOWELL, Pr. A. N. Sc. VII, June, 1854, 92.

4128. Colorado river, California. A. Schott.

31. DIPSOSAURUS DORSALIS, Hallowell.

PLATE XXXII, FIGS. 7—13.

Crotaphytus dorsalis, B. & G. Pr. A. N. Sc. VI, Aug. 1852, 126.
Dipsosaurus dorsalis, HALL, Pr. A. N. S. VII, June, 1854, 92.

4178. Colorado desert. A. Schott.

32. CALLISAURUS VENTRALIS, Baird.

Homalosaurus ventralis, HALLOWELL, Pr. A. N. Sc. VI, October, 1852, 179.

2981. Colorado desert. A. Schott.

33. HOLBROOKIA MACULATA, Girard.

Holbrookia maculata, GIRARD, Pr. Am. Assoc. for 1850–1851, 201

2807. San Antonio to El Paso. J. H. Clark.

34. HOLBROOKIA PROPINQUA, B. & G.

Holbrookia propinqua, B. & G., Pr. A. N. Sc. VI, Aug. 1852, 126.

2671. Indianola to San Antonio. J. H. Clark.

35. HOLBROOKIA APPROXIMANS, Baird.

Holbrookia approximans, BAIRD, Pr. A. N. Sc. Dec. 1858.

SP. CH.—Similar in size and general character to *H. maculata.* Tail shorter than body. Two small vertical indigo black patches on the side of the belly, entirely visible from below, with a light blue areola. Central point of belly about opposite the middle point between the two patches. No light stripe on side of neck. Upper parts and sides gray, sprinkled with whitish. Head broad, very convex above. Hind foot about one-third the head and body.

Tamaulipas. Lt. Couch.

36. HOLBROOKIA TEXANA, B. & G.

PLATE XXX.

Cophosaurus texanus, TROSCHEL, Wiegmann's Archiv für 1850, 1852, 389.
Holbrookia texana, B. & G. Pr. A. N. Sc. VI, Aug. 1852, 125.

2659. San Pedro, Texas. J. H. Clark.—2782. Colorado desert, California. A. Schott.

37. HOLBROOKIA AFFINIS, B. & G.

Holbrookia affinis, B. & G. Pr. A. N. Sc. VI, Aug. 1852, 125.

2667. Sonora. J. H. Clark.

38. TAPAYA HERNANDEZII, Girard.

Tapaya hernandezii, GIRARD, Herp. U. S. Ex. Ex. 1858, 395.

SP. CH.—Head large, depressed; vertex broad, slightly inclined anteriorly, snout protruding. Cephalic plates small; occipital and temporal spines small and conical. External ears simple, granular. Mental scales very small; the series adjoining the sub-maxillar shields larger than the rest. Several folds under the throat minutely granular. Abdominal scales small, sub-rhomboid, acuminated. Terminal pores small, not continuous across the interfemoral region. Color blackish brown above, with a double or quadruple series of black, light margined spots. Beneath yellowish, unicolor, or else obscurely maculated.—(GIRARD.)

No. 195. Between Guadalupe mountains. Pecos. A. Schott.—Chihuahua. J. H. Clark.

39. TAPAYA ORNATISSIMA, Girard.

Phrynosoma orbiculare, HALL. Sitgreaves' Exped. 1853, 125; pl. viii, ix.
Tapaya ornatissima, GIRARD, Herp. U. S. Ex. Ex. 1858, 396.

No. 205. Pecos to Rio Grande, Texas. Dr. Kennerly.—No. 206. Janos to San Luis Springs. Dr. Kennerly.

40. PHRYNOSOMA CORNUTUM, Gray.

Agama cornuta, HARL. J. A. N. Sc. IV, II, 1825, 299.
Phrynosoma cornutum, GRAY, Syn. Rept. Griff. Cuv. IX, 1831, 45.

No. 121. Indianola, and 122, San Antonio. J. H. Clark.—No. 126. Eagle Pass, Texas. A. Schott.—No. 124. Los Nogales. Dr. Kennerly.

41. PHRYNOSOMA REGALE, Girard.

PLATE XXVIII, FIGS. 1—3.

Phrynosoma regale, GIRARD, Herp. U. S. Ex. Ex. 1858, 406.

SP. CH.—Vertex and occipital regions quite depressed. Temporal and occipital spines flat and acute, constituting a continuous series, very much inclined backwards. Labial plates proportionally well developed, unequal, and rugose. Scales under the chin small, rounded, sub-convex; largest series sub-pyramidal and acuminated. Pectoral scales moderate, and acuminated also. Lower series of sub-pyramidal scales at the periphery of the abdomen obsoletely developed. Abdominal scales sub-rhombic, not acuminated, and slightly, though distinctly, keeled. Femoral pores small and closely set together; the series from either side not continuous on the interfemoral region, over which they somewhat extend. Post-anal scales very minute. Ground color brownish olive above, back, limbs, and tail transversely maculated with black; beneath yellowish, with small black spots over the middle of the abdomen.—(GIRARD.)

From the valleys of the Gila and Colorado rivers.

No. 161. Gila and Colorado desert. A. Schott.

42. DOLIOSAURUS MC'CALLI, Girard.

PLATE XXVIII, FIGS. 4—6.

Anota mc'callii, HALLOW. Pr. A. N. Sc. VI, 1852, 182.
Doliosaurus mc'callii, GIRARD, Herp. U. S. Ex. Ex. 1858, 400.

SP. CH.—Cephalic plates moderate. Occipital spines rather elongated and slender. Sub-maxillar shields very large; posterior ones spinous. No external auricular aperture. Scales under the chin very small, sub-equal, with two distant longitudinal series of somewhat larger ones. Pectoral scales large, carinated, and acuminated. A triple series of sub-pyramidal scales at the periphery of the abdomen; middle one alone well developed. Abdominal scales small, slightly carinated. Femoral pores extending somewhat over the interfemoral region, though the series from either side are not continuous. Yellowish olive above, with a dorsal black line, and a double series of rounded spots on either side of the back, uniting into one along the tail; whitish yellow beneath, unicolor.—(GIRARD.)

No. 162. Gila and Colorado deserts. A. Schott.

43. DOLIOSAURUS MODESTUS, Girard.

Phrynosoma modestum, GIRARD, Stansbury's Expl. Great Salt Lake, 1852, 361, 365; pl. vi, figs. 4—8.
Doliosaurus modestus, GIRARD, Herp. U. S. Ex. Ex. 1858, 409.

No. 164. San Antonio to El Paso.—No. 166. Sonora. J. H. Clark.—No. 168. Los Nogales.—No. 170. Janos. Dr. Kennerly.

44. CNEMIDOPHORUS GRAHAMII, B. & G.

PLATE XXXII, FIGS. 1—6.

Cnemidophorus grahamii, B. & G. Pr. A. N. Sc. VI, Aug. 1852, 128.

3045. San Antonio to El Paso. J. H. Clark.—3044. Los Nogales. Dr. Kennerly.

45. CNEMIDOPHORUS INORNATUS, Baird.

Cnemidophorus inornatus, BAIRD, Pr. A. N. Sc. Dec. 1858.

SP. CH.—Scales on the gular fold smaller than those on the breast anteriorly, and scarcely larger than those on the middle of the chin. Scales of back tubercular and elevated. Hind feet about two-fifths the head and body. General color light greenish olive, paler beneath. No lines on the body.

3032. Pesquieria Grande. New Leon. Lieutenant Couch.

46. CNEMIDOPHORUS OCTOLINEATUS, Baird.

Cnemidophorus octolineatus, BAIRD, Pr. A. N. Sc. Dec. 1858.

SP. CH.—Gular fold as in the last. Hind foot not two-fifths the head and body. Scales of back depressed. General color light greenish olive, paler beneath. Back with eight equi-distant and approximated light lines.

3009. Pesquieria Grande. New Leon. Lieutenant Couch.

47. CNEMIDOPHORUS PERPLEXUS, B. & G.

Cnemidophorus perplexus, B. & G. Pr. A. N. Sc. VI, Aug. 1852, 128.

3020. San Pedro, Texas. J. H. Clark.—3008. Tucson. Dr. Kennerly.

48. CNEMIDOPHORUS TIGRIS, B. & G.

PLATE XXXIII.

Cnemidophorus tigris, B. & G. Pr. A. N. Sc. VI, Ap. 1852, 69.
? *Cnemidophorus marmoratus*, B. & G. Pr. A. N. Sc. VI, Aug. 1852, 128.
Cnemidophorus undulatus, HALLOW. Pr. A. N. Sc. VII, June, 1854, 94.

3024. San Antonio to El Paso. J. H. Clark.—3036. Laredo. A. Schott.

49. CNEMIDOPHORUS GRACILIS, B. & G.

PLATE XXXIV, FIGS. 7—14.

Cnemidophorus gracilis, B. & G. Pr. A. N. Sc. VI, Aug. 1852, 128.

3019. Castanuelas, Coahuila. Lieutenant Couch.

CNEMIDOPHORUS GULARIS, B. & G.

PLATE XXXIV, FIGS. 1—6.

Cnemidophorus gularis, B. & G. Pr. A. N. Sc. VI, Aug. 1852, 128.
Cnemidophorus guttatus, HALLOW. Pr. A. N. Sc. VII, Oct. 1854, 192.

3010. Eagle Pass, Tex. A. Schott.—2989. Indianola, Tex. J. H. Clark.

50. HELODERMA HORRIDUM, Wiegm.

PLATE XXVI.

Heloderma horridum, WIEGMANN, Isis, 1829, 627.—IB. Herp. Mex. 1834.

2971. Sierra de la Union. Sonora. A. Schott.

51. GERRHONOTUS WEBBII, Baird.

PLATE XXIV, FIGS. 1—10. Young.

Gerrhonotus webbii, BAIRD, Pr. A. N. Sc. Dec. 1858.

SP. CH.—Tail 2½ times the head and body. Scales strongly carinated. Dorsal scales in 48 transverse rows. Body encircled by 26 rows of scales, of which 12 are ventral. Hind feet longer than from snout to ear. Above leaden olivaceous brown, lighter beneath. Back with ten or twelve blackish bars, bordered in front by brownish or reddish yellow.

3205. San Diego.—3078. San Diego to El Paso. Dr. Webb.

52. GERRHONOTUS NOBILIS, Bg.

PLATE XXV, FIGS. 1—8.

Elgaria nobilis, B. & G., Pr. A. N. Sc. VI, Aug. 1852, 129.

3076. Coppermines of N. M. J. H. Clark.

53. GERRHONOTUS INFERNALIS, Baird.

Gerrhonotus infernalis, BAIRD, Pr. A. N. Sc. Dec. 1858.

SP. CH.—Dorsal scales carinated, in 16 longitudinal rows; ventral in 12. Nasal plate in contact with the 2d labial only. Tail twice as long as head and body. 51 transverse rows of scales from head to tail. Color clear light olive, with 8 cross bars of dusky. Beneath yellowish, marbled faintly with dull olive. Head plain.

3090. Devil's river, Tex. Dr. Kennerly.

54. GERRHONOTUS OLIVACEUS, Baird.

Gerrhonotus olivaceus, BAIRD, Pr. A. N. Sc. Dec. 1858.

SP. CH.—No single frontal. A series of three pairs of plates between the vertical and rostral, becoming successively smaller. Two post-nasals; one loral. 39 transverse rows of scales on back from head to tail. 12 longitudinal rows above; the 6 central strongly carinated. Color dark olive green, with a series of faint dusky bars. Beneath greenish white.

3096. Near San Diego, Cal. A. Schott.

55. PLESTIODON GUTTULATUS, Hallow.

Lamprosaurus guttulatus, Hallowell, Pr. A. N. Sc. VI, Dec. 1852, 206.
Plestiodon guttulatus, Hallowell, Pr. A. N. Sc. VII, Dec. 1857, 215.

3167. San Elziario, Tex. Dr. Kennerly.

56. ANOLIS CAROLINENSIS, Cuv.

2988. Victoria, Tex. Dr. Kennerly.

57. PHYLLODACTYLUS TUBERCULOSUS, Wiegm.

Plate XXIII, Figs. 1—8.

Phyllodactylus tuberculosus, Wiegm. Nova Acta, K. L. C. Acad. XVII, i, 241.

No. 3208. Durango, Mex. Dr. T. H. Webb.

57. SPHAERIODACTYLUS NOTATUS, Baird.

Plate XXIV, Figs. 29—37.

Sphaeriodactylus notatus, Baird, Pr. A. N. Sc. Ph. Dec. 1858.

Sp. Ch.—Scales on back and sides large, equal, strongly carinated; those on belly smaller, smooth, hexagonal. Above light brownish yellow, uniformly dotted above with reddish brown, most distinct on the head, least so on the belly.

Hab.—Key West, Fla., Professor Agassiz and Prof. W. H. B. Thomas. Type No. 3215.

This species, though not an inhabitant of the regions traversed by the Boundary Commission, is introduced for the purpose of illustrating, by its figures, more fully the characters of *Stenodactylus* and *Phyllodactylus.*

58. STENODACTYLUS VARIEGATUS, Baird.

Plate XXIII, Figs. 9—27.—Plate XXIV, Figs. 11—19.

Stenodactylus variegatus, Baird, Pr. A. N. Sc. Dec. 1858.

Sp. Ch.—Head very broad. Hind foot contained six times in head and body. Above brownish yellow, ith irregular small blotches of light reddish brown, sometimes in broad transverse bands. Edges of eyelids and whole under surface opaque white.

3217. Colorado desert. A. Schott.—3211. Live Oak Creek, Tex. Dr. Kennerly.

59. PLESTIODON OBSOLETUS, B. & G.

Plate XXV, Figs. 9—16.

Plestiodon obsoletum, B. & G. Pr. A. N. Ṣc. VI, Aug. 1852, 129.

3133. Western Texas. J. H. Clark.—3117. El Paso. J. H. Clark.

60. PLESTIODON TETRAGRAMMUS, Baird.

Plestiodon tetragrammus, Baird, Pr. A. N. Sc. Dec. 1858.

Sp. Ch.—One post-nasal plate; post-frontal and inter-nasals separated by the post-nasal. Five supra-orbitals. Dorsal scales of equal width. Light olive green above; sides with two yellowish lines, separated by six rows of darker olive scales. Upper labials pure yellowish. Body encircled by about 28 rows of scales. No dorsal stripe.

3139. Salado river. Dr. Kennerly.—3124. Matamoras, Mex. Lt. Couch.

61. LYGOSOMA LATERALE, Dum. Bib.

Scincus lateralis, SAY. Long's Exp. II, 1823, 324.
Lygosoma lateralis, DUM. BIB. Erp. Gen. VI, 1839, 719.

3126. Indianola, Tex.—3146. San Pedro, Tex. J. H. Clark.

62. ANNIELLA PULCHRA, Gray.

Anniella pulchra, GRAY, Annals & Mag. Nat. Hist. X, 1852, 440.

3190. Coast mountains, near San Diego. A Schott.

OPHIDIA.

63. CROTALUS ATROX, B. & G.

PLATE I.

Crotalus atrox, B. & G. Catal. N. A. Serpents, Jan. 1853, 5.

Nos. 258, 256, 4226, 271. Throughout Texas. Schott, Clarke, Kennerly.
Nos. 466, 467, 468. Bottoms of Gila and Colorado. A. Schott.

64. CROTALUS CONFLUENTUS, Say.

Crotalus confluentus, SAY. Long's Exped. II, 1823, 48.—B. & G. Catal. N. A. Serpents, Jan. 1853, 8.

No. 4228. Between San Antonio and El Paso. J. H. Clark.—No. 4247. Cimeron. J. H. Clark.

65. CROTALUS MOLOSSUS, B. & G.

PLATE II.

Crotalus molossus, B. & G. Catal. N. Am. Serp. Jan. 1853, 10.
Crotalus ornatus, HALLOW. Pr. A. N. Sc. VII, Oct. 1854, 192.

No. 485. Fort Webster. Copper Mines, N. M. J. H. Clark.

66. CROTALUS TIGRIS, Kennicott.

PLATE IV.

SP. CH.—Body slender; head small, very much depressed, narrow behind; nose remarkably broad and obtuse; whole outline of head nearly quadrangular. Superciliaries and frontals smooth; space between superciliaries very wide; four frontal plates, six post-frontals. Two rows of scales between sub-orbital chain (which is complete) and the labials. Labials 14 above, 13–14 below. Dorsal rows 21–23; very slightly carinated. Dorsal scales broad, rounded behind. Color yellowish ash above, with rather small indistinct dorsal brown blotches anteriorly; two posterior thirds of body banded with brown.—(KENNICOTT.)

Deserts of Gila and Colorado.
No. 471. Sierra Verde and Pozo Verde. A. Schott.

67. CROTALUS CERASTES, Hallowell.

PLATE III.

Crotalus cerastes, HALLOWELL, Pr. A. N. Sc., June, 1854, 95.

SP. CH.—Head small, angles rounded; nose obtuse, much depressed; rostral as broad as high; nostril in the middle of a single large plate. Lateral edge of superciliary plate elongated into a horn-like process directed upward over the eye. Two rows of scales between the sub-orbital series (which is complete of large scales,) and the labials. Upper labials 11–13, lower 12–13. Dorsal rows of scales 21, slightly carinated; each scale along the middle of the back with a tubercular swelling toward the centre. Crown tubercular. Entire head and upper parts of a light yellowish, with a dorsal series of small indistinct blotches, below which are several irregular rows of isolated brown dots. A narrow brown stripe extends from the orbit back over the angle of the mouth.—(KENNICOTT.)

No. 482. Colorado river, California. A. Schott. Deserts of Gila and Colorado.

68. CROTALOPHORUS CONSORS, B. & G.

Crotalophorus consors, B. & G. Catal. N. A. Serp., Jan. 1853, 12.

SP. CH.—Twenty-five rows of dorsal scales all carinated, except the two first on either side. Color olivaceous brown, with seven series of deeper blotches, one dorsal, and three on each side; the dorsal series rather small and transverse. Yellowish white line from nostril passes between the eye and pit, and along the upper labials to the angle of the mouth.—(KENNICOTT.)

No. 512. Indianola, Texas. J. H. Clark.

69. CROTALOPHORUS EDWARDSII, B. & G.

PLATE V, FIG. 1.

Crotalophorus edwardsii, B. & G. Catal. N. Am. Serp., Jan. 1853, 15.

SP. CH.—Twenty-three rows of dorsal scales, two first rows smooth. Vertical plate sub-pentagonal, tapering to an acute point posteriorly. Color light yellowish brown, with chestnut blotches lighter than *C. tergeminus* or *C. consors*. Lateral blotches proportionally small. Yellowish line from nostril to angle of mouth as in *C. consors*. No vertebral reddish line.—

No. 506. Sonora. J. H. Clark.—511. Brownsville, Texas. J. H. Clark.

70. TOXICOPHIS PUGNAX, B. & G.—Moccasin.

PLATE VI.

Toxicophis pugnax, B. & G. Catal. N. Am. Serpents, Jan. 1853, 20.

SP. CH.—No loral plate. Second labial pressed out of place and with its apex alone on the edge of the mouth. Twenty-five dorsal rows. Above light olive-brown with transverse dark zigzag bands. Cheeks uniform light colored.—(KENNICOTT.)

4262. Indianola, Texas. J. H. Clark.—4263. Eagle Pass, Texas. A. Schott.

71. ANCISTRODON CONTORTRIX, B. & G.—Copperhead.

Boa contortrix, LINN. Syst. Nat. I, 1766, 273.
Agkistrodon contortrix, B. & G. Catal. Serp. 1823, 17.

567. Indianola to San Antonio. J. H. Clark.—571. Sabinal, Texas. J. H. Clark.

72. ELAPS TENER, B. & G.—Coral Snake.

PLATE VII, FIG. 1.

Elaps tener, B. & G. Catal. N. Am. Serp. Jan. 1853, 22.

SP. CH.—Head narrow and elongated for the genus, continuous with the neck and body. Body fawn-colored (in alcohol) annulated with black and yellow. The yellow annulations separating the black and fawn color are each 2 to 2½ scales wide. Vertical and occipital plates narrow and elongated. Rostral wide and high; prefrontals sub-quadrangular.—(Kennicott.)

1119. San Pedro, Texas. J. H. Clark.—1127. Eagle Pass, Texas. A. Schott.—1132. Indianola, Dr. Kennerly.

73. DIPSAS SEPTENTRIONALIS, Kennicott.

Plate VIII, Fig. 1.

Sp. Ch.—Body moderately slender, very much tapering anteriorly and posteriorly. Tail slender, about one-fifth the total length. Head ovoid, somewhat depressed, very large; twice as wide posteriorly as the neck. Crown flattened, concave behind the eyes; temporal regions much swollen. Snout obtusely pointed, scarcely more depressed than the vertical region. Vertical plate pentagonal elongated; broader in front, concave on the sides. Occipitals triangular, nearly as broad as long. Superciliaries small. Rostral broader than high. Nasals much larger than pre-frontals; two-thirds as high as long; emarginate above to receive the exterior edge of pre-frontals. Loral smaller than upper pre-orbital; as high as wide. Three pre-orbitals; the upper more than three times as large as either of the two lower, its inner angle produced to the vertical, separating the superciliary and post-frontal. Two post-orbitals; upper largest. Eight upper labials; sixth and seventh four times as large as either of the three anterior ones. Ten lower labials; fifth and sixth largest. Dorsal scales in 21 to 23 rows, narrow, acute posteriorly; first lateral row much widest. Body above, with broad, lustrous, brownish black half rings on a light yellowish ground. The black rings six to eight scales wide in the middle, narrowing very much, or even rounded off, laterally, sometimes not extending quite to the abdomen. Light intervals, one or two scales wide, on the vertebral region, widening to three or four times as much near the abdomen. Abdomen and lower surface of head uniform light yellowish. An irregular light occipital ring. Upper labials and anterior part of head brownish—lighter than the dark dorsal bands.—(Kennicott.)

The specimen figured has the anterior sub-caudal scutellae entire, but this is a monstrosity, the sub-caudal scutellae being normally all divided in this genus.

No. 4267. Matamoras, Tam. Lieut. Couch.—No. 2288. Brownsville, Tex. Van Vliet.

74. EUTAENIA PROXIMA, B. & G.

Coluber proximus, Say, Long's Exped. I, 1823, 187.
Eutainia proxima, B. & G. Catal. N. Am. Serp. Jan. 1853, 25.

Sp. Ch.—Body stouter than *E. saurita* or *E. faireyi*. Brown or blackish above. Three longitudinal stripes; the dorsal ochraceous yellow or brown, the lateral greenish white or yellow on the third and fourth rows of scales. Dorsal stripe covering one and over two half rows of scales. Lateral row of scales usually the color of the abdomen. Total length about three and one half times that of the tail. Dorsal rows 19.—(Kennicott.)

Texas.

75. EUTAENIA ORNATA, B. & G.

Plate IX.

Eutainia parietalis, B. & G. Catal. N. Am. Serp. Jan. 1853, 28. (Not *Coluber parietalis*, Say.)

745. Indianola, Tex. J. H. Clark.—438. Lower Rio Grande, Tex. J. H. Clark.—768. Near San Antonio. Dr. Kennerly.

76. EUTAENIA MARCIANA, B. & G.

Eutainia marciana, B. & G. Catal. N. Am. Serpents, Jan. 1853, 36.

SP. CH.—Head triangular, short, but very broad posteriorly. Nose pointed; upper labials 8; 6th largest; 7th nearly as large. Body rather stout, sub-cylindrical; tail short. Prominent color light brown; a vertebral paler line and one lateral on each side on the second and third lateral rows more or less indistinct. Three series of square black spots on each side, of about 56–60 in each series, from occiput to anus. Sides of head black, with a crescentic patch of yellowish posterior to the labial plates. Three and sometimes four black vittæ radiating from the eye across the jaws. A double white spot with a black margin on the suture of occipital plates.

1417. Indianola. J. H. Clark.—1418. Eagle Pass, Tex. A. Schott.—1419. San Pedro, Rio Grande, Tex. J. H. Clark.

77. NERODIA WOODHOUSII, B. & G.

Nerodia woodhousii, B. & G. Catal. N. Am. Serp. Jan. 1853, 42.

SP. CH.—Form of *Nerodia erythrogaster*, but the head broader behind and more flattened above. 25 dorsal rows, all carinated. Three rather large post-orbitals; ante-orbitals narrow; loral large. Three series of quadrangular dark blotches on a brownish clay colored ground. The middle series separated by narrow white lines; the lateral by intervals of the ground color wider than themselves; the three series perfect to the head. A double yellow occipital spot. A yellow spot between the superciliary and vertical plate. A black line indistinctly seen from the posterior point of the eye to the angle of the mouth. Abdomen unspotted.—(KENNICOTT.)

1308. Indianola, Tex. J. H. Clark.

78. REGINA GRAHAMII, B. & G.

PLATE VII, FIG. 2. Young.

Regina grahamii, B. & G. Catal. N. Am. Serpents, Jan. 1853, 47.

SP. CH.—Dull dark brown, with a dorsal light brown line, margined on each side by a narrow indistinct black line. A broad yellowish stripe on the first, second, and third rows, margined above with an indistinct black line on the fourth and fifth rows, and below by a distinct narrow black line on the lower fourth of the first lateral row and the extreme ends of the abdominal scutellae. Abdomen yellowish, tinged posteriorly in the old by olive, with a single central row of small sub-triangular black spots posteriorly, which disappear on the anterior third of the body, and are sometimes indistinct in young specimens. Dorsal rows of scales 19.—(KENNICOTT.)

1376. Rio Salado, Tex. J. H. Clark.

79. REGINA CLARKII, B. & G.

PLATE X. Adult.—PLATE XI, FIG. 2. Young.

Regina clarkii, B. & G. Catal. N. Am. Serp. Jan. 1853, 48.

SP. CH.—Yellowish brown above, with four longitudinal bands of deeper brown. Abdomen dull yellow, with two clouded brown bands dotted with black. One ante-orbital. Dorsal rows of scales 19, all carinated.

1392. Indianola. J. H. Clark.—1377. Indianola. Dr. Kennerly.

80. HETERODON COGNATUS, B. & G.

Heterodon cognatus, B. & G. Catal. N. Am. Serp. Jan. 1853, 54.

SP. CH.—Body shorter and stouter than in *H. platyrhinos*, head strikingly larger. Vertical plate longer than occipitals. Dorsal rows of scales 23–25; outer smooth; next scarcely carinated. Scales of the rest with keels extending to their tips. Scales broader than in preceding. Disproportion between scales of the back, before and behind, not conspicuous. Light chestnut, with 20 yellow blotches from head to anus, and 9 on the tail. Beneath yellow.

1271. Indianola. J. H. Clark.

81. HETERODON NASICUS, B. & G.

PLATE XI, FIG. 1.

Heterodon nasicus, B. & G. Stansbury's Rep. Great Salt Lake, 1852, 352.—IB. Catal. Serp. 1853, 61.

SP. CH.—Vertical broader than long. Rostral excessively broad and high. Azygos plate surrounded behind and on the sides by many small plates (12–15.) Sometimes a second loral. Labials short and excessively high. Dorsal rows of scales 23, exterior alone smooth. A dorsal series of about 50 blotches, with four or five others on each side. Body beneath black. A narrow white line across the middle of the superciliaries; a second behind the rostral. A broad dark patch from the eye to the angle of the mouth, crossing the last two labials.

Fort Webster, and 1260, Sonora. J. H. Clark.—1249. Eagle Pass. A. Schott.—1262. Copper Mines. J. H. Clark.

82. PITYOPHIS BELLONA, B. & G.

Churchillia bellona, B. & G. Stansbury's Report, 1852, 350.
Pituophis bellona, B. & G. Catal. N. Am. Serp. Jan. 1853, 66.
Pituophis affinis, HALLOWELL, Pr. A. N. Sc. VI, 1852, 181.

SP. CH.—Head broad behind, tapering to the snout. Snout rather pointed, but less so than *P. sayi*; elevated. Crown flattened between the eyes and posteriorly. Vertical plate broad anteriorly; elongated and narrow posteriorly. Ante-orbitals two, sometimes one. Post-orbitals three or four. Frequently a supplemental plate before the vertical. Dorsal rows of scales 29 to 35, the seven outer rows smooth. Head spotted with black; transverse frontal bar extending from one orbit to the other, well marked; the oblique post-ocular vitta rather narrow. Color of the body whitish yellow. A dorsal series of 45 to 65 sub-quadrate blotches from head to anus, transversely elongated posteriorly; three or four smaller series on each side. The lateral blotches are longitudinally elongated near the head, and vertically elongated posteriorly where they form a series of jet black vertical bands. There are 10 or 15 black bands on the tail. Abdomen yellow, with an external series of black spots on each side.—(KENNICOTT.)

1542. 1543. Through Texas. J. H. Clark.—1546. San Bernardino to Rio San Pedro, Mexico. Dr. C. B. Kennerly.

83. ARIZONA[1] ELEGANS, Kennicott.

PLATE XIII.

SP. CH.—Body rather more slender than in *Pityophis sayi* and *P. bellona*, and head narrower; otherwise bearing a general resemblance in form to these and other species of *Pityophis*, especially in the protruding

[1] GENUS ARIZONA, Kennicott.—Size large. Body sub-cylindrical, deeper than wide. Head ovoid rather narrow. Snout obtusely pointed, projecting beyond the lower jaw. Eye small. Cleft of mouth curved. Cephalic plates normal. Pre-frontals elongated longitudinally; post-frontals elongated transversely. Pre-frontals extending down in front of nostrils. Rostral large, and protruding apex turned back between pre-frontals. Nostril between two plates; anterior smallest. A narrow loral as long as the post-frontals. One or two pre-orbitals, two post-orbitals. Dorsal scales all perfectly smooth, in about 29 to 31 rows. Post-abdominal scutella entire; sub-caudal all biped.

In its general aspect this genus somewhat resembles *Pityophis*, from which it differs widely, however, in its single pair of post-frontals, smooth scales, and different nasals, loral, pre-frontals, &c.—(KENNICOTT.)

and recurved nostril. The tail forms nearly one-sixth of the total length. Head depressed anteriorly, arched, and much more elevated posteriorly, where it is not much wider than high, nor much wider than the neck. Vertical plate sub-pentagonal, broad in front, tapering, and very acute posteriorly. Occipitals large, perfect, longer than the vertical. Superciliaries small. Pre-frontals of greater longitudinal extent than post-frontals, separated for more than two-thirds of their length by the apex of the nostril, the anterior angle extending down in front of the pre-nasal to below the level of the nostril. Post-nasal more than twice as large as pre-nasal. Loral very narrow, as long as both nasals together, and longer than the post-frontals. One ante-orbital, sometimes a second very small one below; two post-orbitals of about equal size. Two narrow and much elongated temporal shields just behind the post-orbitals, entirely filling the space between the seventh upper labial and the occipital; behind this, small scale-like temporal shields. Eight upper labials, seventh twice as large as any other. Fourteen lower labials, seventh largest. Dorsal scales in 29 to 31 rows all perfectly smooth; central rows not much smaller; outer row largest, but not as high as long.

Body whitish yellow above, with a dorsal series of transversely quadrate light olive brown blotches and two smaller lateral series on each side. Abdomen uniform clear whitish. The dorsal series of blotches are indistinctly edged with blackish; they cover three or four scales longitudinally and twelve or thirteen rows transversely, and are separated by regular intervals of one and a half scales of the ground color. The lateral blotches become more or less indistinct in age from a dark suffusion over the ground color. Each dorsal scale occupying a dark blotch is edged with lighter. In the young, the head above is light brown, with a blackish bar across the post-frontals and through the eyes to the angles of the mouth; there is also a dark blotch below the eye and some smaller ones on the crown posteriorly. In older specimens these markings become obsolete, leaving the head uniform light glossy olive brown.—(KENNICOTT.)

1722. Rio Grande. A. Schott.—4266. Between Arkansas and Cimarron. J. H. Clark.

84. SCOTOPHIS LINDHEIMERI, B. & G.

Scotophis lindheimeri, B. & G. Catal. N. Am. Serp. Jan. 1853, 74.

SP. CH.—Head broader behind and more pointed on the snout than in *S. alleghaniensis;* snout more narrow and elevated. Vertical plate as broad anteriorly as long; 27 to 29 dorsal rows, central rows distinctly carinated. A dorsal series of dark leaden brown or amber brown quadrangular blotches, with an elongated lateral series on each side. Intermediate space brownish yellow. Abdomen nearly all yellow.—(KENNICOTT.)

Indianola, Texas. J. H. Clark.

85. SCOTOPHIS EMORYI, B. & G.

PLATE XII.

Scotophis emoryi, B. & G. Catal. N. Am. Serp. Jan. 1853, 157.

SP. CH.—Head narrow, elongated, widening suddenly behind the eyes. Snout elongated, narrow, elevated, and obtuse; outline anterior to the eyes sub-quadrangular. Crown flattened. Eyes very large. Vertical plate much longer than wide, narrow posteriorly, much wider in front. Superciliaries narrow. Loral large, trapezoidal, acutely angled behind. Dorsal rows 25 to 27; carination of scales very faint, barely perceptible on the central rows. Above ash-grey with a dorsal series of transverse brown blotches, on each side of which are two others of smaller size; indistinct traces of a third. A frontal brown vitta passing back through the eye and crossing the angle of the mouth on to the side of the neck. Two blotches on the upper labials which are not margined with black.—(KENNICOTT.)

2257. Howard Springs, Texas. J. H. Clark.—1716. Pecos to Rio Grande. Dr. C. B. Kennerly.

86. OPHIBOLUS BOYLII, B. & G.

Ophibolus boylii, B. & G. Catal. N. A. Serp. Jan. 1853, 82.
Coronella balteata, Hallowell, Pr. A. N. Sc. Jan. 1853, 236.

Gila desert. A. Schott.

87. OPHIBOLUS SPLENDIDUS, B. & G.

Plate XIV.

Ophibolus splendidus, B. & G. Catal. N. A. Serp. Jan. 1853, 83.

Sonora. J. H. Clark.—1851. Tucson, Sonora. A. Schott.

88. OPHIBOLUS SAYI, B. & G.

Coronella sayi, Holbrook, N. Am. Herp. III, 1842, 99.
Ophibolus sayi, B. & G. Catal. N. Am. Serp. 1853, 84.

1699. Indianola, Texas. J. H. Clark.

89. GEORGIA OBSOLETA, B. & G.

Plate XV.

Georgia obsoleta, B. & G. Catal. N. Am. Serp. Jan. 1853, 158. (Not *Coluber obsoletus*, of Say.)

1862, 1864. Eagle Pass and Lower Rio Grande, Texas. A. Schott and J. H. Clark.

90. BASCANION FLAVIVENTRIS, B. & G.

Coluber flaviventris, Say, Long's Exped. II, 1823, 185.
Bascanion flaviventris, B. & G. Catal. N. Am. Serp. Jan. 1853, 96.

Indianola to San Antonio. J. H. Clark.

91. MASTICOPHIS SCHOTTI, B. & G.

Plate XVIII.

Masticophis schotti, B. & G. Catal. N. Am. Serp. 1853, 160.

1972. Eagle Pass, Texas. A. Schott.

92. MASTICOPHIS ORNATUS, B. & G.

Plate XVII.

Masticophis ornatus, B. & G. Catal. N. Am. Serp. Jan. 1853, 102.

1971. San Antonio to El Paso. J. H. Clark.—1970. Howard Springs, Texas. A. Schott.

93. MASTICOPHIS TESTACEUS, B. & G.

Plate XVI.

Coluber testaceus, Say, Long's Exped. II, 1853, 48.
Psammophis flavigularis, Hallow. Pr. A. N. Sc. VI, 1852, 178.
Masticophis flavigularis, B. & G. Catal. N. Am. Serp. 1853, 99.

Sp. Ch.—Dorsal rows 17. Tail one-fourth the length of body. In alcohol light dull yellow, tinged with brown above. Beneath, two longitudinal series of blotches distinct anteriorly; when the epidermis is removed the whole animal appears of a soiled white. In life there is a tinge of rose color, and in some specimens the whole body is said to be nearly of a brick red.

Texas generally.

1990. Indianola. A. Schott.—2002. Presidio del Norte. J. H. Clark.—2118. Brownsville, Texas. J. H. Clark.

94. SALVADORA GRAHAMIAE, B. & G.

Plate V, Fig. 2.

Salvadora grahamiæ, B. & G. Catal. N. Am. Serp. Jan. 1853, 104.

2081. Sonora. J. H. Clark.—2080. Presidio del Norte. J. H. Clark.—2082. Sonora. A. Schott.—2083. Northern Sonora. Dr. C. B. Kennerly.

95. LEPTOPHIS MAJALIS, B. & G.

Leptophis majalis, B. & G. Catal. N. Am. Serp. Jan. 1853.

1435. Castroville, Texas. Dr. C. B. Kennerly.—Indianola, Texas. J. H. Clark.—143. Eagle Pass, Texas. A. Schott.

96. SONORA SEMIANNULATA, B. & G.

Plate XIX, Fig. 3.

Sonora semiannulata, B. & G. Catal. N. Am. Serp. Jan. 1853, 117.

3109. Sonora. J. H. Clark.

97. RHINOCHILUS LECONTII, B. & G.

Plate XX.

Rhinochilus lecontii, B. & G. Catal. N. Am. Serp. Jan. 1853, 120.

2014. Pecos, Texas. A. Schott.

98. LAMPROSOMA OCCIPITALE, Hallowell.

Plate XXI, Fig. 1.

Rhinostoma occipitale, Hallowell, Pr. A. N. Sc. VII, June, 1854, 95.
Lamprosoma occipitale, Hallow. Pr. A. N. Sc. VIII, Dec. 1856, 310.

Sp. Ch.—Body moderately stout, tapering slightly toward the neck and considerably toward the tail, which is about one-sixth of the total length, slender, and narrowing suddenly at its base. Vertical proportionally large, tapering posteriorly. Superciliaries scarcely two-thirds as long as the vertical. Occipitals nearly as broad as long. Anterior edge of pre-orbitals concave, to receive the convex upper edge of the nasal. Loral elongated; exceedingly narrow. Orbitals small; pre-orbital quadrangular, but little higher than wide; upper post-orbital much the largest. Upper labials eight; seventh largest. Lower labials seven; fourth largest. Dorsal scales rather short and broad, disposed in fifteen rows; those of the lateral rows about as high as long; the central rows not very much smaller; the first lateral row no larger than the second and third. In the alcoholic specimen the ground color above is light greenish yellow, tinged with dull pinkish along the vertical region; below whitish yellow, with narrow black rings entirely around the body. In life, the whole ground color above may have been pinkish. The black rings are two scales wide, and separated by regular intervals of five or six scales of the ground color throughout. The first three anterior rings do not reach the abdomen, but behind this to the tip of the tail all pass completely around the body. They are narrowest on the first lateral rows of scales and widest on the middle of the abdomen and vertebral region. On the head is an occipital crescentic blotch, passing from one eye to the other back through the middle of the occipitals.—(KENNICOTT.)

The specimens described are with some hesitation referred to *Lamprosoma occipitale.* Dr. Hallowell's type of this species differs in having only transverse dorsal bands, which do not reach the abdomen, except on the tail, in place of the perfect annulations; and these bands are only one and one-half scales wide and separated by three and a half or four scales of the ground color; the occipital crescentic blotch is broader, but does not extend to the eyes. Should these characters be permanent, the specimens described above must be referred to another species, which might with propriety be called *Lamprosoma annulatum.*

No. 2105–6. Colorado desert. A. Schott.

99. LAMPROSOMA EPISCOPUM, Kennicott.

PLATE VIII, FIG. 2.

SP. CH.—Form tapering very little toward the neck—which is not much narrower than the head—and moderately tapering toward the tail. The tail forms about one-fifth of the total length. Head rather depressed; crown flattened posteriorly. Snout broad, rounded, and depressed. Vertical a third longer than wide; scarcely tapering behind; acute posteriorly, and usually slightly concave on the sides. Superciliaries and occipitals short and narrow. Nasal elongated; nostril very small in the centre of the plate. Loral elongated; not half as large as the ante-orbital, which is itself small and vertically elongated. Post-orbitals of nearly equal size. Rostral subpentagonal; the apex acute and turned back upon the crown, entering slightly between the pre-frontals. Seven upper labials; fourth, fifth, and sixth largest and nearly equal in size, seventh very small. Lower labials seven; the fourth very much the largest. The dorsal scales in fifteen rows; they increase regularly in size from the central rows, which are much the smallest, to the first lateral row, which is higher than long. The color of the entire upper parts of head and body is uniform light olive brown, tinged with green; but on close inspection each scale is seen to be very minutely mottled with black toward the centre, and upon stretching the skin the base of each scale is black. The abdomen is uniform whitish green. In a specimen from Rio Seco the exposed base of each scale is black, giving the body somewhat of an indistinctly mottled appearance even when the skin is not stretched. The colors become lighter after soaking long in alcohol, and the black at the bases of the scales becomes more or less effaced.—(KENNICOTT.)

2042. Eagle Pass. A. Schott.—2045. San Antonio to Rio Grande. Dr. Kennerly.

100. DIADOPHIS REGALIS, B. & G.

Diadophis regalis, B. & G. Catal. N. Am. Serp. Jan. 1853, 115.

SP. CH.—Form stout. Head very short, broad behind. Eye very small. Vertical plate broad, tapering posteriorly. Dorsal scales in 17 rows. Body above uniform greenish ash; beneath light yellow, with scattered black spots. No occipital ring.—(KENNICOTT.)

2062. Sonora. J. H. Clark.—2064. Eagle Springs, Texas. J. H. Clark.

101. DIADOPHIS DOCILIS, B. & G.

PLATE XXI, FIG. 3.

Diadophis docilis, B. & G. Catal. N. Am. Serp. Jan. 1853, 114.

SP. CH.—In form, the most slender of the genus. Head rather broad behind; snout narrow; crown arched posteriorly. Eyes small. Vertical plate pentagonal, elongated, tapering but little posteriorly.

Dorsal scales proportionally small, disposed in 17 rows, the outer row scarcely largest. Color above uniform ash gray; beneath light yellow, sparsely spotted all over with black. Head dark brown above; below yellow, thickly mottled with black. As in all the western species, the color of the abdomen deepens into reddish towards the tip of the tail. A yellowish white occipital ring $2\frac{1}{2}$ to 3 scales in width.—(KENNICOTT.)

2074. Devil's river, Texas. J. H. Clark.—2066. Tucson, Sonora. A. Schott.—2078. Pecos to Rio Grande. Dr. C. B. R. Kennerly.

102. TAENIOPHIS IMPERIALIS, B. & G.

PLATE XIX, FIG. 1.

Taeniophis imperialis, (B. & G.) GIRARD in Gilliss' Chile, II, 1855, 215. (Named, not described.)

SP. CH.—Form tapering anteriorly and posteriorly. Tail continuous with the body. Head proportionally narrow posteriorly, and broad on the nose; snout rounded. Rostral gibbous, twice as broad as high, the apex below the anterior frontals. Loral as high as long, and more than half as large as the anterior frontals, which are themselves about one-third the size of the post-frontals. The vertical is pentagonal, narrow, and elongated, scarcely wider anteriorly, the acute posterior point entering between the occipitals; one ante-orbital large, vertically elongated, broader above; two post-orbitals small, upper largest. Dorsal scales smooth, in 19 rows; central scales narrow and acute, outer rows much broader, especially the first. Body above deep purplish black, with two dorsal stripes of yellowish brown from head to tip of tail, and separated by a narrower vertebral line of the ground color. Head black above, with two narrow yellow lines from the nostrils to the sides of the occiput, crossing the upper angle of the orbit. Upper labials and under part of head yellowish, minutely mottled with black. Ground color of the back extending on to the ends of the abdominal scutellae. Middle of abdomen uniform light yellowish in the alcoholic specimen; said to be bright red in life.—(KENNICOTT.)

2060. Brownsville. Texas. Capt. Van Vliet.

103. TANTILLA GRACILIS, B. & G.

Tantilla gracilis, B. & G. Catal. N. Am. Serp. Jan. 1853, 132.

2040. Indianola, Texas. J. H. Clark.—2038. Eagle Pass, Texas. A. Schott.

104. TOLUCA[1] LINEATA, Kennicott.

PLATE XXI, FIG. 2.

SP. CH.—Body exceedingly stout. Tail about one-sixth of the total length. Head short, wedge-shaped, scarcely wider than the neck. Snout much depressed, acutely pointed, and projecting beyond the lower jaw. Crown arched throughout. Vertical plate very large, sub-hexagonal, the anterior extremity elongated in a narrow process to the anterior frontals, thus widely separating the post-frontals. Occipitals nearly as broad as long. Superciliaries small. Rostral proportionally large, the apex obtusely pointed, and its centre forming the acute point of the nose. Nasal pentagonal, much elongated, pointed posteriorly. One small sub-pentagonal ante-orbital, as long as high; two post-orbitals, upper slightly largest; seven upper labials, first much smaller than the second and succeeding ones; lower labials, six. Dorsal scales in 17 rows. The scales of the first lateral rows are higher than long; those of the central rows narrower; the outer row largest. Color above uniform light brownish ash, with three imperfect longitudinal blackish stripes, each on a single row of scales.

[1] GENUS TOLUCA, Kennicott.—Size small. Body very stout, sub-cylindrical, deeper than wide. Tail short and thick. Head short and broad, wedge-shaped, almost continuous with the body. Snout pointed and projecting. Cephalic plates normal. Vertical large, sub-hexagonal, the elongated anterior angle separating the post-frontals. Occipitals shorter than vertical. Rostral turned back upon the crown, and occupying the entire point of nose. A single elongated nasal; no loral. Dorsal scales smooth. Post-abdominal scutella divided; sub-caudal all divided.

The vertebral stripe the most distinct; the lateral stripes obsolete on the fourth lateral row on each side. The stripes are formed by scales with the bases black, and the tips of the ashy ground color; the scales involved by the vertebral stripe have their tips even lighter than the ground color. Each dorsal scale appears, on close examination, to be minutely mottled with blackish, and when the skin is stretched the covered base of each scale is seen to be black. Head brownish ash, minutely mottled with blackish; labials lighter. Abdomen uniform greyish white.—(KENNICOTT.)

2103. Valley of Mexico. Dr. T. Potts.

105. RENA DULCIS, B. & G.

Rena dulcis, B. & G. Catal. N. A. Serp. Jan. 1853, 143.

San Pedro to Comanche Springs. J. H. Clark.

BATRACHIA.

105. BUFO AMERICANUS, Leconte.

PLATE XXXIX, FIGS. 1—4.

Bufo americanus, (LEC. MSS) HOLBROOK, N. Am. Herp. V, 1842, 17; pl. iv.

SPEC. CHAR.—Head above, grooved; a ridge from the snout to the occiput, hence at almost right angle to the tympanum. The ridge from either side is slightly diverging from before backwards. Skin upon the head thick, and adhering to the skull. Parotids sub-reniform. Tympanum well developed. A sub-gular vocal sac in the male sex. Limbs large and stout. First finger larger than the second. A large carpal corneous disk, and a small one. No membranous fold at the inner lower edge of the tarsus. Toes semi-palmated. Two metatarsal tubercles, the internal large and spade-shaped, the external small and sub-conical. Large papilla on the back. Greenish or yellowish brown above, with scattered black patches or spots. A vertebral fuscous vitta or streak. Two black patches under the eye, occasionally spotted about the nostrils and over the jaws. Beneath dirty yellowish.—(GIRARD.)

Atlantic States.

106. BUFO PUNCTATUS, B. & G.

PLATE XXXIX, FIGS. 5—7.

Bufo punctatus, B. & G. Pr. A. N. Sc. VI, Oct. 1852, 173.

SPEC. CHAR.—Upper surface of the head flattened, with a sub-orbicular row of granules. Tip of the snout prominent. A blunt granular ridge along the sides of the upper surface. Parotid glands rounded or sub-triangular. Tympanum well developed and eventually elliptical. Eyes large. Tongue elliptical. First finger longer than the second. Two carpal disks. No membranous fold along the inner edge of the tarsus. Toes very slightly webbed at the base. Upper surface of the body covered with minute granules. Two small metatarsal tubercles. Above yellowish brown, with minute black dots over the head, body, and limbs. Granules reddish. Beneath yellowish white, with scattered minute blackish dots under the head and over the chest.—(GIRARD.)

Western Texas. J. H. Clark.

107. BUFO NEBULIFER, Grd.

PLATE XL, FIGS. 1—4.

Bufo granulosus, B. & G. Pr. A. N. Sc. VI, Oct. 1852, 173.
Bufo nebulifer, GIRARD, Pr. A. N. Sc. VII, May, 1854, 87.

SPEC. CHAR.—Upper surface of head deeply excavated or concave; concavity bordered with conspicuous ridges. Skin very thin and firmly adhering to the skull. Parotid small; eyes and tympanum rather large. Tongue moderate, broadest posteriorly. Upper jaw emarginated. Inner finger much longer than the second, which is longer than the fourth. Two well developed carpal disks. No membranous fold along the tarsus. Toes semipalmated; two metatarsal tubercles. Palm of hands and sole of feet tuberculous. Yellowish brown with a dorsal broad streak of the same tint. An interocular black bar on either side of the dorsal streak extending to the whole length of the body. Sides maculated; upper part of limbs barred. Beneath unicolor in the adult, spotted in the young.—(GIRARD.)

Quite abundant along the shores of the Gulf of Mexico, north and south of the Rio Grande del Norte, (Rio Bravo.)

108. BUFO SPECIOSUS, Grd.

PLATE XL, FIGS. 5—10.

Bufo speciosus, GIRARD, Pr. A. N. Sc. VII, May, 1854, 86.

SPEC. CHAR.—Head moderate; its upper surface smooth and even, else showing slight traces of ridges or crests. Snout sub-truncated and rounded; nostrils sub-terminal. Mouth large; upper jaw slightly emarginated. Tongue small, elongated, free posteriorly upon the fourth of its length. A sub-gular vocal bladder in the male sex. Tympanum of medium size. Parotids sub-rhomboid, of stoutish appearance. Limbs of moderate development. First finger much longer than the second, which is equal to the fourth. A large sub-circular carpal disk. Toes semipalmated. Two metatarsal, spade-shaped processes; the innermost being much the largest. A membranous fold at the inner lower edge of the tarsus. Skin above covered with papillae of moderate development, and with small warts beneath. Color above greenish brown, maculated; no dorsal lighter vitta or streak. Beneath greenish or yellowish white, unicolor.—(GIRARD.)

Lower Rio Grande.

109. BUFO ALVARIUS, Grd.

PLATE XLI, FIGS. 1—6.

SPEC. CHAR.—Upper surface of head nearly plane upon its middle region; orbits bordered by a low and rounded off ridge; its skin being thin and adhering to the skull. Parotids well developed and sub-reniform. Eyes and tympanum rather large also. Tongue elongated, broadest posteriorly. Upper jaw emarginated. Two large carpal callosities. A membranous fold at the inner lower edge of the tarsus. Toes palmated; two metatarsal tubercles. Palms and soles coarsely granular. Upper surface of body exhibiting numerous glandular tubercles; large pustular swelling upon the thighs. Color uniformly dark green.—(GIRARD.)

Valley of Gila and Colorado. A. Schott.

110. BUFO HALOPHILA, B. & G.

PLATE XLI, FIGS. 7—12.

Bufo halophila, B. & G. Pr. A. N. Sc. VI, Feb. 1853, 301.

SPEC. CHAR.—Upper surface of head without either crests or grooves. Skin thick and adhering to the skull. Snout rounded. Parotids of medium size. Eyes moderate. Tympanum small. Tongue lanceolated; broadest posteriorly. Upper jaw emarginated. Inner finger longer than the second. Two carpal disks. A membranous fold along the inner edge of the tarsus. Toes semipalmated; two metatarsal tubercles. A narrow yellowish dorsal streak or vitta. Ground color greenish yellow, with numerous black spots and dots distributed all over the upper surface and sides of the body, head, and limbs. Beneath unicolor, of a dingy yellow.—(GIRARD.)

Coast of California.

111. BUFO INSIDIOR, Grd.

PLATE XLI, FIGS. 13—18.

Bufo insidior, GIRARD, Pr. A. N. Sc. VII, May, 1854, 88.

SPEC. CHAR.—Upper surface of head plane and smooth. Snout sub-acute, protruding. Mouth moderate; upper jaw slightly emarginated. Tongue elongated, tapering towards both extremities. Tympanum inconspicuous. Parotids large and elongated, obliquely situated across the shoulders. Limbs moderate

First finger equal to the second in length. A carpal disk and a tubercle. Toes slightly webbed at their base. Two metatarsal tubercles. No membranous fold at the inner lower edge of the tarsus. Skin papillous above, warty beneath. Above of a bluish slate tint with black markings. Beneath unicolor, of a dingy yellow tint—(GIRARD.)

Chihuahua and Sonora.

112. BUFO WOODHOUSII, Girard.

Bufo dorsalis, HALLOWELL, Pr. A. N. Sc. VI, 1852, 181. (Not of Spix.)
Bufo woodhousii, GIRARD, Pr. A. N. Sc. VII, May, 1854, 86.

Sonora.

113. BUFO COGNATUS, Say.

Bufo cognatus, SAY, Long's Exped. II, 1823, 190.

Texas.

114. BUFO DEBILIS, Girard.

Bufo debilis, GIRARD, Pr. A. N. Sc. VII, May, 1854, 87.

Lower Rio Grande.

115. RANA CATESBIANA, Shaw.

Bull Frog.

Rana catesbiana, SHAW, Gen. Zool. Amphibia, 1802, 106; pl. xxxiii.
Rana pipiens, HARLAN, Sill. Am. Jour. X, 1825, 62.—HOLBROOK, N. Am. Herp. IV, 1842, 77; pl. xviii. (Not of Linnaeus.)

San Antonio, Texas. Dr. Kennerly.

116. RANA MONTEZUMAE, Baird.

PLATE XXXVI, FIGS. 1—6.

Rana montezumae, BAIRD, Pr. A. N. Sc. VII, April, 1854, 61.

SP. CH.—Head as wide as long. Body generally smooth, except pustulation on the sides and sometimes above. No fold of skin, either on the sides or around the tympanum. Tympanum about size of eye; but little variation with sexes. Large vocal vesicles on each side behind jaws. Tongue broad, cornua short. Palatine protuberances close together. Toes not webbed beyond middle of last phalanges on the external sides. Color purplish olive above, grayish below, with crowded spots or vermiculations of whitish all over the body. In young specimens, rounded, areolated, dark blotches on the back and fore leg. Size, that of *Rana pipiens* of Holbrook. (*R. catesbiana.*)

City of Mexico. Major W. Rich.

117. RANA BERLANDIERI, Baird.

PLATE XXXVI, FIGS. 7—10.

SP. CH.—Size large. Body stout, robust. Eye distant not quite $1\frac{1}{2}$ times its diameter from tip of snout, and contained $2\frac{2}{3}$ times in the length of jaw from rictus. Tympanum two-thirds the diameter of the eye. A vocal vesicle on each side of the head. A glandular fold on each side the jaw, and another one broad and depressed on each side of the body. Between these is one pair of ridges along the coccyx;

several pairs more interrupted anterior to it. Skin corrugated and irregular, quite pustular in some specimens. Feet webbed from the bulb of the toes; excavated on the inner edges; last joint of longest free. Femur about half the length of body, shorter than the tibia.

Color above greenish olive, with distant sub-circular blotches of darker, scarcely areolated in the preserved specimens. Beneath yellowish white, with brown mottlings on the throat. An indistinct whitish line on the side of the head, especially in the young; the lateral ridge bronzed.

Southern Texas generally.

118. RANA AREOLATA, B. & G.

PLATE XXXVI, FIGS. 11, 12.

Rana areolata, B. & G. Pr. A. N. Sc. VI, Oct. 1852, 173.

Indianola. J. H. Clark.

119. SCAPHIOPUS COUCHII, Baird.

PLATE XXXV, FIGS. 1—6.

Scaphiopus couchii, BAIRD, Pr. A. N. Sc. VII, April, 1854, 62.

SP. CH.—Outer toe but little shorter than the third. Hand nearly as long as the forearm. Above grayish ash in alcohol, with dark markings. A dark line down the back from each orbit. Beneath white; length about one inch or more. Older specimens apparently more greenish.

Lower Rio Grande. Lieutenant Couch.

120. ACRIS CREPITANS, Baird.

PLATE XXXVII, FIGS. 14—17.

Hylodes gryllus, DEKAY, N. Y. Zool. III, 1842, 70; pl. xxii. (Not of Leconte.)
Acris crepitans, BAIRD, Pr. A. N. Sc. VII, April, 1854, 59.

SP. CH.—Brownish above. The median region of head and body above bright green; a dark triangle between the eyes. Three oblique blotches on the sides, nearly equidistant; the first behind the eye, the last on the flanks and running up on the back; all usually margined with lighter. A narrow white line from the eye to the arm. Beneath yellowish white. Inferior face of thigh plain. Tibia a little more than half the length of the body; foot rather smaller. Head rather obtuse, scarcely longer than broad. Web of hind foot extending to the penultimate articulation of the fourth toe.

Indianola, Texas. J. H. Clark.—New Braunfels. F. Lindheimer.

121. HELOECETES CLARKII, Baird.

PLATE XXXVII. FIGS. 4—9.

Heloecetes clarkii, BAIRD, Pr. A. N. Sc. VII, April, 1854, 60.

SP. CH.—Snout acute, projecting. Extremities somewhat dilated. Tibia half the distance between eye and anus. Foot but little longer, not nearly half the length of body. Above grayish brown or ash, with distinct large circular blotches. A dark band from snout through eye and tympanum down the sides, and a whitish line on the side of jaw. Body about one inch long

Indianola, Texas. J. H. Clark.

122. HYLA SEMIFASCIATA, HALLOW.

Hyla semifasciatum, HALLOW. Pr. A. N. Sc. VIII, Dec. 1856, 306.

Indianola. J. H. Clark.

123. HYLA EXIMIA, Baird.

Plate XXXVIII, Figs. 8—10.

Hyla eximia, Baird, Pr. A. N. Sc. VII, April, 1854, 61.

Sp. Ch.—Smooth above. Tibia not half the total length of body. Hind foot not longer than arm from elbow. Bluish above, with two dark longitudinal stripes; beneath white. A dark band from the eye along the sides, margined above and below by a white line, the latter reaching only to the arm, behind which the outline of the dark band is indistinct. Legs not banded. Body about an inch long.

City of Mexico. Major Rich.

124. HYLA VANVLIETII, Baird.

Plate XXXVIII, Figs. 1—3.

Hyla vanvlietii, Baird, Pr. A. N. Sc. VII, April, 1854, 61.

Sp. Ch.—Nearly smooth above. Tympanum nearly as large as the eye. Tibia half as long as the body, longer than arm from elbow, which in turn exceeds the foot. Ash gray or olive, with an irregular cruciform dorsal blotch. A black spot on the side above the foreleg. A white spot under the eye. Thigh and leg with three transverse bands each. Their inner surfaces (when flexed) scarcely reticulated but spotted with white upon a darker ground. Inside of tibia uncolored. Body two inches long.

Brownsville, Texas. Captain Van Vliet.

125. HYLA AFFINIS, Baird.

Plate XXXVIII, Figs. 4—7.

Hyla affinis, Baird, Pr. A. N. Sc. VII, April, 1854, 61.

Sp. Ch.—Body rough. Tympanum two-thirds the size of eye. Tibia not quite half the length of the body, but reaching more than half way from anus to centre of eyes. Color ash grey or green, with numerous rounded dorsal blotches. Three transverse bands on each thigh and leg. No vermiculation on anterior and posterior faces of hind legs, nor on lower part of sides. A light spot under the eye. Web of hand extending only to the third joint of the second finger. Arm from elbow less than tibia, but longer than hind foot. About 1½ inches long.

Northern Sonora. J. H. Clark.

126. AMBLYSTOMA PROSERPINA, B. & G.

Plate XXXV, Figs. 7—14.

Amblystoma proserpina, B. & G. Pr. A. N. Sc. VI, Oct. 1852, 173.

Salado river, Texas.

127. AMBLYSTOMA TEXANUM, Baird.

Plate XXXV, Fig. 15.

? *Salamandra texana*, Matthes, Allg. Deut. Nat. Zeitung, I, 1855, 266.

Lower Rio Grande.

128. SIREN LACERTINA, L.

Lower Rio Grande, Texas.

ALPHABETICAL INDEX.

LIST OF PLATES.

FIGS. 9—18. *Stenodactylus variegatus*, Baird. Female. No. 3217.—Figs. 10, 11, 12, head from different views; fig. 13, inguinal region; fig. 14, a hind toe; and fig. 16, a finger from below, (not quite accurately represented;) fig. 15, hand from below; fig. 17, skin of back; and 18, skin of belly. All magnified except fig. 9.

FIGS. 19—27. *Stenodactylus variegatus*, Baird. Male. No. 3213—References as above; figure of skin of belly wanting. The under surfaces of the toes and fingers, figs. 24, 26, not quite accurate. All magnified except fig. 19.

PLATE XXIV, FIGS. 1—10. *Gerrhonotus webbii*, Baird. Young.—Fig. 1, animal; figs. 2. 3, 4, views of head; fig. 5, upper surface of left hand. fig. 6, tip of a finger, highly magnified; fig. 7, upper surface of left hind foot; fig. 8. portion of belly; fig. 9, portion of the side, showing the lateral fold. All magnified except fig. 1.

FIGS. 11—19. *Stenodactylus variegatus*, Baird. Young.—Fig. 11, animal; figs. 12, 13, 14, views of head; fig. 15, left hand from above; fig. 16, tip of finger; fig. 17, left hind foot from above; fig. 18, portion of back; fig. 19, portion of belly. All magnified except fig. 11.

FIGS. 20—28. *Plestiodon guttulatus*, Hallowell. Young.—Fig. 20, animal; figs. 21, 22, 23, views of head; fig. 24, left hand from above; fig. 25, left foot from above; fig. 26, its longest toe, much magnified; fig. 27, scales of back; and fig. 28, of belly. Fig. 20, natural size of specimen; the others magnified.

FIGS. 29—37. *Sphaeriodactylus notatus*, Baird.—Fig. 29, animal; figs. 30, 31, 32, views of the head; fig. 33, left hand from above; fig. 34, a digit from below; fig. 35, left foot from above; fig. 36, scales of back; fig. 37, scales of belly. Fig. 29, natural size; the others magnified.

PLATE XXV, FIGS. 1—8. *Gerrhonotus nobilis*, Baird.—Fig. 1, animal; fig. 2, head from above; fig. 3, eyelids; fig. 4, head and arm from below; fig. 5, anal region; fig. 6, scales of back; fig. 7, scales of side; fig. 8, scales of belly. Figs. 3, 6, 7, and 8, magnified.

FIGS. 9—16. *Plestiodon obsoletus*, B. & G.—Fig. 9, animal; fig. 10, head from above; fig. 11, eyelids; fig. 12, (right hand side,) head from below; fig. 12, (left hand side,) arm from below; fig. 14, inguinal region and hind leg from below; fig. 15, scales of back; fig. 16, scales of belly. Figs. 11, 15, 16, magnified.

PLATE XXVI, FIGS. 1—10. *Heloderma horridum*, Wiegm. No. 2791.—Fig. 1, animal; fig. 2, head from below; fig. 3, auditory aperture; fig. 4, anal region; fig. 5, fore finger from the side; fig. 6, fore finger from below; fig. 7, hind toe from the side; fig. 8, hind toe from below; fig. 9, skin of back; fig. 10, skin of belly.

PLATE XXVII, FIGS. 1—12. *Euphryne obesus*, Baird.—Fig. 1, animal; fig. 2, head from above; fig. 3, head from below; fig. 4, anal region; fig. 5, fore finger from below; fig. 6, fore finger from the side; fig. 7, hind toe from below; fig. 8, hind toe from the side.

PLATE XXVIII.—FIGS. 1—3. *Phrynosoma regale*, Girard.

FIGS. 4—6. *Doliosaurus m'callii*, Girard.

PLATE XXIX.—FIGS. 1—3. *Sceloporus poinsettii*, B. & G.

FIGS. 4—6. *? Sceloporus spinosus*, Wiegm.

PLATE XXX.—FIGS. 1—2. *Holbrookia texana*, B. & G. Male.

FIGS. 3—7. *Holbrookia texana*, B. & G. Female.—Fig. 3, animal; fig. 4, head from above; fig. 5, palpebral region; fig. 6, inferior surface of body; fig. 7, dorsal scales, much magnified.

PLATE XXXI, FIGS. 1—8. *Crotaphylus wislizenii*, B. & G. No. 2685.—Fig. 1, animal; fig. 2, head from above; fig. 3, head from the side; fig. 4, inferior surface of body; fig. 5, a fore finger from the side; fig. 6, a hind toe from the side; fig. 7, dorsal scales; fig. 8, a femoral pore. All magnified, except figs. 1 and 4.

PLATE XXXII, FIGS. 1—6. *Cnemidophorus grahamii*, B. & G. No. 3046*b*.—Fig. 1, animal; fig. 2, head from above; fig. 3, palpebral region; fig. 4, under surface of body; fig. 5, scales of back; fig. 6, femoral pore; figs. 3, 5, and 6 magnified.

FIGS. 7—13. *Dipsosaurus dorsalis*, Hallow. No. 2699.—Fig. 7, animal; fig. 8, side of head; fig. 9, head from above; fig. 10, head from beneath; fig. 11, anal region, under surface of hind leg; fig. 12, side view of a fore finger; fig. 13, side view of a hind toe. All magnified except fig. 1.

PLATE XXXIII, FIGS. 1—8. *Cnemidophorus tigris*, B. & G. No. 3061. California.—Fig. 1, animal; fig. 2, head from above; fig. 3, head from the side; fig. 4, under surface of body; fig. 5, side view of a fore finger; fig. 6, side view of a hind toe; fig. 7, scales of side; fig. 8, a femoral pore. All magnified, except fig. 1.

N. B. The division of the loral plate into two, as seen in fig. 3, is unusual, the narrow linear one superposed on the larger being almost peculiar to the specimen figured.

PLATE XXXIV, FIGS. 1—6. *Cnemidophorus gularis*, B. & G.—References as in preceding cases, fig. 3 being the palpebral region, and fig. 6 the scales of sides and belly.

FIGS. 7—14. *Cnemidophorus gracilis*, B. & G. No. 3034.

PLATE XXXV, FIGS. 1—6. *Scaphiopus couchii*, Baird. Adult.

FIGS. 7—14. *Amblystoma proserpina*, B. & G.—Fig. 7, animal; figs. 8, 9, 10, views of head; fig. 11, open mouth, showing the tongue; fig. 12, roof of mouth magnified; fig. 13, palm of hand; fig. 14, sole of foot.

FIG. 15. *Amblystoma texana*, Baird.

PLATE XXXVI, FIGS. 1—4. *Rana montezumae*, Baird. Adult.

FIGS. 5, 6. *Rana montezumae*, Baird. Young.

FIGS. 7—10. *Rana berlandieri*, Baird.

FIGS. 11, 12. *Rana areolata*, B. & G.

PLATE XXXVII, FIGS. 1—3. *Batrachyla longipes*, Baird. Mexico.

FIGS. 4—9. *Heloecetes clarkii*, Baird.

FIGS. 10—13. *Hylarana fusca*, Baird.

FIGS. 14—17. *Acris acheta*, Baird.

PLATE XXXVIII, FIGS. 1—3. *Hyla vanvlietii*, Baird.

FIGS. 4—7. *Hyla affinis*, Baird.

FIGS. 8—10. *Hyla eximia*, Baird.

FIGS. 11—13. *Hyla vociferans*, Baird.

PLATE XXXIX, FIGS. 1—4. *Bufo americanus*, Lec.

FIGS. 5—9. *Bufo punctatus*, B. & G.

PLATE XL, FIGS. 1—4. *Bufo nebulifer*, Girard.

FIGS. 5—11. *Bufo speciosus*, Girard.

PLATE XLI, FIGS 1—6. *Bufo alvarius*, Girard.

FIGS. 7—12. *Bufo halophila*, B. & G.

FIGS. 13—18. *Bufo insidior*, Girard.

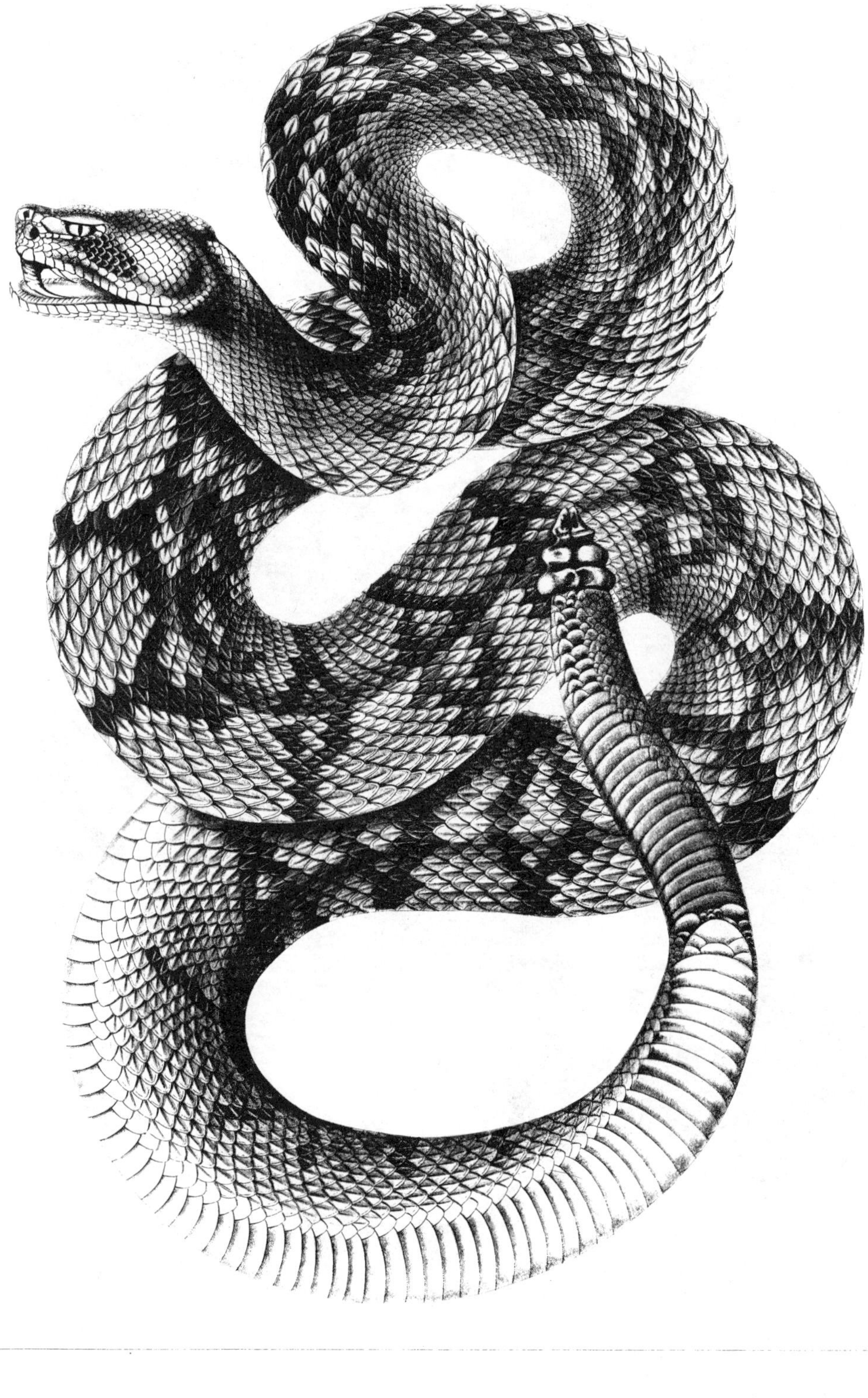

J. H. Richard del. W. H. Dougal sc.

J. H. Richard del. W. H. Dougal sc.

Fig. 2.
Fig. 1.

Fig. 1
Fig. 2

Fig. 2.

Fig. 1.

J. H. Richard del. W. H. Dougal sc.

J. H. Richard del. W. H. Dougal sc.

J. H. Richard del. W. H. Dougal sc.

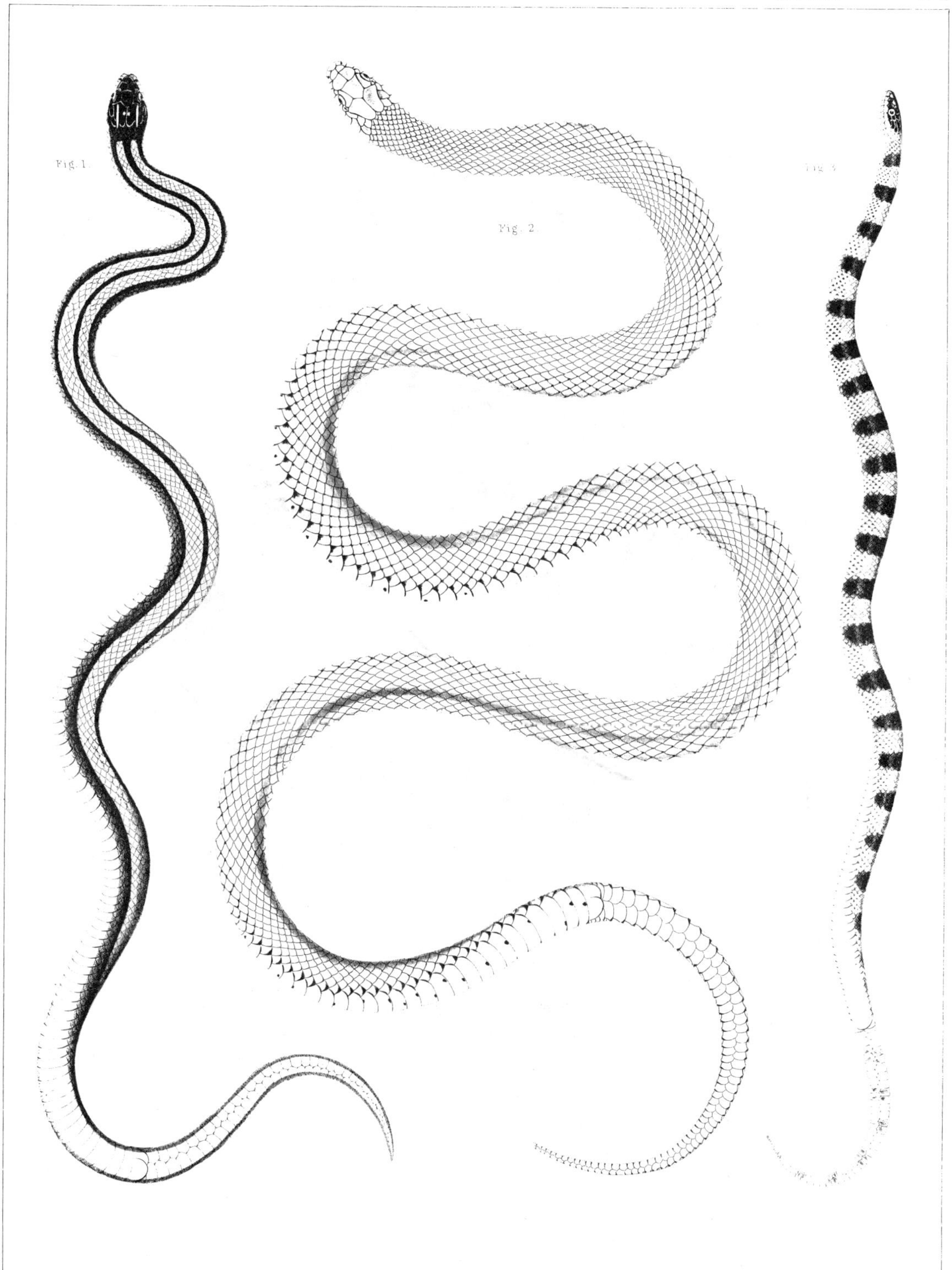
Fig. 1.
Fig. 2.
Fig. 3.

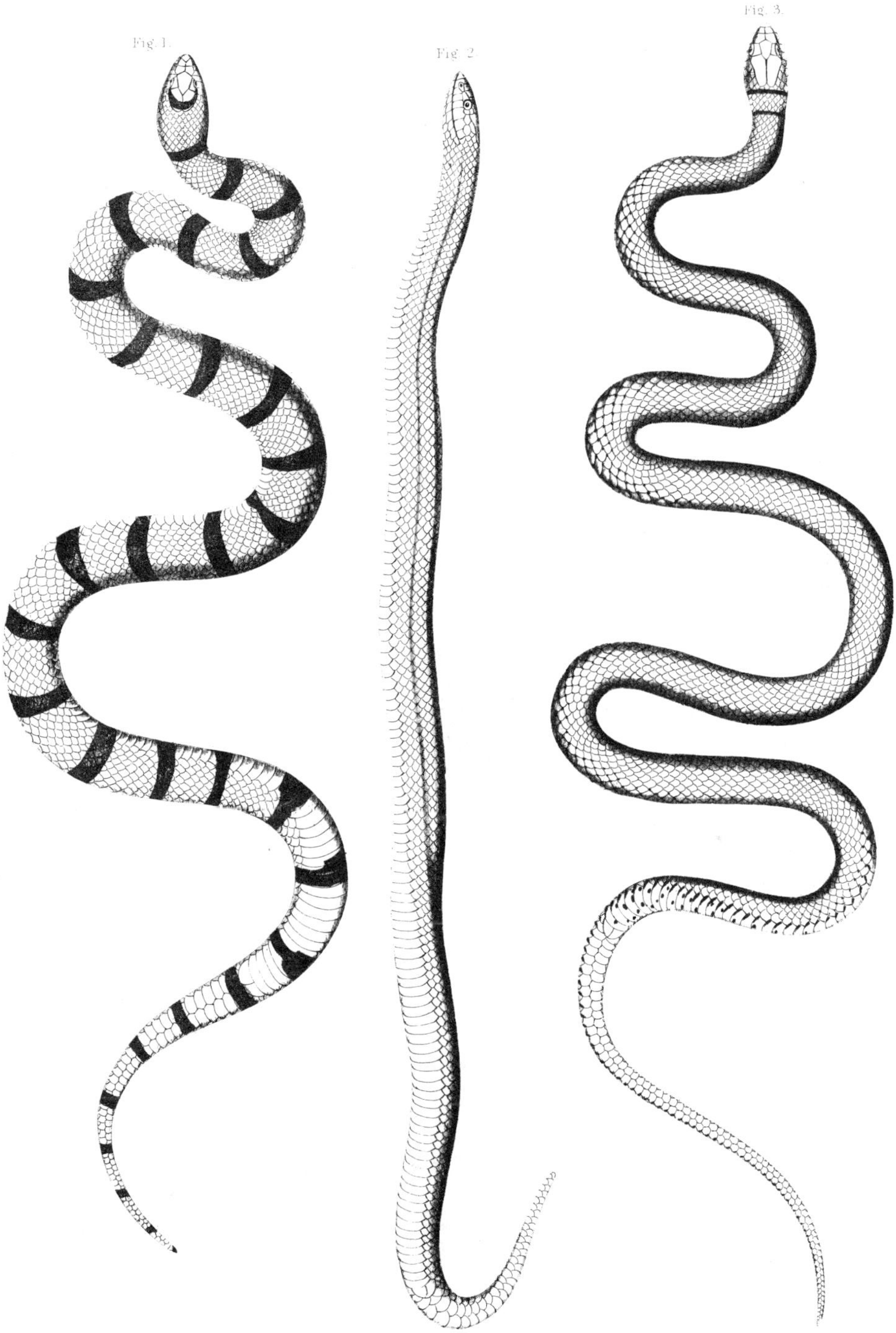

J.H. Richard del. W. H. Dougal sc.

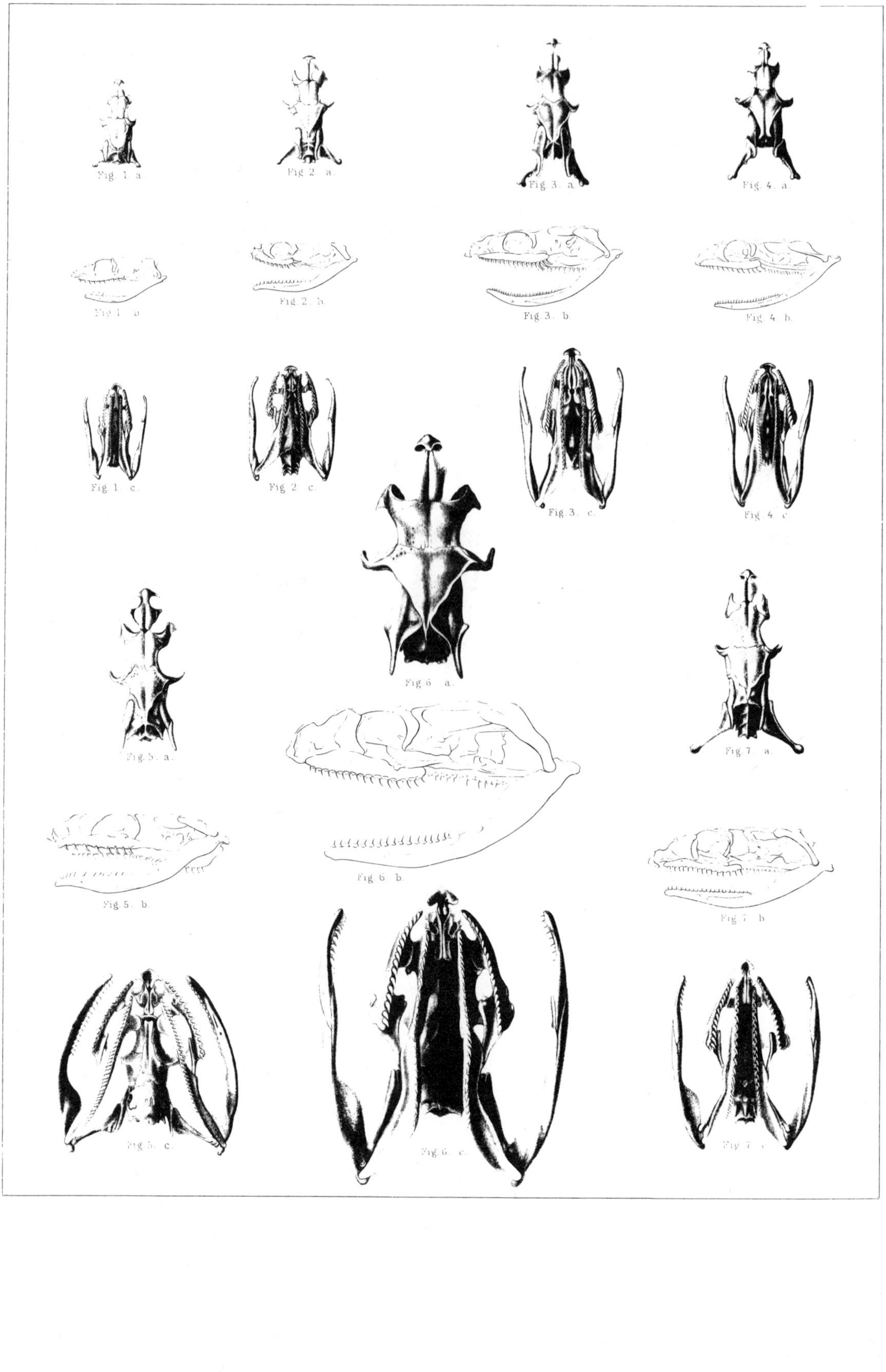
Fig. 1. a.
Fig. 2. a.
Fig. 3. a.
Fig. 4. a.
Fig. 1. b.
Fig. 2. b.
Fig. 3. b.
Fig. 4. b.
Fig. 1. c.
Fig. 2. c.
Fig. 3. c.
Fig. 4. c.
Fig. 6. a.
Fig. 5. a.
Fig. 7. a.
Fig. 6. b.
Fig. 5. b.
Fig. 7. b.
Fig. 5. c.
Fig. 6. c.

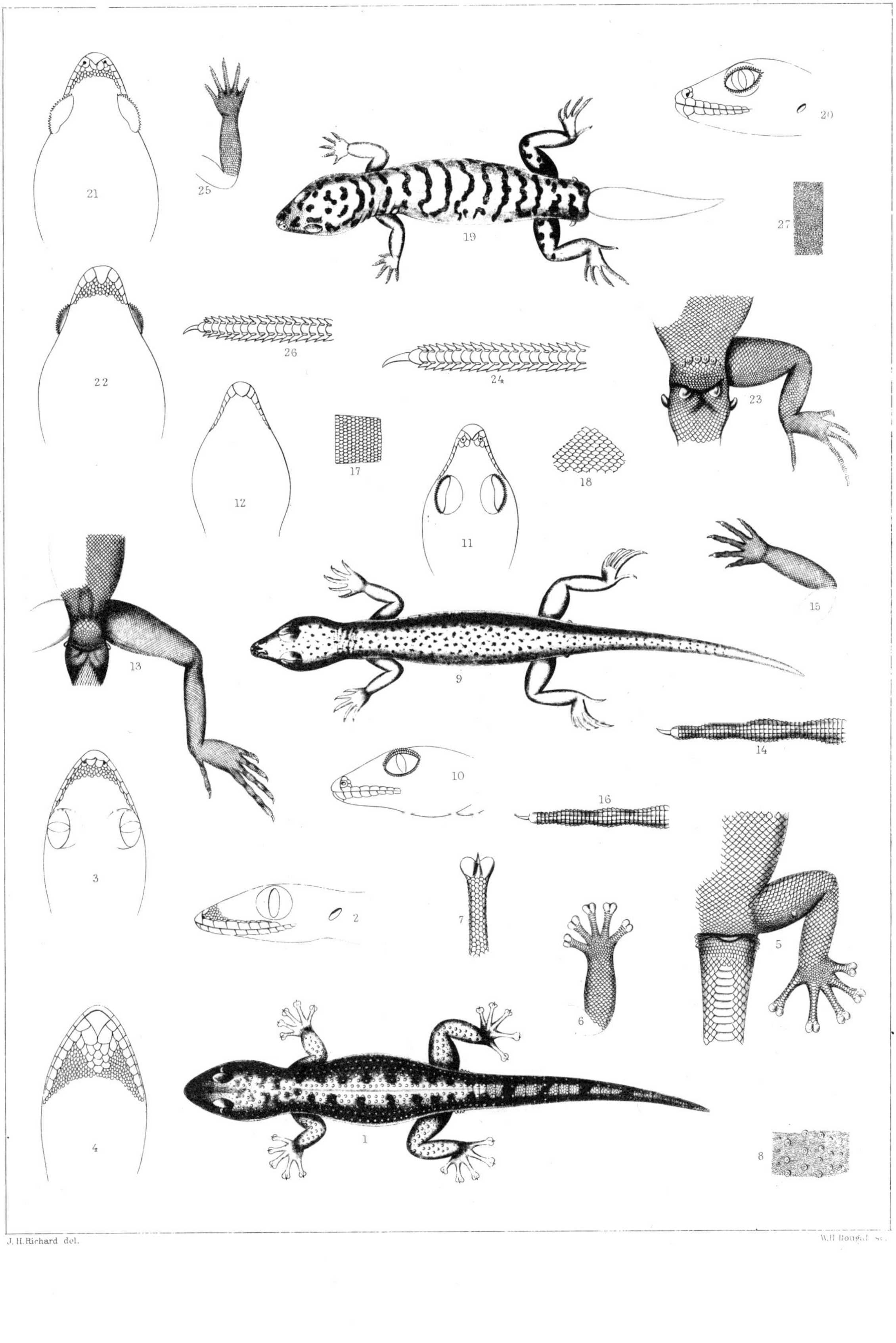

J. H. Richard del. W. H. Dougal sc.

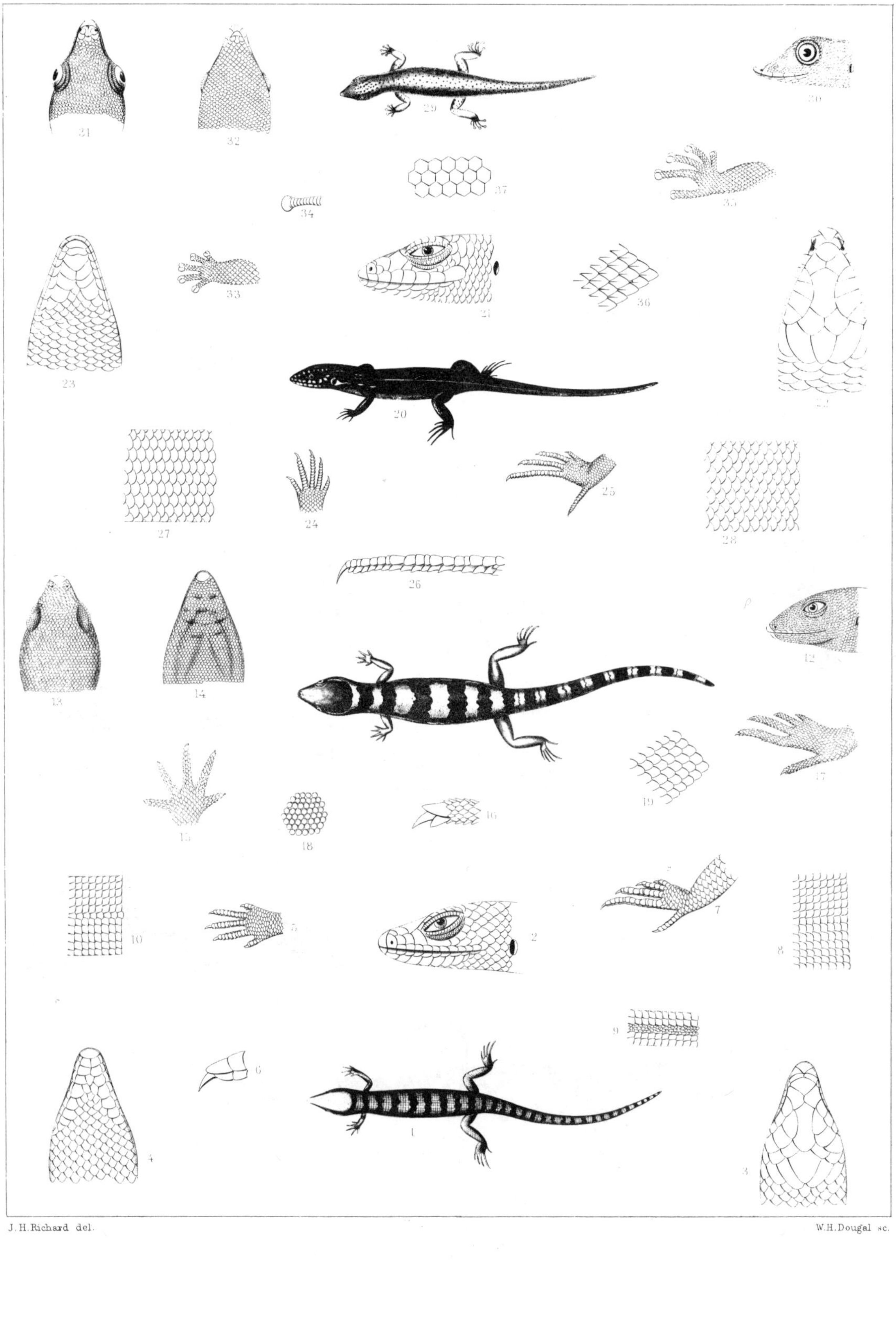

J. H. Richard del. W. H. Dougal sc.

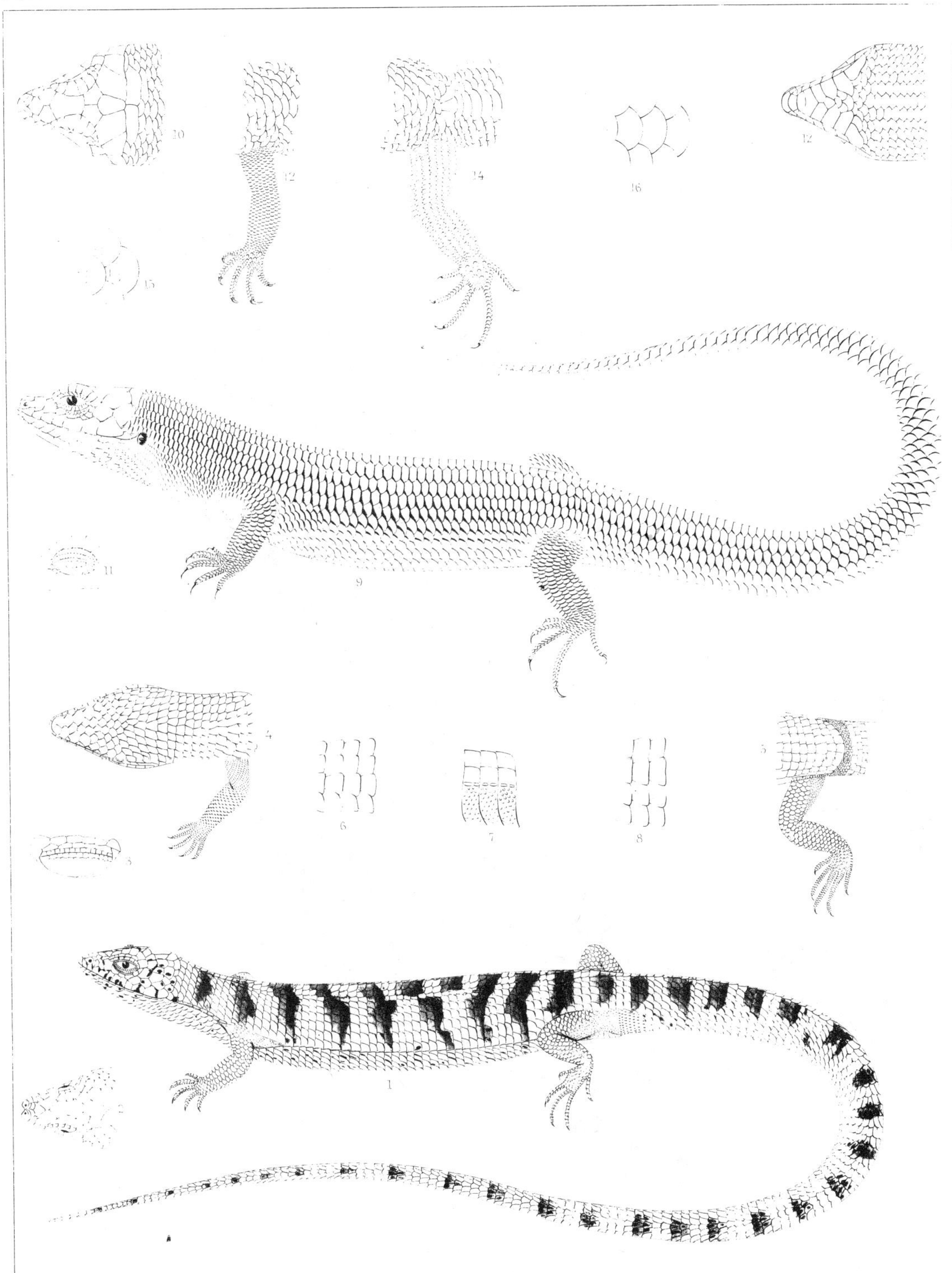

J. H. Richard del. W. H. Dougal sc.

J. H. Richard del. W. H. Dougal sc.

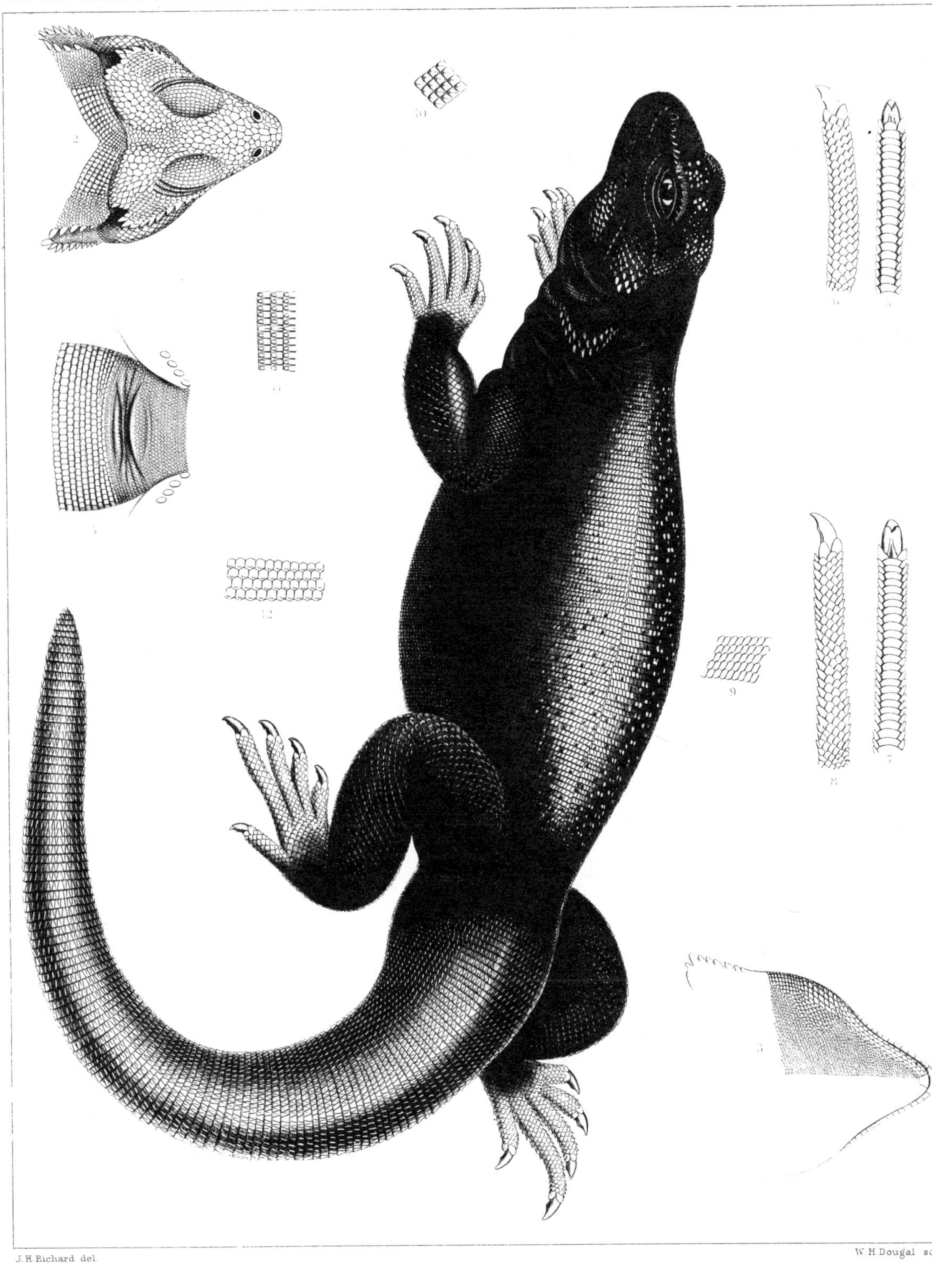

J. H. Richard del. W. H. Dougal sc.

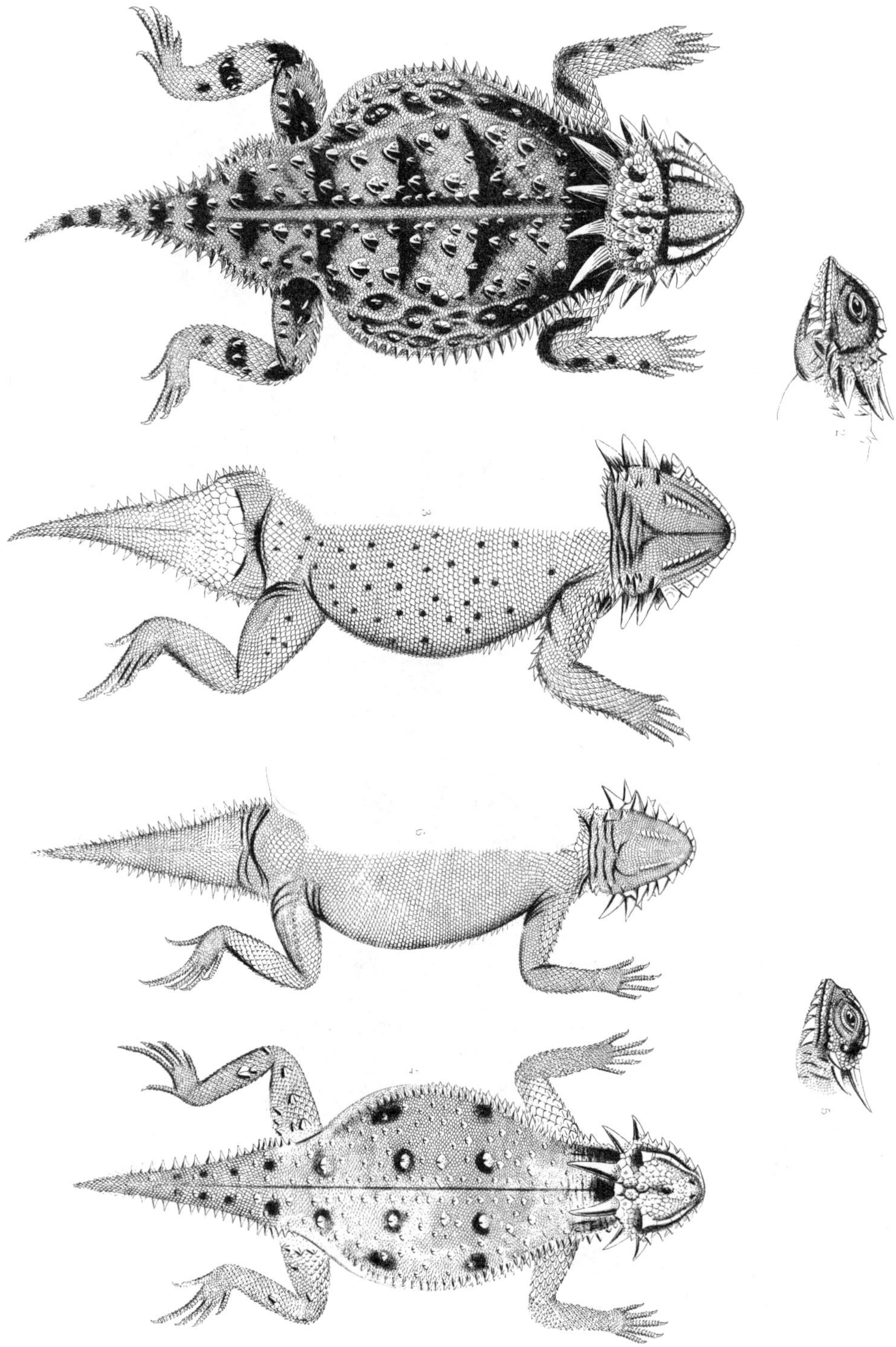

J. H. Richard del. W. H. Dougal sc.

J. H. Richard del. W. H. Dougal sc.

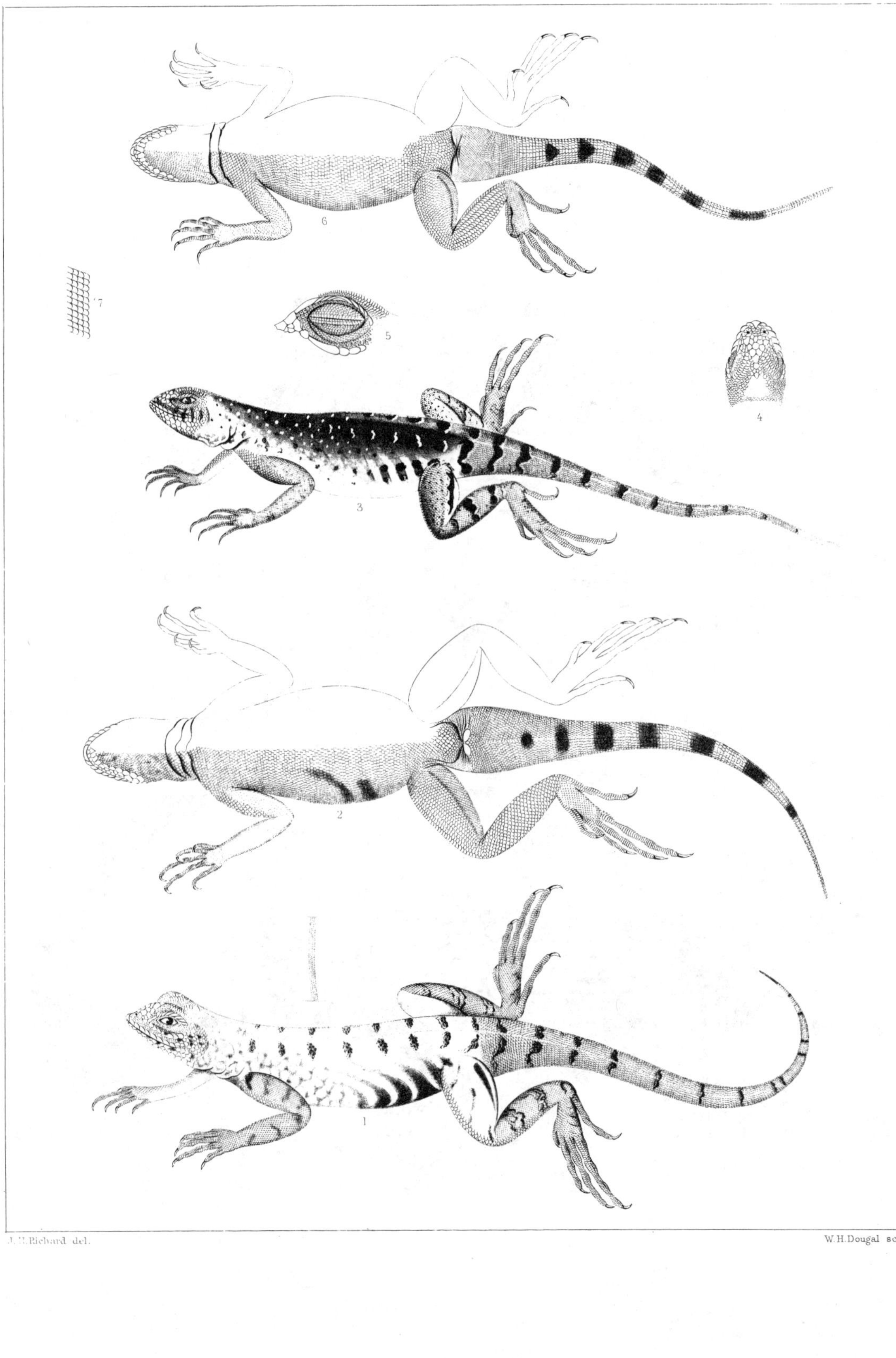

J. H. Richard del. W. H. Dougal sc.

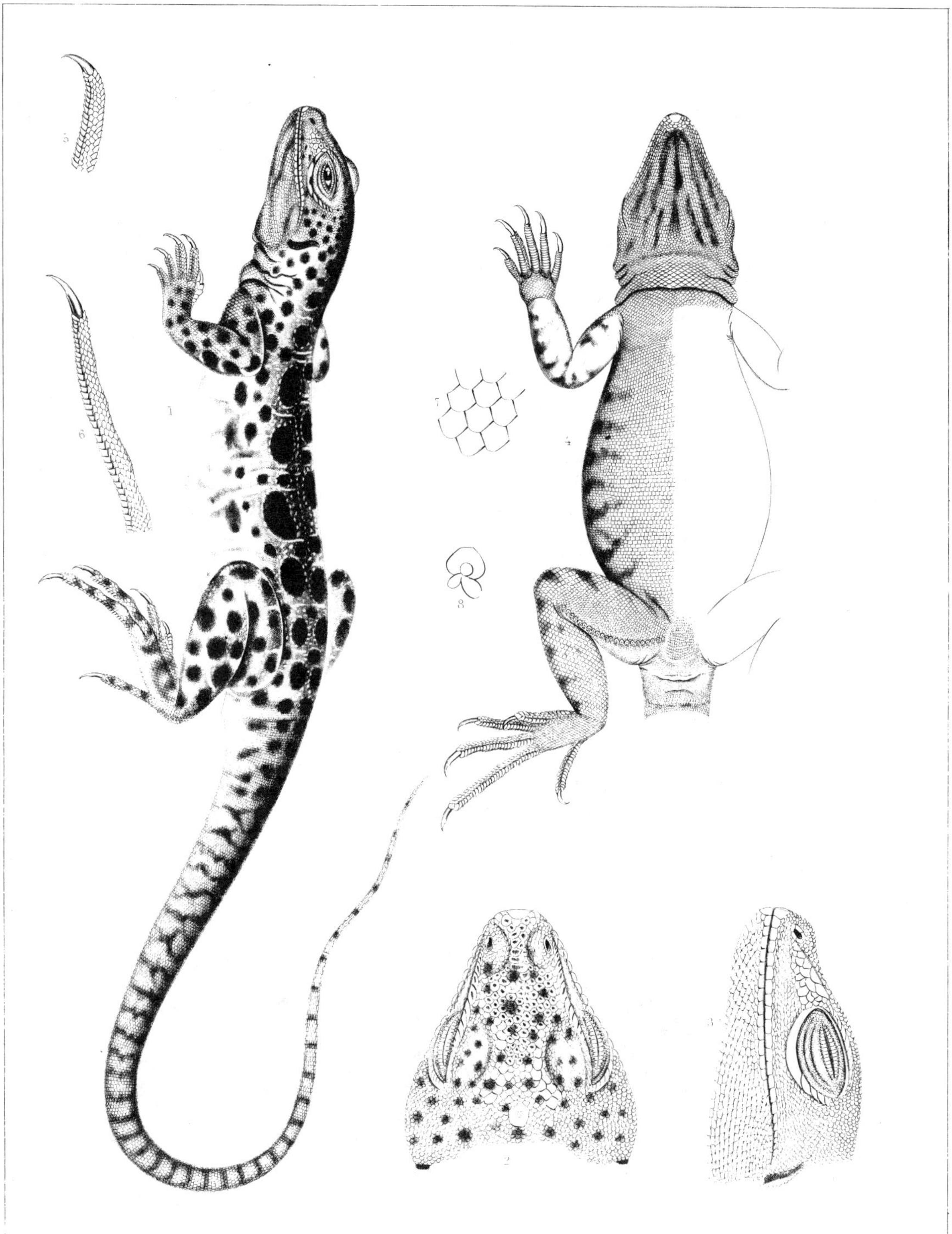

J. H. Richard del. W. H. Dougal sc.

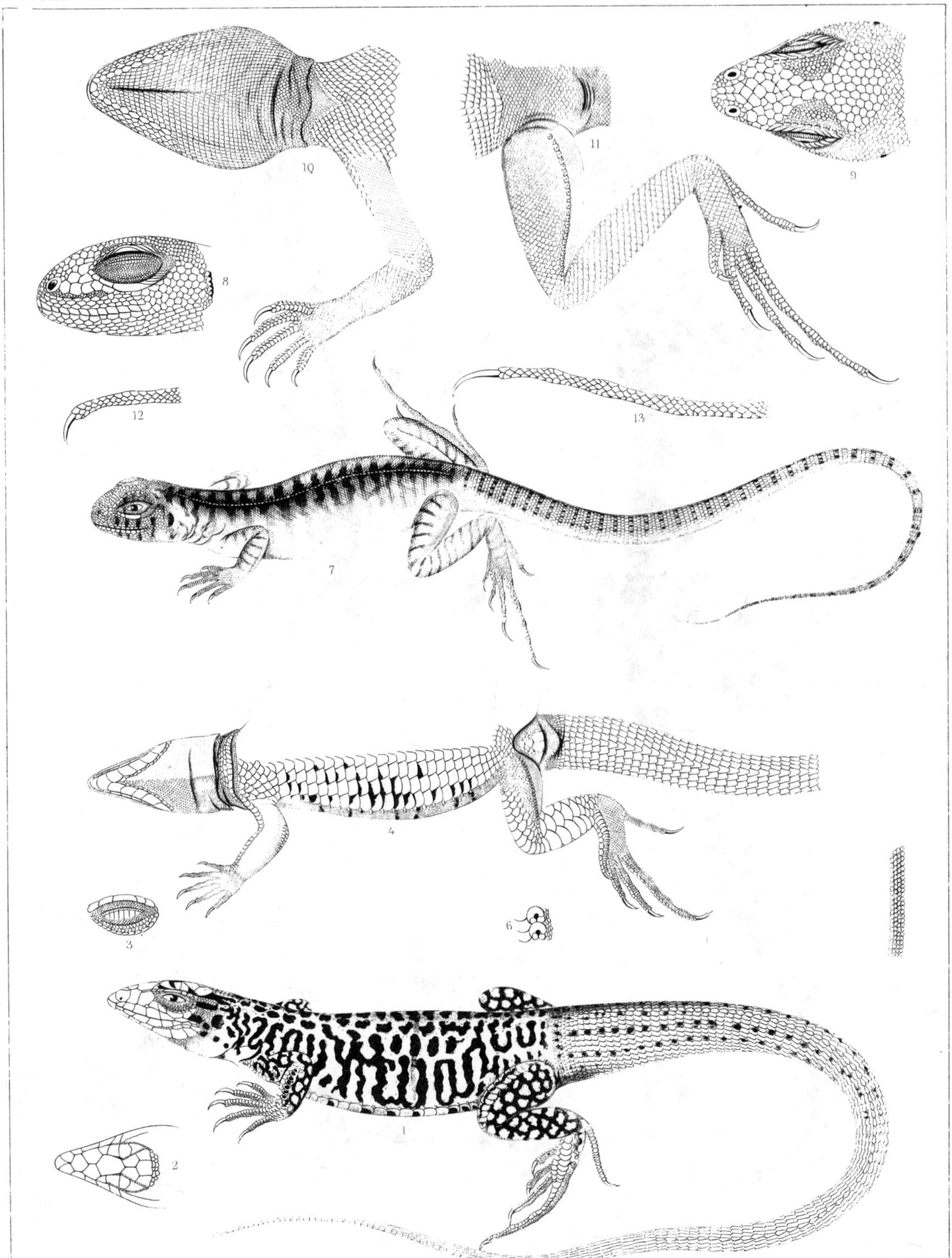

J. H. Richard del. W. H. Dougal sc.

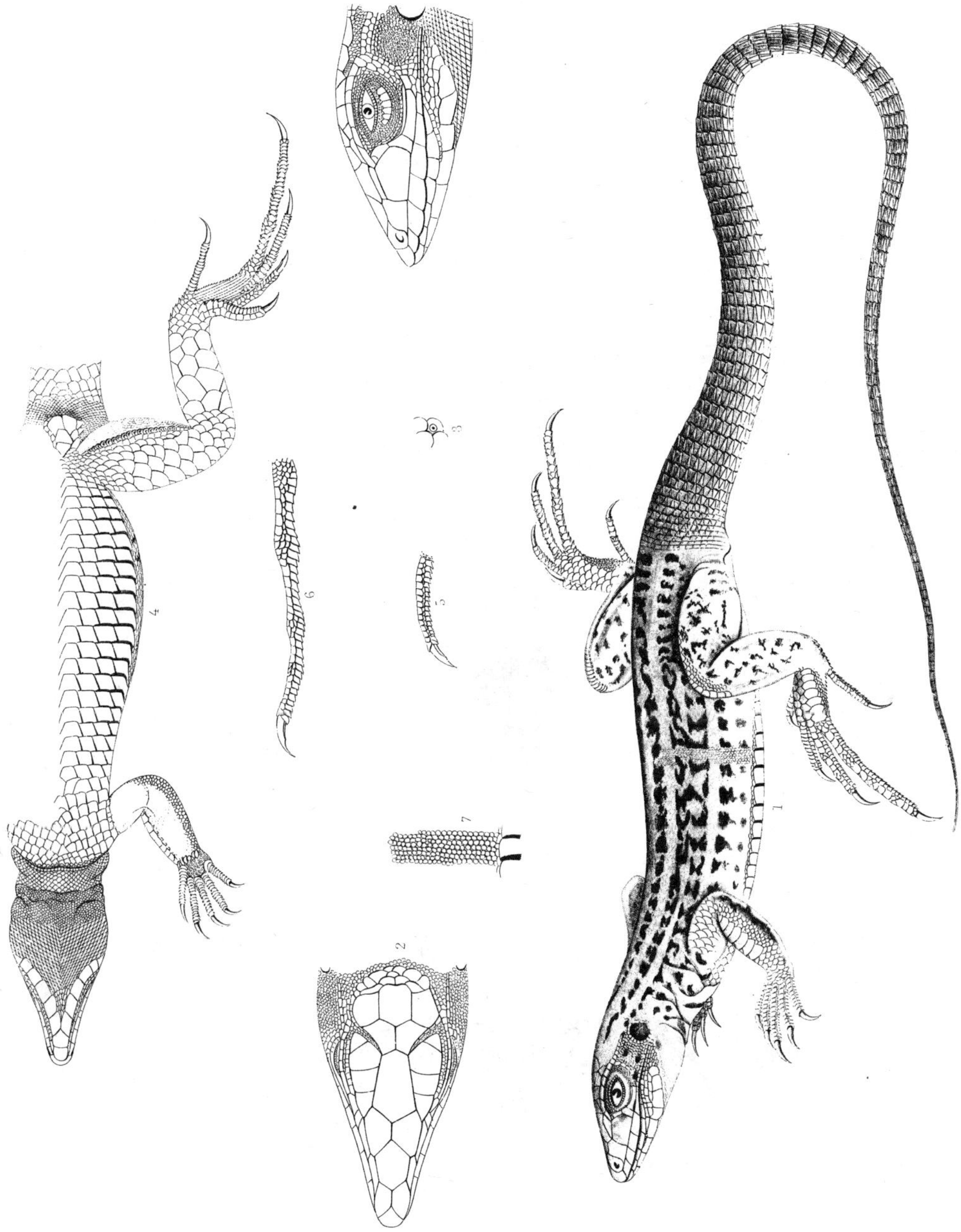

J. H. Richard del. W. H. Dougal sc.

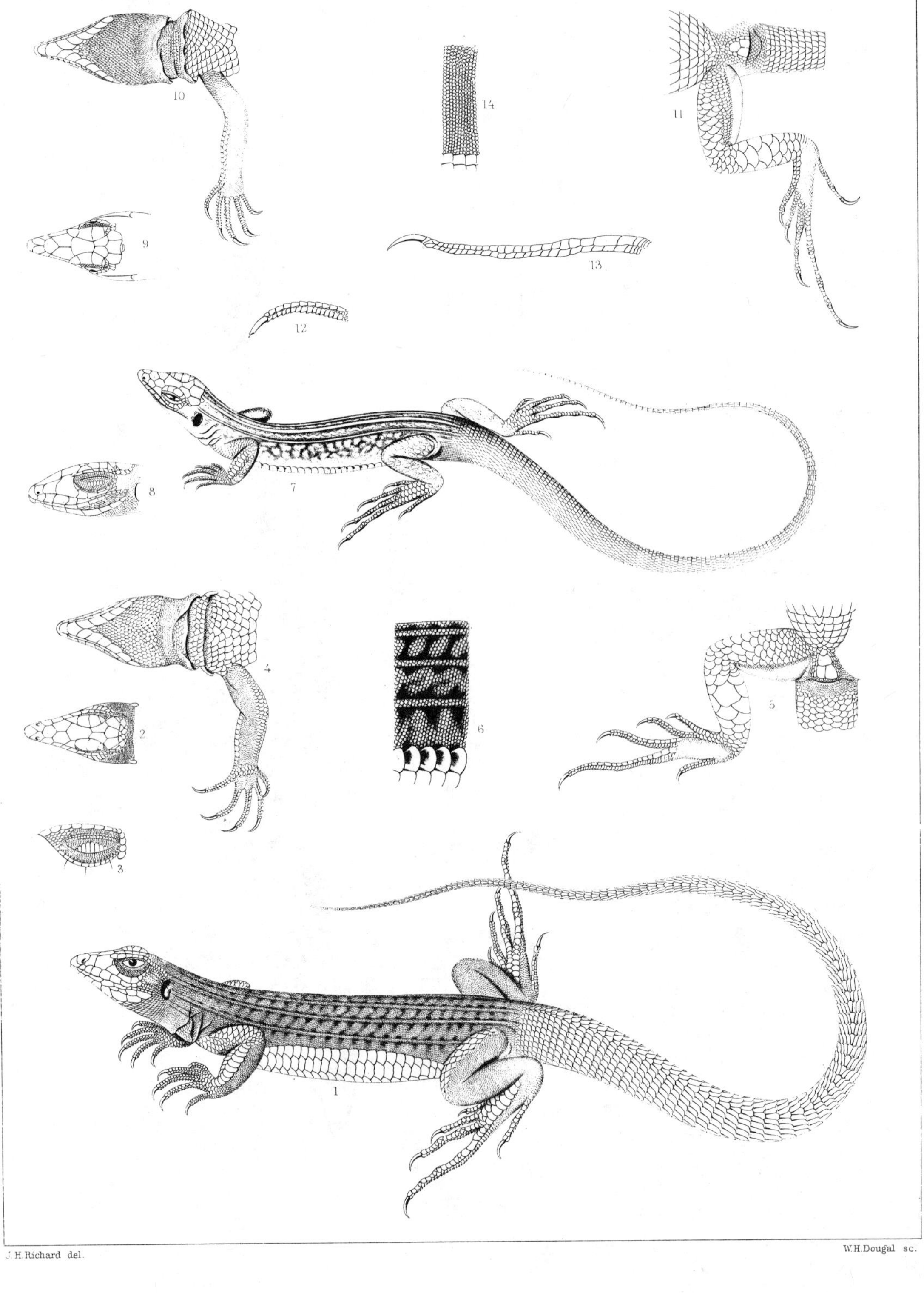

J.H.Richard del. W.H.Dougal sc.

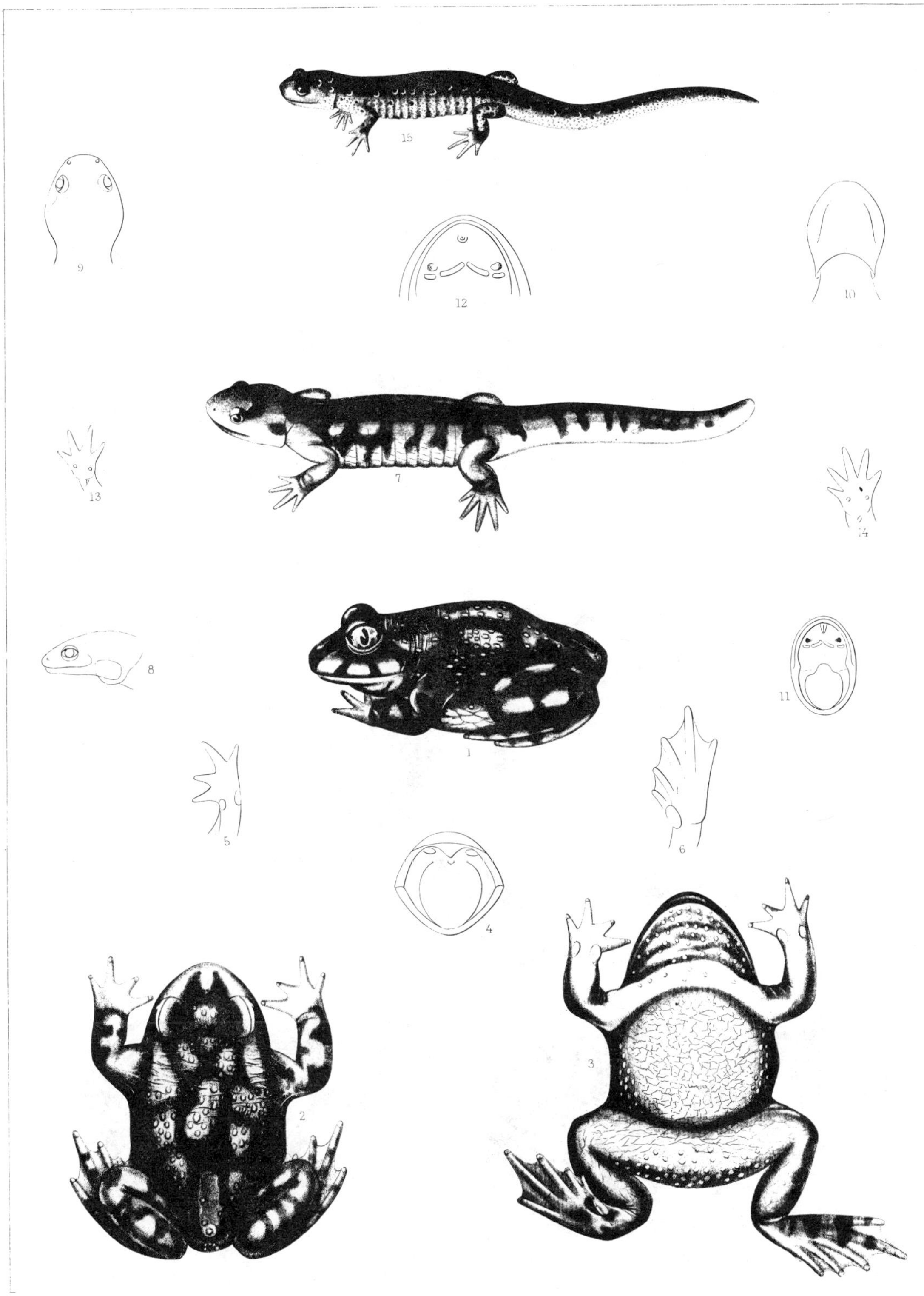

J. H. Richard del. W. H. Dougal sc.

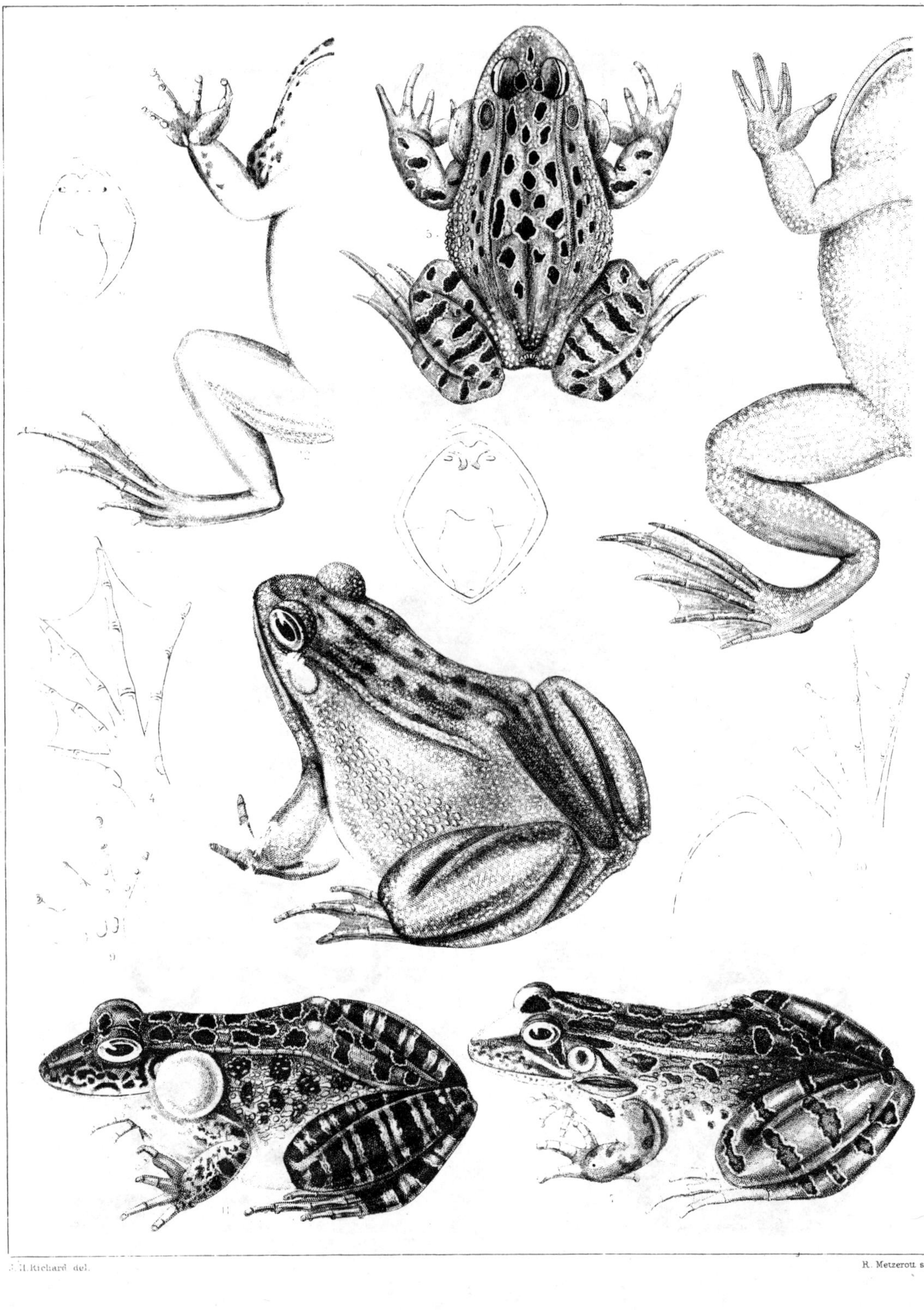

J. H. Richard del. R. Metzerott sc.

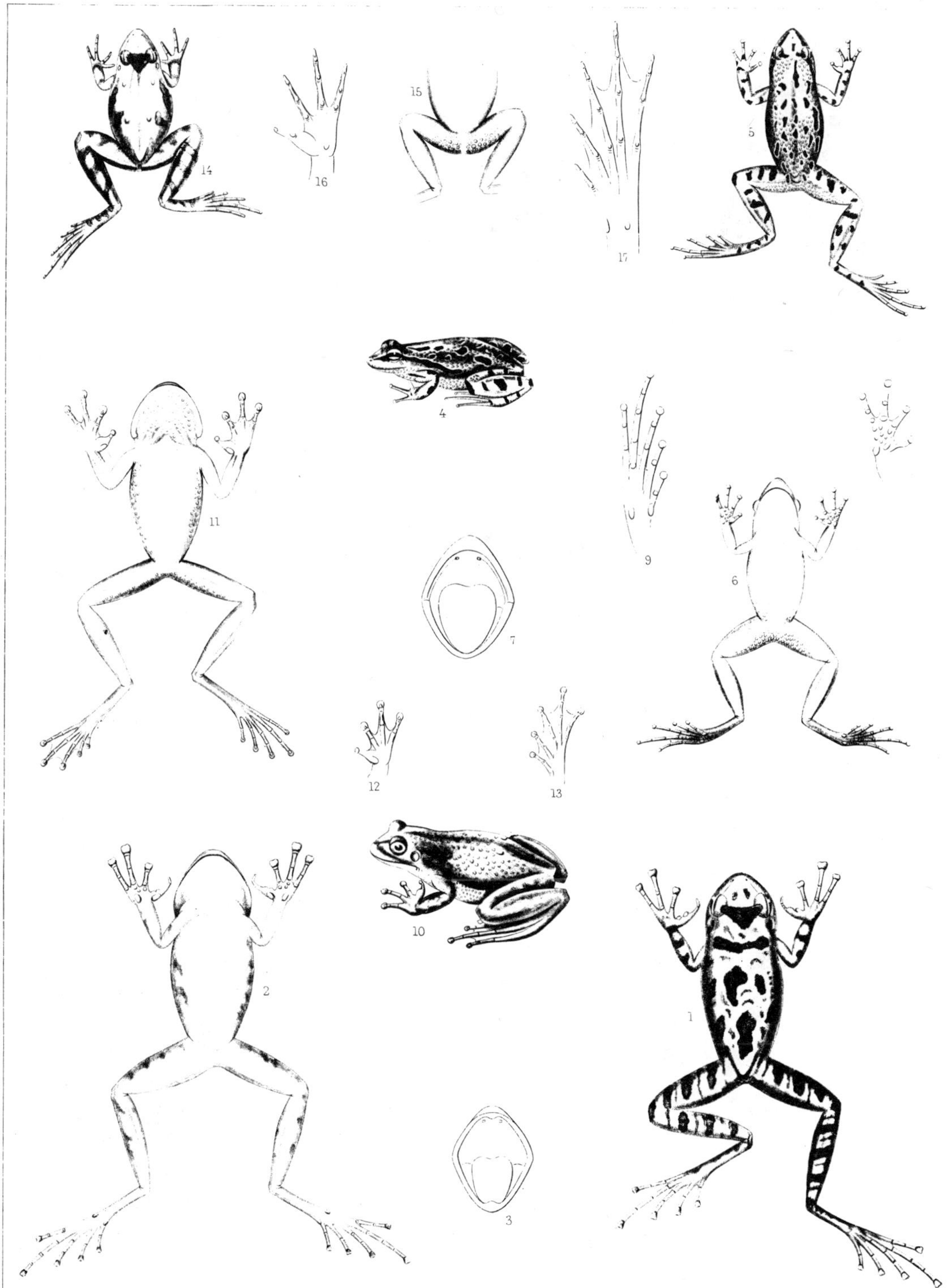

J. H. Richard del. W. H. Dougal sc.

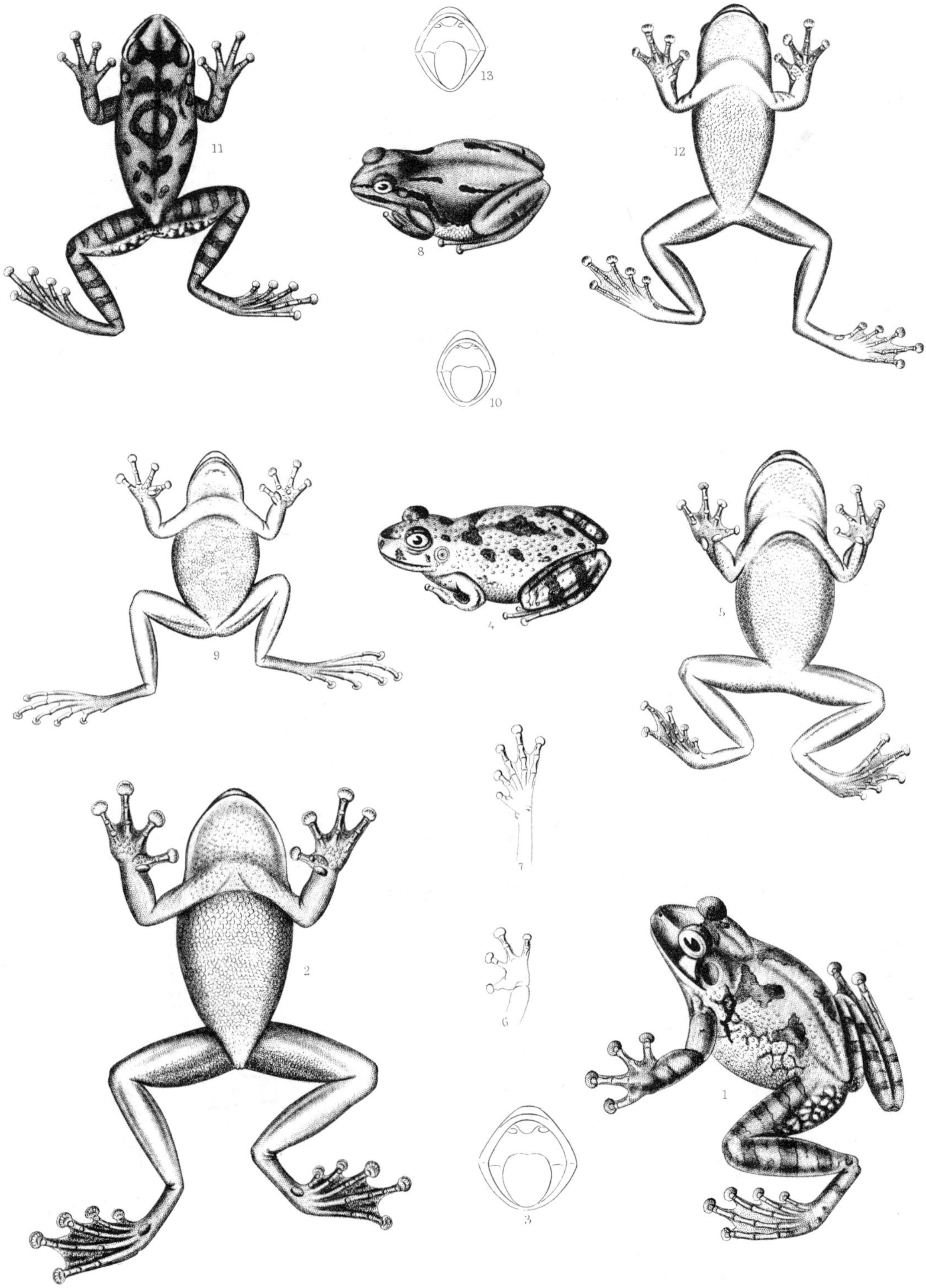

J. H. Richard del. W. H. Dougal sc.

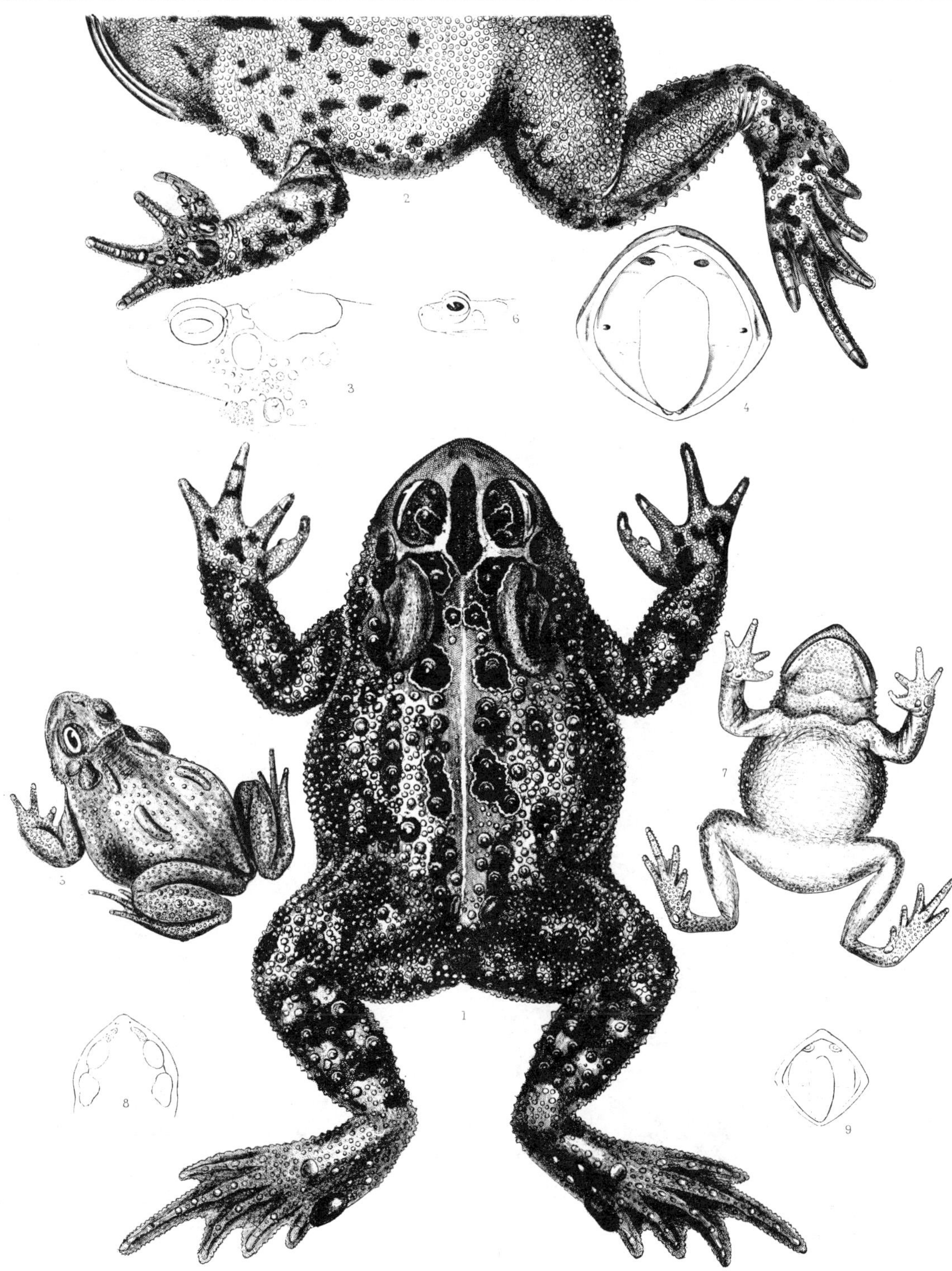

J. H. Richard del. R. Metzeroth sc.

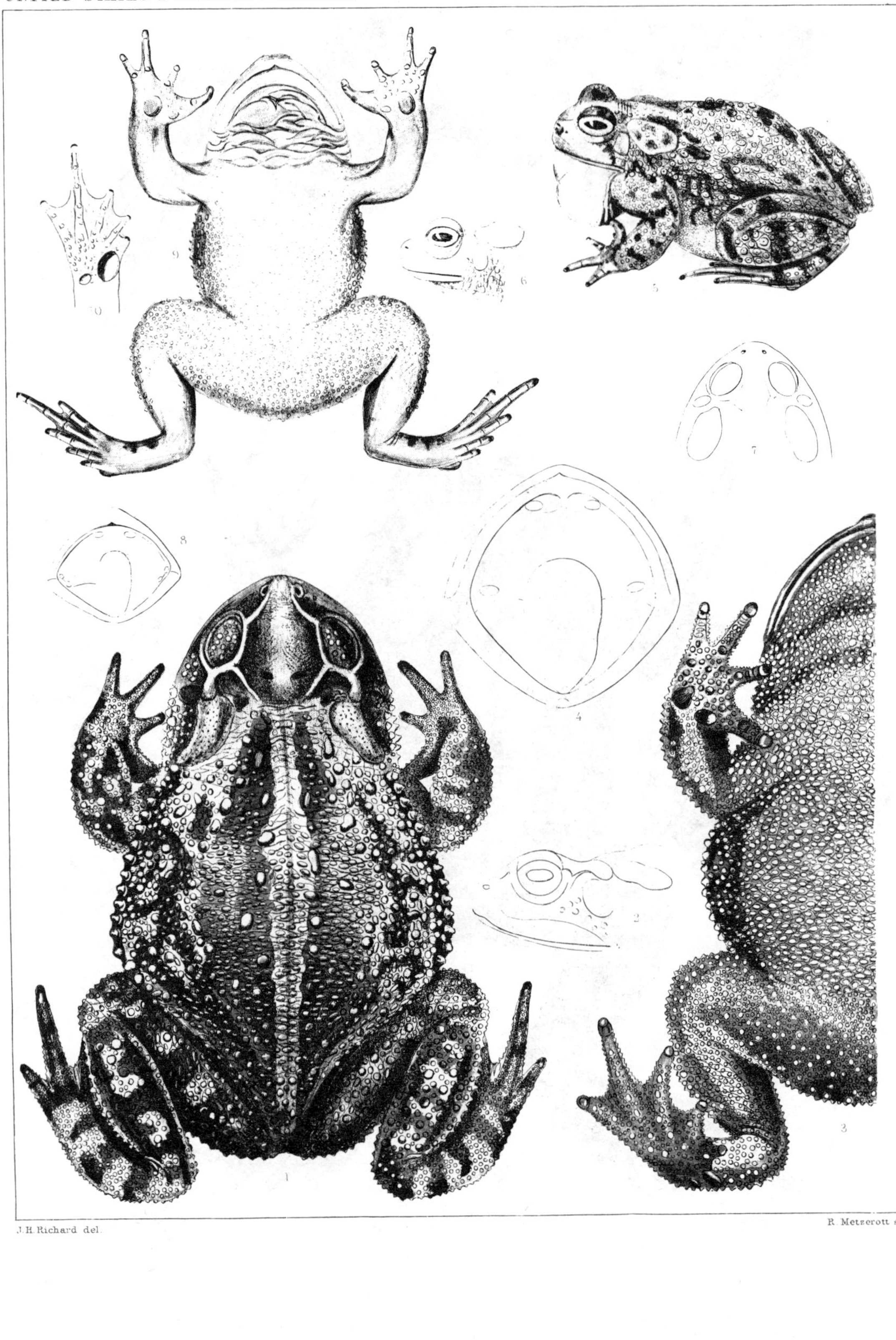

J. H. Richard del. R. Metzerott sc.

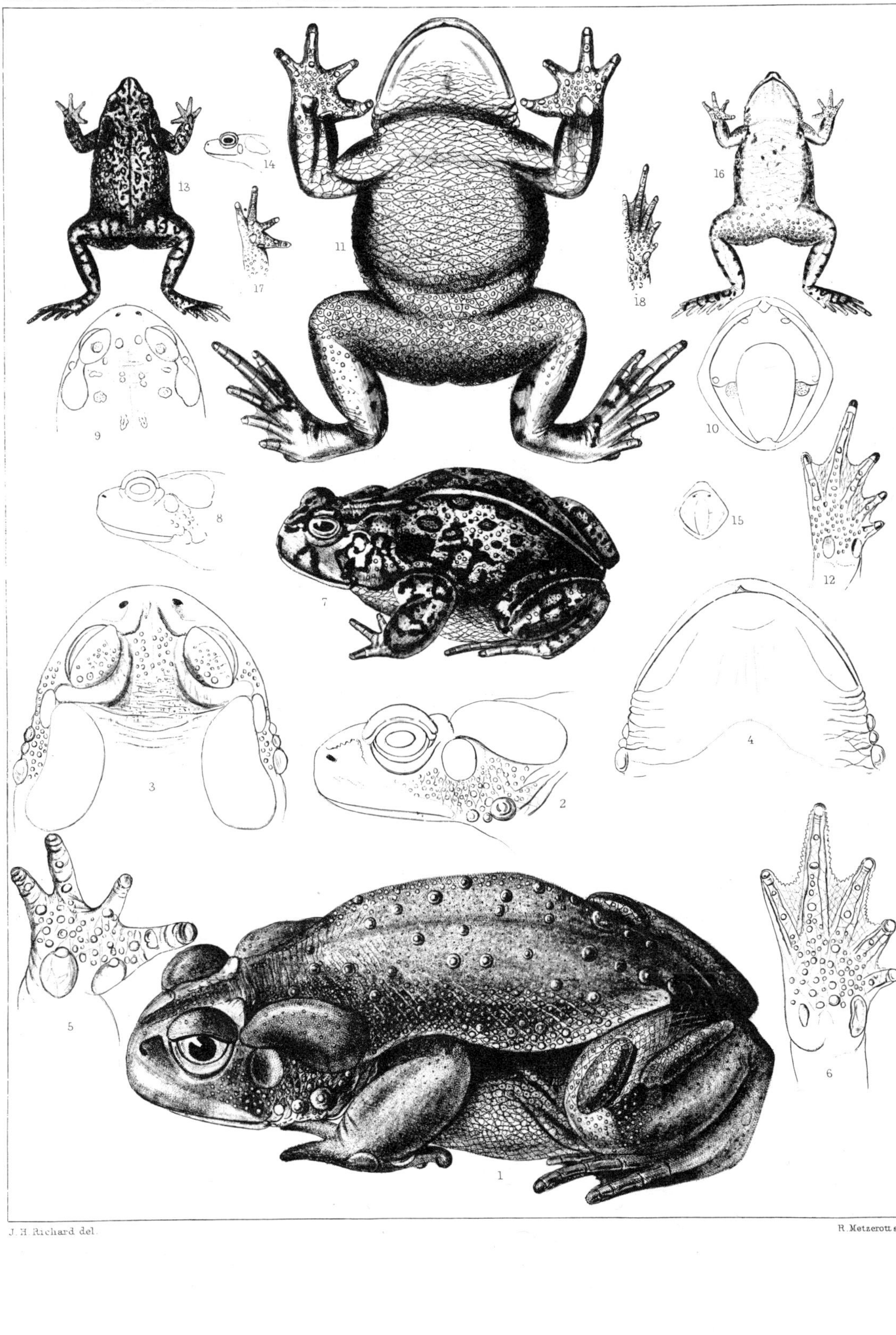

J. H. Richard del. R. Metzerott sc.

DESCRIPTION DE QUELQUES ESPÈCES DE REPTILES DE LA CALIFORNIE

H[enri Marie] D[ucrotay] De Blainville

DESCRIPTION

DE QUELQUES ESPÈCES

DE

REPTILES DE LA CALIFORNIE,

PRÉCÉDÉE

DE L'ANALYSE D'UN SYSTÈME GÉNÉRAL D'ERPÉTOLOGIE ET D'AMPHIBIOLOGIE;

PAR M. H. D. DE BLAINVILLE.

M. P. E. Botta, actuellement voyageur du Muséum d'histoire naturelle de Paris, pendant un premier voyage de circumnavigation qu'il fit dans les années 1827, 1828 et 1829, à bord d'un bâtiment de commerce nommé *le Héros*, commandé par M. Duhautcilly, eut l'occasion de faire un grand nombre d'observations d'histoire naturelle sur des animaux de toutes les classes, en même temps qu'il eut soin de recueillir ceux que sa position lui permettait de conserver. La Faculté des sciences fit l'acquisition du plus grand nombre des objets rapportés, les autres furent cédés au Muséum d'histoire naturelle ou à M. le prince Masséna, sous la condition expresse qu'ils ne pourraient être publiés que par M. Botta ou avec son agrément. J'étais en effet dans l'intention de l'engager à publier lui-même le résultat fort intéressant sous tous les rapports de ses recherches; j'avais même dans cette intention fait faire un certain nombre de figures par un dessinateur fort habile, M. de Bièvre, qui a succombé depuis à une maladie de poitrine. Mais, d'une part, n'ayant pu continuer à supporter les frais des dessins et des gravures par suite de la nouvelle secousse révolutionnaire de 1830, et de l'autre, M. Botta, à son retour d'un nouveau voyage en Syrie, en Égypte et en Abyssinie, en ayant presque aussitôt entrepris un troisième dans l'Arabie heureuse, vers laquelle il se dirige en ce moment, je me suis décidé, à sa sollicitation et dans l'intérêt de la science, à publier moi-même les matériaux les plus importans de son voyage. Je commence aujourd'hui par les reptiles, parce que c'est la partie dans laquelle il me semble y avoir plus de choses nouvelles, et que d'ailleurs, cela me permettra de faire connaître mon système de classification de ces animaux, ainsi que celui des amphibiens, tel que je l'ai exposé, en 1834, à la Faculté des sciences dans mon cours de philosophie zoologique, et depuis les travaux importans sur ces deux classes d'animaux que nous devons surtout aux zoologistes étrangers.

Il est en effet peu de classes d'animaux qui, depuis le commencement de ce siècle jusqu'aujourd'hui, aient été le sujet de travaux plus nombreux et plus importans que celles qui renferment les animaux remplissant dans la série l'intervalle compris entre les oiseaux et les poissons. Autant ces animaux ont été long-temps négligés, autant ils sont aujourd'hui étudiés, comme il me serait aisé de le démontrer par l'analyse historique que je pourrais donner, mais qui, réduite même à ce qu'elle doit être, serait beaucoup trop longue pour être reprise dans un mémoire de la nature de celui-ci.

Je me bornerai donc comme le résultat du travail que j'ai fait à ce sujet, à énoncer les assertions historiques suivantes :

(1740.) Linné réunit le premier sous un nom commun (*amphibia*) les animaux de ces deux classes, mais dans un ordre peu convenable.

(1756.) Brisson les réunit également sous un nom commun, évidemment plus convenable, celui de *reptilia*, et définit la classe unique qu'il en forme de manière à faire entrevoir une bien meilleure distribution que eelle de Linné, d'après la considération de l'état de la peau mise avant celle des membres.

(1768.) Laurenti adopte la manière de voir de Brisson pour le nom, et même jusqu'à un certain point pour les subdivisions intérieures qu'il porte à quatre, les tortues, dont il ne devait pas parler, puisqu'il ne traite, comme l'indique le titre de son ouvrage, que des reptiles supposés à tort ou à raison venimeux, les lézards, les serpens et les grenouilles. Mais en outre il est conduit par une étude plus approfondie de ces animaux à établir un assez grand nombre de subdivisions génériques fort bonnes, et adoptées depuis, en indiquant même les parties différentielles.

(1790.) Merrem, en introduisant la considération du système squameux de la tête chez les serpens, pour la distinction des espèces, et en assignant des noms à chacun des scutelles qui le constituent, introduit un élément de grande importance dans la classification des reptiles, parce qu'il est aussi constant que facile.

(1799.) M. Brongniart, en exécutant la définition de Brisson, perfectionne évidemment le système erpétologique de Laurenti, en lui donnant une meilleure disposition, et en employant des dénominations plus convenables.

Latreille (1801), Oppel (1811), commencent un nouveau perfectionnement que je développe et régularise peut-être (1816) en formant de ces animaux deux classes distinctes, reptiles et amphibiens, d'après la considération de l'ensemble de l'organisation traduite par des caractères extérieurs, et en établissant des coupes secondaires de même valeur dans chacune de ces classes.

Enfin Wagler (1830) et Wagner (1834) étendent la considération du système squameux et du système dentaire de manière à permettre la caractéristique des espèces et leur groupement en sections d'une manière beaucoup plus certaine et plus facile, ce qu'imitent maintenant tous les erpétologistes.

Comme résultat de ces améliorations successives apportées à l'erpétologie, on voit qu'aujourd'hui et depuis long-temps la position de ces animaux dans la série n'est plus contestée, et qu'ils doivent remplir la lacune comprise entre les oiseaux et les poissons. La disposition générale est également généralement adoptée, c'est-à-dire qu'on est convenu de commencer par les tortues et de finir par les cœcilies, ou au moins par les sirènes, celles-là étant évidemment les plus rapprochées des oiseaux, et celles-ci des poissons, comme le montre l'ensemble de l'organisation, aussi bien que les caractères extérieurs.

Quant à leur distribution intérieure, on peut également dire que les véritables principes de la zooclassie nécessitent leur séparation en deux classes aussi distinctes par les caractères anatomiques, que par ceux que l'on regarde comme plus essentiellement zoologiques ou extérieurs, chacune d'elles offrant des degrés d'organisation ou ordres qui se correspondent assez, et la dégradation marchant assez bien dans chacune avec le développement du tronc aux dépens des membres; les tortues ou chéloniens, les emydosauriens ou crocodiles, les sauriens et les ophidiens dans la première classe, les grenouilles ou batraciens, les salamandres ou pseudosauriens, les cœcilies ou pseudophidiens dans la seconde : aussi ces deux classes, sous les noms de reptiles et d'amphibiens, sont-elles presque généralement adoptées par les zoologistes actuels.

Dans l'intérieur de chacun des ordres, les zoologistes sont encore assez d'accord pour la distinction des familles qui correspondent au reste assez bien aux genres de Linné; mais ils le sont beaucoup moins pour l'ordre à établir entre elles. Nos principes sont ici les mêmes que dans toutes les autres parties de la zoologie : cet ordre devant être déterminé de manière à ce que les animaux placés à la tête de la division, quelle qu'elle soit, soient ceux qui se rapprochent davantage de la précédente et que ceux qui sont à la fin passent à la suivante.

Ainsi parmi les tortues ou chéloniens l'ordre à suivre n'est pas comme je l'ai cru long-temps, celui qui des espèces les plus aquatiques passerait à celles qui sont le plus terrestres, ou *vice versâ;* mais de celles qui se rapprochent le plus des oiseaux à celles qui sont plus voisines des crocodiles.

Il en est de même des saurophiens, chez lesquelles la dégradation est généralement indiquée des espèces qui se meuvent le plus avec leurs membres, le moins avec le tronc, à celles chez lesquelles la locomotion est tellement et si complétement troncale, que les membres après quelques oscillations finissent par disparaître entièrement. Quant aux genres ou sous-genres qui entrent dans la composition de ces familles, les mêmes principes doivent être admis pour la disposition des espèces que l'on pourra ensuite grouper en plus ou moins grand nombre de petites divisions. On ne doit cependant les élever à la dignité de genres, que lorsque avec des différences dans l'organisation, traduites par des caractères extérieurs, il s'ensuit quelques différences constantes et notables dans les mœurs et les habitudes des animaux.

C'est en suivant ces principes que nous sommes parvenu au système général de classification des reptiles et des amphibiens dont nous allons exposer une analyse raisonnée avant que de passer à la description des espèces de la Californie.

TYPE I. A. VERTÉBRÉS (*Osteozoa*).

SOUS-TYPE II. — O. Vivipara.

CLASSE III. PTERODACTYLES (*PTERODACTYLIA*).

Corps court, pourvu en avant d'une tête à mâchoires longues et garnies de dents coniques, espacées, en arrière d'une queue conique assez courte, et sur les côtés, de deux paires de membres, les thoraciques ayant le doigt externe beaucoup plus long que les autres et sans ongle, les pelviens fort élevés et à cinq doigts égaux onguiculés.

OBSERV. Cette classe ne contient encore que le genre ptérodactyle, connu seulement à l'état fossile, et que nous pensons n'être ni un mammifère de la famille des chauves-souris, comme Sœmmering l'a pensé, ni même un reptile proprement dit, comme G. Cuvier l'a dit, mais un être faisant le passage des oiseaux aux reptiles, et dont le système épidermique n'était peut-être pas squameux.

CLASSE IV. REPTILES (*REPTILIA*).

Animaux ovipares (sans mamelles), couverts d'une peau fortement épidermée (sans poils ni plumes), rampant plus ou moins sur le sol, soit avec le tronc ou les membres, soit à l'aide du tronc seulement.

Observ. Ces animaux, par leur organisation, forment évidemment un type distinct que l'on ne peut rapporter ni aux oiseaux ni aux amphibiens, et dont les différences principales portent sur la considération de la tête, de la colonne vertébrale et des membres.

Les reptiles, sous le rapport des mœurs, offrent aussi de très-grandes différences, soit avec les autres animaux, soit même entre eux.

Les grandes différences que les reptiles présentent sous le double rapport de l'organisation et des mœurs et des habitudes, fournissent de nombreux élémens pour leur classification; aussi peut-on en trouver d'excellens même dans la disposition de l'écaillure ou pholidose et dans la forme des parties qui la constituent.

Ordre I. CHELONIENS (*CHELONIA*).

Corps court, plus ou moins orbiculaire, plus ou moins inflexible ou immobile par la réunion des os du tronc et la solidification du derme, pourvu, en avant, d'un cou long et extensible, portant une tête courte, à mâchoires également courtes, dépourvues de dents, en arrière d'une queue conique, plus ou moins alongée, et sur les côtés de deux paires de membres très-diversiformes.

Observ. L'organisation des tortues est si particulière qu'elle se laisse même assez difficilement rapporter à celle des autres ostéozoaires; cependant, en l'étudiant convenablement, c'est évidemment avec les oiseaux, d'une part, et avec les crocodiles de l'autre, qu'il y a plus de rapprochemens à faire.

Le point le plus important à considérer, c'est leur séjour ou leur habitation dans la mer, dans l'eau douce, dans les marais, ou à la surface de la terre, parce qu'il en est résulté des modifications importantes dans la solidité de leur carapace, et surtout dans la forme de leurs extrémités.

La distribution méthodique des tortues nous paraît devoir porter sur la distinction et la séparation des doigts qui sont d'autant plus mobiles ou d'autant plus séparés, que les espèces se rapprochent davantage des crocodiles, caractère qui est concomitant avec le séjour, mais non pas d'une manière absolue. Ainsi, il nous semble que les tortues de mer, par la forme de leurs membres et surtout des antérieurs, dont les doigts sont indistincts et immobiles entre eux sous la peau, se rapprochent davantage des manchots, et que les trionyx, par la grande liberté des doigts réunis cependant par une membrane, et dont trois seulement sont pourvus d'ongle, se rapprochent évidemment des crocodiles. C'est d'après cela que nous partageons les tortues de la manière suivante :

FAM. I. LES T. DE MER (*CHELONEA*).

Corps très-aplati, ou déprimé, pourvu de membres disposés en larges nageoires par l'aplatissement et la réunion des doigts entièrement cachés sous la peau.

G. *dermochelys*, *chelonœa*, *caretta*.

FAM. II. LES T. DE TERRE (*TESTUDINEA*).

Corps très-bombé, très-solide, à pieds courts cylindriques, tronqués, pourvus de doigts courts et distincts au moins par des ongles gros et courts.

OBSERV. Cette division comprend chez les erpétologistes récens un assez grand nombre de genres fort peu importans, établis sur la considération de la mobilité de quelques parties de la carapace ou du plastron et nommés *testudo*, *kinnyxis* ou *cinyxis*, *pyxis*, *chersus*, etc.

FAM. III. LES T. D'EAU DOUCE (*EMYDA*).

Corps ovale, médiocrement bombé, solide, résistant, porté sur des pieds terminés par des doigts distincts, plus ou moins palmés par une membrane intermédiaire, et pourvus d'ongles longs et aigus, en aussi grand nombre que de doigts.

OBSERV. L'organisation des tortues de cette famille ne présente réellement aucune différence bien importante avec celle des tortues de terre, vers lesquelles on passe d'une manière presque insensible; aussi, les subdivisions génériques que les erpétologistes y ont établies, et qui deviennent tous les jours plus nombreuses, sont-elles assez peu importantes. Elles portent essentiellement sur la mobilité des pièces du sternum et sur le système d'écaillure des bords de la carapace, sur le nombre des doigts, etc.

Cette famille peut être subdivisée en trois genres, comprenant chacun plusieurs sous-genres.

Emys comprenant les sous-genres *emys*, *terrapena*, *cistudo*, *cinosternon*.

Chelemys comprenant les sous-genres *sternothera*, *pentonyx*, *platemys*, *chelodina*.

Chelys subdivise en *chelys* proprement dit, *saurotypus* et *chelydra*.

FAM. IV. LES T. MOLLES (*AMYDA*).

Corps suborbiculaire, très-déprimé, flexible au moins dans toute sa circonférence, et pourvu de pieds, dont les doigts sont très-distincts, largement palmés, avec des ongles longs et aigus en moins grand nombre qu'eux.

OBSERV. Cette famille s'éloigne déjà des autres tortues, parce que les pièces du squelette sont moins élargies, moins soudées entre elles, et que la peau qui revêt la carapace n'est solidifiée par des pièces osseuses que dans le milieu du dos, et nullement sur ses bords, ce qui leur donne une sorte de mollesse d'où est provenu le nom sous lequel on les désigne.

Quelques espèces ont encore des rudimens de pièces solides cutanées en avant et en arrière du bord de la carapace, tandis que d'autres n'en ont aucune trace; c'est là-dessus que sont établis les genres *emyda* et *trionyx*.

ORDRE II. PLESIOSAURIENS (*PLESIOSAURIA*).

Corps ovale, assez alongé, mou, au moins dans ses parties supérieures, pourvu en avant d'un très-long cou, portant une très-petite tête à mâchoires courtes, armées de dents, en arrière d'une petite queue conique, et sur les côtés de deux paires de membres entièrement pinniformes, et formés de doigts non distincts sans ongles et entièrement cachés sous la peau.

OBSERV. Cet ordre ne contient encore qu'un seul genre qui n'est connu qu'à l'état fossile. D'après l'étude que j'ai faite de son squelette, il me semble que cet animal doit être considéré comme faisant le passage des cheloniens ou tortues aux emydo-sauriens ou crocodiles.

Ordre III. EMYDOSAURIENS (*EMYDOSAURIA*).

Corps plus ou moins alongé, un peu déprimé, cataphracté en dessus, pourvu, en avant, d'une tête aplatie, armée de mâchoires longues, très-fortes et dentées, se continuant en arrière en une queue très-longue, très-puissante, plus ou moins comprimée, et porté sur des membres courts, étalés à (4-5) doigts plus ou moins palmés, et garnis d'ongles en moins grand nombre qu'eux.

Observ. Cet ordre, que la plupart des erpétologistes séparent des véritables sauriens dont l'organisation est toute différente, s'en distingue en effet par l'immobilité de toutes les parties de la mâchoire supérieure, entre autres de l'os carré, comme chez les tortues, par l'existence d'un véritable pénis unique, comme dans celles-ci, et par presque tous les autres points de l'organisation.

Les espèces qui entrent dans ce degré d'organisation ont été réparties dans différens genres ou sous-genres, d'après la considération peu importante de la forme et de la longueur du museau, ce qui produit les alligators ou crocodiles proprement dits, les caïmans et les gavials; mais on a pu, avec plus de raison, en séparer des animaux fossiles, qui offrent en effet des caractères bien plus tranchés, et que M. Geoffroy a nommés *teleosaurus* et *stenosaurus*.

Ordre IV. SAUROPHIENS (*SAUROPHIDIA*).

Corps très-diversiforme, de plus en plus cylindrique et alongé, couvert de squames plus ou moins dissemblables, pourvu, en avant, d'une tête petite à mâchoires dentifères, en arrière d'une queue très-variable dans sa longueur; membres également très-variables en nombre, et même tout-à-fait nuls.

L'anus transversal.

L'organe mâle de la génération double ou séparé en deux parties distinctes.

Observ. Lorsqu'on étudie la série des espèces de reptiles qui entrent dans cet ordre, on est obligé de reconnaître qu'elles se nuancent, pour ainsi dire, de la première à la dernière, de telle sorte qu'il est presque impossible d'établir une séparation un peu tranchée entre les espèces qui ont des membres et celles qui en sont dépourvues, c'est-à-dire entre les sauriens et les ophidiens. Aussi, avons-nous cru devoir prendre pour dénomination commune un mot composé de ces deux noms, au lieu de celle de bispeniens que nous avions d'abord employée, et qui, quoique juste, était peut-être trop significative.

Les seuls caractères communs à tout cet ordre, sont d'avoir : 1° la peau couverte de productions épidermiques en forme de tubercules, d'écailles ou de plaques plus ou moins étendues ; 2° l'anus transverse et bordé en avant par une disposition operculiforme du système squameux ; 3° l'organe excitateur mâle nettement séparé en deux parties indépendantes, l'une à droite et l'autre à gauche.

Dès-lors, on conçoit combien il est facile de répartir les espèces dans de grands genres linnéens, ou petites familles assez distinctes.

Sous-ordre. — Sauriens (*Sauria*).

Corps partagé d'une manière assez distincte en ses cinq parties, la tête, le cou, la poitrine, le ventre ou abdomen, la queue, et pouvu constamment de membres complets ou rudimentaires.

Le cœur très-antérieur, le poumon à deux lobes égaux ou subégaux.

L'anus couvert d'un opercule formé de plusieurs scutelles, ainsi que l'opercule surcilier.

Observ. Ce sous-ordre, facile à distinguer pour la plus grande partie des espèces, qui sont en effet pourvues de deux paires de membres complets le devient de moins en moins à mesure qu'on s'approche des derniers genres, chez lesquels ces organes semblent osciller non seulement dans leurs divisions terminales, mais même dans leur nombre, avant de finir par disparaître complétement à l'extérieur du moins, car à l'intérieur il existe presque toujours une ceinture osseuse antérieure et postérieure.

Chez les zoologistes modernes, les grands genres de Linné sont des familles que l'on peut parfaitement adopter, afin de mettre de l'ordre dans la disposition des espèces, ordre qui, en même temps qu'il permet de les reconnaître plus aisément, dessine aussi d'une manière convenable la série.

On peut aussi trouver dans le système d'écaillure des caractères assez tranchés; ainsi, au-dessus des orbites, l'espèce d'opercule qui recouvre les yeux et celui de l'anus sont toujours composés de plusieurs scutelles chez les sauriens et jamais chez les ophidiens.

Fam. I. GECKOS (*GECKONES*) (G. *Gecko* L.).

Corps assez court, subdéprimé, nettement séparé par un rétrécissement colliforme, de la tête également déprimée, à museau court et parabolique, et prolongé en une queue médiocre et conique.

Membres courts, très-étalés, terminés par cinq doigts également courts, et plus ou moins dilatés en dessous par une disposition particulière de la peau.

Narines petites, rondes et plus ou moins terminales.

Yeux à paupières peu fendues, à pupille verticale.

Tympan assez enfoncé.

Peau couverte partout d'écailles fort petites, tuberculées en dessus, un peu plus grandes et imbriquées en dessous.

Ongles très-aigus, et plus ou moins rétractiles.

Dents maxillaires, petites, égales, entières, implantées.

Dents palatines nulles.

Langue large, épaisse, entière, non exsertile.

Observ. Les sauriens qui constituent ce genre ont véritablement quelque chose de particulier dans les mœurs et les habitudes, et même jusqu'à un certain point dans l'organisation qui les distingue des autres espèces.

La considération de la disposition des doigts propre à les faire adhérer aux plans verticaux et même horizontaux, a conduit M. Cuvier à une très-bonne distribution des espèces, qu'il est convenable d'adopter, sans qu'il soit absolument nécessaire d'en faire autant à l'égard des genres ou subdivisions que ce zoologiste ou d'autres depuis lui ont établies.

Je subdiviserai les espèces de geckos ainsi qu'il suit :

A. *Espèces dont les doigts sont dilatés dans toute leur étendue.*

Les Geckos proprement dits, formant le genre *platydactylus* (Cuv.), subdivisés ensuite suivant l'intégrité de la plaque sous-digitale, ou sa division par un sillon médian, le nombre des ongles des doigts, et autres caractères encore plus insignifians en *phelsuma*, *platydactylus*, *tarentola*, *gecko*, *ptychozoon*, *pteropleura*, *thecadactylus*.

B. *Espèces dont les doigts sont dilatés seulement dans la moitié basilaire de leur longueur.*

Les Demi-Geckos (*Hemidactylus* Cuv.).

C. *Espèces dont les doigts sont dilatés seulement à l'extrémité.*

Les Tiers-Geckos (*Ptyodactylus* Cuv.).

Subdivisés en *ptyodactylus*, *uroplatus*, *crossurus*, *sphœrodactylus*, d'après des considérations d'une moindre importance, et entre autres, de la forme conique ou comprimée de la queue.

D. *Espèces dont les doigts ne sont pas élargis, mais seulement striés en travers et denticulés.*

Les Quart-Geckos (*Stenodactylus* Cuv.) subdivisés suivant que les écailles de la queue ne sont pas ou sont verticillées, en *stenodactylus*, *eublepharis* et *pachydactylus*.

E. *Espèce dont les doigts ne sont ni élargis ni même striés.*

Les Subgeckos (*Gymnodactylus* Cuv.) subdivisés suivant que la queue est ronde ou très-aplatie, en *gymnodactylus* proprement dit, en *goniodactylus* et en *phyllurus*.

Fam. II. LES CAMÉLÉONS (G. *CHAMÆLEO* L.)

Corps très-comprimé, caréné en dessus, comme en dessous.

Tête grande, comprimée, tétraèdre, à angles très-saillans, plus ou moins prolongés à la nuque.

Queue fort longue, arrondie et prenante.

Membres très-longs, très-grêles, élevés, à cinq doigts subégaux, réunis jusqu'aux ongles en deux paquets opposés, 3 - 2 en avant, 2 - 3 en arrière.

Yeux à voile unique, très-peu fendu.

Tympan non visible.

Narines fort petites, rondes, subterminales.

Peau entièrement couverte de très-petites écailles graniformes, plus grandes et plus saillantes dans la ligne médiane.

Point de pores fémoraux.

Ongles très-courts et très-aigus.

Dents maxillaires, entièrement implantées ou soudées dans les mâchoires.

Point de dents palatines.

Langue épaisse, charnue, en forme de disque, portée à l'extrémité d'un long pédoncule vermiforme et extensible.

Observ. Quoique cette famille ne soit composée que d'un seul genre, elle ne mérite pas moins d'être distinguée de toutes les autres, par un grand nombre de points de son organisation extérieure et intérieure

Le frontal antérieur et le postérieur se réunissent au rebord supérieur de l'orbite, de manière à ce que le véritable frontal est seulement interorbitaire, comme cela a lieu dans les chelonées; le pariétal réunit ses deux branches postérieures en une seule, pour aller sur l'occiput se joindre à la pointe du squameux, ce qui constitue une sorte de pyramide sur la nuque; les fosses nasales sont latérales et fort petites; le prœmaxillaire est extrêmement petit, terminé par une dent unique, tranchante, un peu comme dans le fouette-queue.

Les dents sont extrêmement petites, triangulaires, comprimées, espacées, croissant un peu de la première à la dernière, et si solidement immergées dans les mâchoires qu'elles semblent denticuler le bord de celles-ci.

La place de ce genre a considérablement varié dans les systèmes d'erpétologie; mais, dans notre manière de voir, il doit suivre celui des geckos, par la ressemblance du système d'écaillure, ainsi que par celle des ongles.

Fam. III. LES AGAMES (*AGAMA* Daud.).

Corps de forme assez variable, déprimé, comprimé, et même fusiforme.

Tête bien distincte par un étranglement colliforme très-profond.

Queue en général longue et plus ou moins comprimée.

Membres assez rapprochés, médiocres, et assez étalés, pourvus de cinq doigts longs, grêles, fort inégaux, surtout en arrière.

Narines subterminales.

Yeux grands, à voile unique, plus ou moins fendu, sans trace de paupière intérieure.

Tympan large, superficiel, très-rarement nul.

Peau couverte d'écailles, le plus souvent carénées et épineuses, à peu près semblables partout.

Ongles comprimés, arqués et aigus.

Dents maxillaires entières, un peu comprimées, tellement enchâssées dans les mâchoires, qu'elles semblent en faire partie, et dont quelques-unes des antérieures plus longues que les autres semblent des espèces de canines.

Palatines nulles.

Langue charnue, épaisse, à peine échancrée, non extensible.

Observ. Le crâne et la face des agames sont toujours fort courts et assez comprimés; la mâchoire inférieure plus longue que la supérieure; l'espace interorbitaire est assez large, ainsi que les os du nez; l'orbite est incomplet.

La distribution des espèces de ce grand genre, ou de cette famille, que Linné et Gmelin comprenaient dans la division des iguanes de leur genre *lacerta*, n'est pas facile; aussi, la plupart des subdivisions génériques que les erpétologistes modernes y ont introduites dans ces derniers temps, sont-elles assez peu importantes, et sans influence un peu marquée sur les mœurs. Elles reposent sur la considération :

1° De la forme et de la disposition du système squameux;

2° De la présence et de l'absence d'une crête médio-dorsale;

3° De la présence et de l'absence des pores fémoraux;

4° De la présence ou de l'absence des dents palatines.

G. Cuvier a tranché la difficulté, en regardant comme iguanes toutes les espèces de sauriens n'ayant pas la tête scutellée, ni la langue extensible, qui n'ont pas de dents palatines, celles qui en sont pourvues étant pour lui des iguanes. Je crois qu'il est préférable d'avoir plus d'égard à l'écaillure, comme d'une bien plus grande valeur; aussi, ce genre nous paraît-il, à cause de cela même, devoir suivre la famille des geckos.

Je me suis efforcé de disposer les espèces de cette famille, de manière à ce que les premières soient les plus semblables aux geckos, et les dernières aux iguanes; elle est partagée en trois ou quatre genres seulement.

A. *Espèces dépourvues de crêtes dorsale et caudale.*

Les Agames (*Agama*), qui par des caractères de troisième importance, tirés de la forme et des proportions des différentes parties du corps, ainsi que de la peau écailleuse, plus ou moins épineuse, ou carénée, sont réparties dans les sous-genres *phrynocephalus*, *stellio*, *phrynosoma*, *platynotus*, *tapelus*, *agama*, *ecphymotis*, *tropidolepis* et *amphibolurus*.

B. *Espèces pourvues d'une crête dorsale, formée d'écailles.*

Les Lophyres (*Lophyrus* Dum.), réparties ainsi d'après la considération de la queue ronde ou comprimée, des écailles plus ou moins unies dans les sous-genres *hypsibatus*, *galeotes*, *lophyrus*, *ophryessa*, *gooniocephalus*, *lyrocephalus*.

C. *Espèces sans crête dorsale, et à queue verticillée par des anneaux d'écailles fort épineuses.*

Les Fouette-queue (*Uromastyx*).

D. *Espèces pourvues d'une crête dorsale, et de dents maxillaires appliquées, plus ou moins denticulées, sans dents palatines.*

Les Agamiguanes (*Agamiguana*) qui, beaucoup plus rapprochées des iguanes par la forme des dents maxillaires, sont réparties d'une manière assez fixe dans les quatre sous-genres, nommés *physignathus*, *brachylophus*, *istiurus* et *amblyrhynchus* par les erpétologistes les plus récens.

Fam. IV. LES DRAGONS (G. *DRACO* L.).

Corps ovale ou renflé dans son milieu, un peu déprimé.

Tête courte, assez grosse, quadrangulaire, à crête surcilliaire fort saillante.

Queue longue, conique et aiguë.

Membres longs et grêles, terminés par cinq doigts inégaux.

Narines sulatérales.

Yeux à paupières bien fendues.

Oreilles sans tympan visible.

Peau plus ou moins dilatée, soit sur les flancs, soit autour du cou, et couverte de très-petites écailles imbriquées, carénées sur la queue et sur les membres, et quelquefois un peu plus grandes sur la tête.

Ongles assez aigus et peu courbés.

Dents maxillaires, immergées dans les mâchoires, dont quelques-unes plus longues que les autres, et quelquefois trilobées.

Observ. La forme de la tête des dragons, proprement dits, rappelle assez celle des agames, mais plus courte, à cause de plus de brièveté dans le museau; l'orbite est grand et entièrement fermé; l'échancrure nasale est petite, oblique et terminale.

Les espèces, qui composent cette petite famille, se laissent très-aisément grouper en sections génériques, au moyen de particularités fort remarquables. En effet, les espèces de véritables dragons ont sur les flancs une large membrane soutenue par une disposition particulière des côtes asternales; tandis que les *chlamydosaurus*, nouveau genre établi par M. Gray, ont cette expansion bornée aux parties latérales de la tête, que les *callysaurus*, genre que nous établissons d'après un joli saurien de la Californie, n'ont plus qu'un simple pli sur les flancs, et que les *sitana* de G. Cuvier ont une large expansion médiane sous la poitrine.

Fam. V. LES IGUANES (*IGUANA* Daud.).

Corps un peu diversiforme, mais seulement comprimé ou fusiforme.

Tête bien distincte, sans étranglement colliforme, aussi marquée que chez les agames, et plus généralement quadrangulaire.

Queue fort longue, et plus ou moins comprimée.

Membres assez rapprochés et étalés, terminés par cinq doigts grêles, alongés et très-inégaux, surtout en arrière.

Yeux à paupières bien fendues.

Tympan large et superficiel.

Peau assez souvent dilatée en goître sous la gorge, et couverte d'écailles fort petites, presque semblables en dessus comme en dessous, et fort rarement carénées.

Des pores fémoraux ou non.

Dents maxillaires nombreuses, très-comprimées, dentelées ou trilobées, appliquées contre les mâchoires, et non implantées.

Dents palatines sur un seul rang.

Langue charnue, épaisse, non extensible, et à peine échancrée.

Observ. Le crâne et la face sont en général courts et tétraèdres; l'espace interorbitaire étroit; l'ouverture nasale ovalaire et médiocre; l'orbite et l'arcade zygomatique complets.

La distribution des espèces de cette famille n'est réellement pas facile, à moins que de faire comme G. Cuvier, et de n'avoir qu'à l'existence des dents palatines.

Toutefois, comme ce caractère se prononce de plus en plus, à mesure qu'on s'éloigne des premières familles de sauriens, et qu'il est presque constant chez les ophydiens, il nous semble que les iguanes doivent être placés après les agames. Du reste, la distribution des espèces s'est faite à peu près d'après la considération des mêmes parties, l'absence ou la présence des pores fémoraux, d'une crête médio-dorsale, et sur la disposition de l'écaillure de la tête qui se régularise de plus en plus, et enfin sur quelques autres particularités assez peu importantes.

La distribution que j'ai adoptée est la suivante :

+ *Sans pores fémoraux.*

A. *Espèces dont le corps est pourvu d'une crête soutenue par les apophyses épineuses prolongées, la tête triangulaire squameuse.*

Les Basilics (*Basiliscus*).

B. *Espèces dont les doigts sont dilatés et garnis de rangées d'é-*

cailles sous l'avant-dernière phalange, avec des ongles très-arqués et très-aigus.

Les Anolis (*Anolis* Daud.) qui, en ayant égard à la présence et à l'absence d'une crête dorsale, ainsi qu'au degré de dilatation des doigts, se rangent dans les sous-genres *xiphosurus*, *dactyloa*, *anolis*, *norops* et *draconura*.

C. *Espèces dont la tête est singulièrement dilatée, le dos denticulé, ainsi que le bord des doigts.*

Les Subiguanes (*Œdicoryphus*).

D. *Espèces dont le dos est garni d'une crête paléacée.*

Les Iguanes (*Iguana*), et qui par la considération de la forme comprimée de la queue ou du front cornu, ou de l'écaillure épineuse et verticillée de la queue, ou d'un prolongement singulier de la nuque constituent les genres *corythophanes*, *hypsilophus*, *metapoceros*, *cyclura* et *chamaleopsis*.

E. *Espèces sans crête, et dont le corps et la queue fort grêles, sont couverts d'écailles lisses.*

Les Marbrés (*Polychrus* Cuv.), et qui ayant des pores fémoraux à moitié cuisse ou dans toute son étendue, ou en étant totalement dépourvues, constituent les sous-genres *polychrus*, *leiolepis*, *laemanctus*.

F. *Espèces sans crête dorsale, et dont la queue est couverte d'écailles fortement épineuses, disposées en verticilles.*

Les Échinés ou (*Urocentron*), réparties d'après la présence ou l'absence des dents palatines dans les sous-genres *oplurus*, *tropidurus*, *strobilurus*, et *doryphorus*.

FAM. VI. LES SAUVES-GARDES (*MONITOR*).

Corps en général alongé et fusiforme.

Tête étroite à face alongée.

Queue fort longue, et plus ou moins comprimée.

Membres robustes, étalés, pourvus de cinq doigts, inégaux, et fort longs surtout en arrière.

Narines grandes, longitudinales, latérales, et très-remontées.

Yeux grands, à trois paupières mobiles.

Tympan large et superficiel.

Peau couverte d'écailles tuberculiformes en dessus, et d'espèces de plaques ou scutelles carrées en dessous.

Pores fémoraux nuls.

Ongles aigus, arqués et puissans.

Dents maxillaires, coniques, plus ou moins arquées, appliquées ou lâchement implantées.

Dents palatines nulles.

Languc grande, extensible, profondément fourchue à sa pointe, et pourvue d'un fourreau à sa base.

OBSERV. La tête osseuse des sauriens de cette famille présente une forme assez particulière, en ce qu'elle est fort étroite, fort alongée; l'échancrure des narines très-grande et très-remontée; l'espace interorbitaire est fort étroit, et l'orbite incomplet.

Les dents ont aussi une forme particulière, en ce qu'elles sont toutes coniques, et plus ou moins en crochet.

Toutes les espèces sont de l'ancien continent; car il ne nous paraît pas probable que l'héloderme soit réellement de cette famille. Elles atteignent toutes une assez grande taille.

Quoiqu'assez peu nombreuses, elles nous semblent devoir former une petite famille distincte, qui doit être placée immédiatement avant les lézards, avec lesquels la forme de la langue, celle des paupières, et même une certaine dissemblance plus prononcée, entre l'écaillure dorsale et ventrale, établissent des rapports évidens.

Les sauves-gardes ne doivent réellement former qu'un seul genre, pour lesquels nous préférerons le nom de monitor qui leur appartient, à celui de tupinambis qui doit indiquer des animaux américains. Cependant on pourra distribuer les espèces d'après la considération de la queue presque ronde, comprimée avec une double crête, ou très-comprimée dans les sous-genres établis pas Wagler, sous les noms de *psamosaurus*, *polydaedalus* et *hydrosaurus*.

Fam. VII. LES LACERTIENS (*LACERTA* L.).

Corps généralement de plus en plus alongé et anguiforme, cylindrique.

Tête peu ou point distincte, scutellée.

Queue presque toujours fort longue, et confondue avec le tronc.

Narines petites, rondes et terminales.

Yeux à deux ou trois paupières mobiles.

Tympan superficiel et rarement nul.

Peau couverte de plaques ou de scutelles, disposées régulièrement sur tout ou partie de la tête, et sur le reste du corps, d'écailles molles, ou osseuses, plus larges en dessous qu'en dessus.

Des pores fémoraux ou non.

Dents maxillaires appliquées, et souvent denticulées, presque de même grandeur.

Dents palatines sur plusieurs rangs ou nulles.

Langue plus ou moins extensible et bifide.

Observ. Cette famille, beaucoup plus nombreuse en espèces que toutes les autres, présente, dans presque tous les points de son organisation, des indices de dégradation évidente, et surtout dans l'appareil locomoteur; en ce que les membres diminuent peu à peu de force et de développement, s'éloignent de plus en plus, de manière à ce que la locomotion troncale devienne prédominante, et finisse par exister seule, quoiqu'il existe presque toujours des rudimens sous-cutanés de la racine des membres; mais le cœur reste constamment placé tout-à-fait à la partie antérieure du tronc, le cou étant fort court, comme dans tous les sauriens. Il y a toujours deux poumons, mais l'un d'eux diminue avec l'alongement du corps.

D'après ce que nous venons de dire sur les nuances de dégradation qu'offrent les animaux de cette famille, il est aisé d'en conclure que les caractères doivent offrir beaucoup d'oscillations, ce qui a conduit à multiplier considérablement le nombre des genres. On les a surtout établis sur le nombre et la forme des membres, sur le nombre des doigts; et quoique, au premier abord, les caractères d'organes aussi importans semblent avoir une grande valeur; au fond, ils signifient fort peu de chose, parce qu'ils reposent sur des organes oscillans, avant de disparaître tout-à-fait, et dont les modifications n'ont aucune influence sur les mœurs et les habitudes des animaux : aussi sommes-nous bien loin de les adopter. Tout oscille en effet chez ces animaux, les paupières, le tympan, les pieds, les doigts, les pores fémoraux, les dents palatines, les poumons même et le cœur, du moins dans sa position.

Quoi qu'il en soit, il en résulte au moins que cette famille ne peut être qu'ici, quoique plusieurs zoologistes aient regardé les dragones comme voisines des crocodiles. Elle doit, au contraire, être placée immédiatement avant les ophidiens.

Quant à sa distribution intérieure, elle est aussi assez facile du moins pour les espèces qui doivent être mises à la tête. Quant à celles qui doivent la terminer, il y a un peu plus de difficultés; le passage aux ophidiens se faisant par deux lignes.

Voici, au reste, la classification que nous avons cru devoir adopter.

Tribu I. — LES TUPINAMBIS (*Neosaurus*).

Corps lacertiforme, couvert d'écailles molles, très-rarement carénées.

Langue étroite, terminée par deux filamens cornés, sans gaîne à la base.

Point de dents palatines.

OBSERV. Ce sont les plus grandes espèces de lézards, qui jusqu'alors n'ont été trouvés que dans le nouveau continent.

Des premières espèces on passe d'une manière graduelle aux dernières qui se distinguent à peine des lézards proprement dits.

A. *Espèces sans collier proprement dit, et sans pores fémoraux.*

Les DRAGONES (*Dracœna* Cuv.) qui, d'après la considération minutieuse de la forme des dents incisives et maxillaires, et de celle de la queue conique, ou comprimée et à deux crêtes, sont réparties dans les sous-genres *dracœna* ou *thorictis*, *podinema*, *ctenodon* et *crocodilurus* établis par Spix ou par Wagler.

B. *Espèces pourvues d'un véritable collier et de pores fémoraux.*

Les AMÉIVAS (*Ameiva*) que les erpétologistes ont répartis dans les sous-genres *tejus*, ou *cnemidophorus*, *acrantus*, et *trachygaster*, suivant que les écailles sont lisses et granuleuses, avec cinq ou quatre doigts aux pieds, ou qu'elles sont carénées et épineuses en dessous.

Tribu II. — LES LÉZARDS (G. *Lacerta*).

Langue profondément bifide à la pointe, exsertile, et ayant une gaîne à sa base.

Écailles molles, lisses ou carénées.

OBSERV. Les sauriens qui constituent cette tribu, se distinguent de ceux de la précédente par la forme de la langue qui est beaucoup plus exsertile; mais, du reste, ce sont presque tous les mêmes caractères, au point qu'on pourra les réunir sans aucun inconvénient.

Quant à la tribu des scinques, la distinction en est beaucoup plus facile par la nature des écailles.

Les espèces de lézards sont assez difficiles à disposer dans un ordre à la fois méthodique et naturel. Nous pensons cependant qu'elles peuvent l'être de la manière suivante :

A. *Espèces dont les écailles du corps sont disposées en séries obliques, et non verticillées,*

+ *non carénées* (LACERTA),

réparties suivant que les tempes sont écailleuses ou scutellées, qu'il y a ou non des paupières, un collier, des dents palatines, des pores fémoraux, dans les genres *zootoca*, *gymnophthalmus*, *podacis*, *eremias*, *scaptera* et *lacerta.*

+ + *carénées et subverticillées* (TACHYDROMUS),

constituant, d'après les mêmes considérations à peu près, les sous-genres *aspistis*, *psamodromus*, *algyra*, *tachydromus*, *cercosaura*, etc.

B. *Espèces dont les écailles du corps sont verticillées, comme celles de la queue,*

+ *carénées et imbriquées*,

formant, d'après la considération de l'absence des paupières, des doigts au nombre de 5-5, de 4-5, ou même tout-à-fait nuls, les genres ou sous-genres *lepidosoma*, *cricochaleis*, *monodactylus* ou *chamæsaura.*

+ + *non carénées*, *non imbriquées*,

et seulement approchées (*saurochalcis*), partagées d'après les doigts, au nombre de 4-4 onguiculés, de 3-3 sans ongles et même de 1-1 également sans ongle dans les sous-genres *brachypus*, *cophias* et *colobus.*

Tribu III. — Les L. Bipèdes (*Diposaurus*).

Corps assez longuement conique, anguiforme, couvert d'écailles molles et pourvu d'une seule paire de pieds monodactyles.

Observ. Ce groupe de sauriens anguiformes est véritablement assez difficile à placer convenablement quand on a égard à l'ensemble de l'organisation, plutôt qu'au nombre des pieds. En effet, la disposition de l'écaillure rappelle assez bien celle de certaines espèces de serpens; aussi Wagler le place-t-il parmi ses *angues*.

Il me semble cependant qu'il doit être mis ici avant les scinques à cause de la grandeur des écailles, la manière dont elles sont largement imbriquées, sans indices de carène.

Les espèces fort peu nombreuses qui constituent cette tribu, ou mieux peut-être ce genre, sont placées dans les sous-genres *pygopus*, *lialis*, *pygodactylus* et *delma*, suivant qu'il y a au devant de l'anus cinq ou quatre paires de pores, ou qu'il n'y en a pas, le tympan étant visible ou non.

Trib. IV. — Les Scinques (*Scincus*).

Corps très-diversiforme, couvert d'écailles solides, osseuses, polymérées.

Langue large, épaisse, papilleuse, plus ou moins échancrée, non exsertile ni pourvue de gaîne à sa base.

Observ. Cette tribu pourra être aisément considérée comme une petite famille par la considération de son système d écaillure qui est tout-à-fait particulier, et qui ne peut être comparé qu'à ce qui existe dans les poissons, les écailles étant comprises dans le derme lui-même, polymérées et tout-à-fait osseuses. Quant au reste de l'organisation, la différence, avec ce qui a lieu chez les lézards, n'est réellement pas considérable.

La distribution des espèces me paraît assez facile en ayant égard à l'absence ou à l'existence d'un sillon de chaque côté du corps, à l'état lisse ou caréné des écailles, avant que de recourir à la considération des membres qui finissent par disparaître entièrement dans les deux grands genres ou sections que nous établissons dans cette tribu, de la manière suivante :

A. *Espèces dont le corps n'est pas pourvu d'un sillon latéral, et dont les écailles sont toujours lisses,*

+ *quadrupèdes* (G. SCINCUS),

subdivisées, d'après la considération du nombre des doigts, des pores fémoraux, fort rares dans cette tribu, des paupières, quelquefois nulles, des dents palatines, du tympan, de la forme des dents maxillaires, et enfin de l'écaillure de la tête, en un assez grand nombre de sections considérées comme des genres sous les noms de *spondylurus*, *ablepharus*, *tiliqua* ou *cyclodus*, *trachysaurus*, *heloderma*, *euprepis*, *scincus*, *gongylus*, *sphœnops*, *tetradactylus*, *tridactylus*, *lygosoma*, *podophis*, *peromeles*, *saiphos* et *lerista*, établis surtout par Wagler.

++ *bipèdes* (G. BIPES Lacepd.),

distinguées en *scelotes* ou *zygnis* et en *ophioides*, suivant que les pieds sont bifides ou indivis.

+++ *nullipèdes* (G. ANGUIS L.),

réparties par les erpétologistes récens dans les trois genres *anguis*, *acontias*, *typhlosaurus*, suivant que les yeux et le tympan sont visibles, que les yeux seuls sont visibles, ou que ni l'un ni l'autre ne sont visibles.

B. *Espèces pourvues d'un sillon latéral et d'écailles carénées ou non,*

+ *quadrupèdes* (G. CORDYLUS),

rangées, d'après la considération des écailles carénées, épineuses et verticillées, carénées sur le dos seulement, non carénées, verticillées et homogènes, ou dissemblables, dans les genres *cordylus*, *gerrhonotus*, *gerrhosaurus*, *saurophis* et *chalcis*,

++ *bipèdes* (G. PSEUDOPUS).

+++ *nullipèdes* (G. Ophysaurus),
subdivisées malgré leur petit nombre en *ophisaurus* et en *otophis*, suivant que la tête tout entière est scutellée, ou qu'il n'y a que le rostre ou museau.

Sous-ordre. — Les Ophidiens ou Serpens (*Ophidia*).

Corps fort alongé, cylindrique, sans presque aucune distinction de tête, de col, de tronc et de queue, presque toujours entièrement dépourvu de membres même rudimentaires; couvert d'écailles plus ou moins dissemblables, sur différentes parties du corps; yeux constamment sans paupières.

Tympan constamment caché.

Langue large, entière, fixée, ou au contraire longue, étroite, profondément bifide et extensible.

Dents coniques, aiguës, plus ou moins dirigées en arrière, et dans le plus grand nombre des cas aussi bien palatines que maxillaires.

Observ. Quoique ces animaux ne diffèrent réellement de ceux du sous-ordre précédent que par des nuances que nous avons indiquées en parlant de celui-ci, on doit cependant remarquer que les modifications d'organisation qu'ils offrent sont assez nombreuses, et qu'elles portent essentiellement sur l'appareil locomoteur; il est en effet réduit à sa partie essentielle, c'est-à-dire au tronc, ce qui entraîne un beaucoup plus grand nombre de vertèbres que dans toute autre classe d'ostéozoaires, et que toutes ces vertèbres sont costifères depuis l'atlas jusqu'à l'anus, sans qu'il y ait trace de sternum.

Ce grand alongement du tronc porte aussi bien sur le cou que sur la poitrine, l'abdomen et la queue, en sorte que le cœur est très-reculé dans la cavité splanchnique; d'où il est résulté que la trachée-artère est aussi fort longue, et que les poumons se sont réduits à un seul lobe; l'autre, d'abord, très-petit et rudimentaire, disparaissant tout-à-fait chez les espèces les plus ophydiennes.

Un autre caractère anatomique de ce sous-ordre de reptiles consiste en ce que les appendices céphaliques ou mâchoires sont le plus souvent dilatables à leur symphise, et que l'inférieure l'est au plus haut point, parce que non seulement l'os carré est mobile à ses deux extrémités, mais encore l'os squameux entièrement détaché du crâne.

Tous ces caractères, qui ne sont au fond qu'une exagération qui existe déjà plus ou moins dans le sous-ordre précédent, ne laissent aucun doute sur la position des ophidiens dans la serie, ni même sur la disposition qu'on doit donner aux espèces nombreuses qui entrent dans cette division. Il est en effet évident que l'on doit commencer par celles qui s'éloignent le moins des derniers sauriens, et finir par celles qui ont les caractères du sous-ordre portées au maximum, ce qui nous semble avoir lieu pour les serpens à sonnettes.

Quant à la distribution intérieure des espèces et à leur répartition en genres et en sous-genres, nous devons chercher à suivre les principes qui conduiront à traduire la série organique, et nous croyons pouvoir y parvenir en prenant en considération :

1° Le plus ou moins de liberté de l'os carré et par conséquent de dilatabilité des mâchoires;

2° La disposition générale de l'écaillure (*pholydosis*) plus ou mois similaire ou dissimilaire;

3° La disposition générale du système dentaire, et surtout dans le cas d'action vénéneuse;

4° La forme et le degré d'extensibilité de la langue, quoique offrant beaucoup moins de variations;

5° La proportion de la queue.

Voici l'ordre que nous avons adopté d'après ces principes.

Tribu I. — LES S. BIMANES.

Une paire de membres antérieurs.

Genre *chirotes*, dont l'organisation a tant de rapports avec celle des amphisbènes que l'on pourrait les regarder comme des espèces de ce genre pourvues de membres antérieurs.

Tribu II. — Les Serpens.

Le corps sans aucune trace de membres ou tout au plus pourvu vers l'anus d'espèces de crochets rudimens des postérieurs.

Section I. *Dents maxillaires non vénéneuses.*

Fam. I. LES AMPHISBÈNES (*AMPHISBÆNA* L.).

Corps cylindrique, presque également obtus aux deux extrémités, et couvert d'une peau plus ou moins écailleuse, et semblable en dessus comme en dessous.

A. *Espèces dont l'épiderme forme des espèces d'écailles carrées, rugiformes, semblables partout, si ce n'est sur la tête.*

Les amphisbènes proprement dites (G. *Amphisbœna*), qui se subdivisent suivant qu'elles ont des pores au devant de l'anus avec un sillon latéral plus ou moins marqué, ou qu'elles n'ont pas de pores au devant de l'anus, avec la tête couverte de petites ou d'une seule grande plaque, des yeux plus ou moins évidens dans les sous-genres *blanus*, *amphisbœna*, *trogonophis*, *lepidosternon*, *cephalopeltis* et *anops*, qui se nuancent d'une manière presque insensible pour passer aux thyphlops.

B. *Espèces dont le corps vermiforme est couvert d'écailles imbriquées semblables en dessus comme en dessous, et qui sont entièrement privées d'yeux.*

Les Typhlops (G. *Typhlops*). Que les erpétologistes ont répartis dans les trois sous-genres *stenosoma*, *typhline* et *rinophis*, suivant que la tête est scutellée seulement en partie, ou couverte d'une seule grande plaque, ou pourvue d'une sorte de rostre

Fam. II. — LES ROULEAUX (G. *TORTRIX* Oppel.).

Corps en général alongé, cylindrique, presque également obtus aux deux extrémités, mais couvert d'écailles en dessus et de scutelles sur un seul rang en dessous.

Observ. C'est un degré de plus vers les serpens les plus évidens; aussi, les mâchoires sont-elles un peu dilatables, les pièces qui les constituent étant réunies par des ligamens assez courts.

L'écaillure est d'ailleurs de plus en plus semblable à celle des couleuvres.

Pour la distribution des espèces, il faut avoir égard à l'écaillure de la tête, ainsi qu'à la forme et proportion des scutelles sous-ventrales et sous-caudales. C'est par ces considérations de fort peu d'importance, que les erpétologistes modernes ont établi les genres *uropeltis*, *cylindrophis*, *tortrix*, *aspidurus*, *cercaspis* et *scytale*, ou *pseudo-boa*.

Fam. III. LES BOAS (G. *BOA* L.).

Corps alongé, plus ou moins colubriforme, à tête renflée et distincte, à museau obtus; la queue atténuée, conique et de médiocre longueur.

Peau couverte d'écailles en dessus, et d'un seul rang de scutelles sous le corps et sous la queue.

Une paire d'appendices en forme de crochets de chaque côté de la queue et quelquefois nuls.

Observ. Cette famille ou ce grand genre linnéen n'est réellement distincte que lorsqu'on la caractérise rigoureusement par la présence des crochets à l'anus, ou par l'existence d'une seule rangée de scutelles sous la queue. mais alors les premiers genres ne lui appartiennent pas dans le premier cas, puisqu'ils n'ont pas de crochets à l'anus; et, dans le second, il faut en retrancher les pithons qui en sont pourvus, mais qui ont une écaillure toute semblable à celle des couleuvres, et par conséquent deux rangs de scutelles sous la queue.

Nous nous en tiendrons à la définition rigoureuse de Linné, et nous reporterons les pithons dans la famille ou division suivante, et nous diviserons les boas en deux genres seulement, comme il suit.

A. *Espèces qui n'ont pas de crochets à l'anus.*

Les Eryx (G. *Eryx*), subdivisées, quoique fort peu nombreuses en *eryx* proprement dits, en *gongylolophis* et en *erpeton*, suivant que les écailles sont lisses, ou qu'elles sont carénées, dissemblables ou partout semblables, avec une paire de longs barbillons pour cette dernière espèce.

B. *Espèces qui ont des crochets à l'anus, et la queue courte et préhensile.*

Les Boas (G. *Boa* B.), réparties par les erpétologistes modernes, en plusieurs genres fort peu distincts qu'ils nomment *enygris*, *boa*, *epicrates*, *xiphosoma* et *eunectes*, d'après la considération minutieuse de l'écaillure de la tête, se rapprochant de plus en plus de celle des couleuvres, et la profondeur des excavations des scutelles labiales, qui s'effacent aussi peu à peu dans les dernières espèces, par la forme du corps plus ou moins comprimée, et même par l'absence des ergots de l'anus, comme dans les xiphosomes.

Fam. IV. — Les Boa-Couleuvres.

Corps, tête et queue de plus et plus semblables à ce qui a lieu dans les couleuvres, couverts d'écailles et de scutelles, celles-ci en double rang sous la queue.

Observ. Cette petite famille est réellement assez peu naturelle, et ne porte rigoureusement que sur la considération du double rang de scutelles sous la queue. Elle renferme en effet des espèces qui ont des crochets à l'anus, comme dans les boas, et d'autres qui en sont dépourvues, et dont la forme générale du corps a peut-être plus de rapports avec les rouleaux qu'avec les couleuvres ordinaires; cependant nous n'avons pas cru devoir les réunir, d'autant plus que les premières couleuvres auront aussi une forme assez raccourcie, et par conséquent un peu amphisbénique.

A. *Espèces pourvues de crochets à l'anus.*

Les Pythons (G. *Python*), qui, sauf la disposition des scutelles de la queue, et peut-être la forme et la longueur de celle-ci, sont de véritables boas que l'on a encore trouvé à séparer en deux genres, *python* et *constrictor*, suivant que la tête est simplement squameuse ou scutellée.

B. *Espèces dépourvues de crochets à l'anus, et dont le corps est plus ou moins cylindrique et assez court.*

Les Couleuvres-Rouleaux, dont les premières espèces s'éloignent véritablement assez des couleuvres qui vont suivre par quelques particularités de l'écaillure de la tête, et par la grande brièveté de la queue et dont les derniers en diffèrent de moins en moins.

La considération rigoureuse des scutelles de la tête dans leur nombre et dans leurs proportions, celle de la longueur et de la forme de la queue, et enfin l'état lisse ou caréné des écailles du corps ont conduit Wagles à établir les genres *xenopeltis*, *catostoma*, *elapoides* et *calamaria*, qui ont tous un loréum oculaire et quelques autres petits caractères distinctifs.

Fam. V. LES COULEUVRES (*COLUBER* L.).

Corps et queue généralement fort alongés, renflé au milieu, atténué aux deux extrémités et surtout en arrière, couvert d'écailles variables de forme et de grandeur en dessus, mais toujours de grandes plaques en dessous, sur un seul rang avant l'anus, et sur deux en arrière.

Écaillure de la tête constamment composée ainsi; scutelles front. 3-5 ou 4 en deux paires; interorbit. 1; occipit. 2 en 1 paire.

Observ. Les espèces de ce genre actuellement assez bien circonscrit, et malgré tous les retranchemens qu'il a subis, sont encore si nombreuses qu'il a fallu, pour les ranger dans un ordre au moins systématique, si non naturel, avoir égard à des considérations extrêmement minutieuses sur l'écaillure de la tête, considérations qui, depuis Merrem, avaient été malheureusement négligées par les zoologistes qui se sont plus spécialement occupés de ces animaux, ainsi que sur la position des narines et sur la forme et la proportion des dents maxillaires, comme l'a fait Wagles. En suivant ces indications, nous avons disposé les espèces de couleuvres dans l'ordre suivant qui nous paraît commencer par les espèces les plus rapprochées des boas-couleuvres, et finir par celles qui ressemblent davantage aux vipères.

A. *Espèces qui ont 4 scutelles frontales et 1 loréale.*

1) Scutelles oculaires 0-1, formant les genres *ophites* et *brachyurhos.*

2) Scutelles oculaires 1-2, constituant les genres *homalosoma*, *oligodon*, *rhinosoma*, *zacholus*, *erythrolamprus*, *dipsas*, *pareas*, *ophis*, *oxyropus*, *lycodon*, *rhinobothrium*, *liophis*, *macrops*, *telescopus*, *dendrophis*, *gonyosoma*, *chlorosoma*, *philodryas*, *herpetodrys*, *psamophis*, et *coluber*, caractérisés par quelques particularités du corps cylindrique ou filiforme, de la tête peu ou très-distincte, du museau plus ou moins prolongé, de la forme et de la proportion des écailles, lisses ou carénées, des dents égales et fort petites, inégales et plus longues en avant ou en arrière, cultriformes ou sulcifères en avant ou en arrière, de la grandeur ou de la petitesse des yeux, etc.

3) Scutelles oculaires 1-3 : formant les genres *tropidonotus*, *leionotus* et *xenodon.*

4) Scutelles oculaires 2-2 : réparties d'après des considérations analogues et fort peu importantes dans les genres *coronella*, *zamenis*, *coelopeltis*, *chrysopelea*, *rhinechis*, *thamnodynastes* et *dryophilax.*

5) Scutelles oculaires 2-3 : ne comprennent que deux espèces

que nous regardons comme nouvelles, et qui nous ont été rapportées, l'une de la Californie, l'autre du mont Liban par M. Botta.

6) Scutelles oculaires 3-4 : le C. *variegatus* seulement.

B. *Espèces qui ont 4 scutelles frontales comme les précédentes, et point de loréale.*

1) Avec scutelles oculaires 1-1.

C. *planiceps* (voyez plus loin sa description).

2) Pourvues de 1-2 scutelles oculaires : réparties suivant qu'elles ont les écailles lisses ou carénées, et d'autres caractères en général peu importans dans les genres *pseudechis*, *clœlia*, *oxybelis*, *dryophis*, *spilotes*, *leptophis*, *dasypeltis*.

3) Pourvues de 1-3 scutelles oculaires.

C. pondecerianus.

4) Pourvues de 3-2 scutelles oculaires (G. *tragops* ou *drynus*).

5) Pourvues de 3-2 scutelles oculaires (G. *langaha* ou *rinhurus*).

6) Pourvues de 3-3 scutelles oculaires (G. *periops*).

C. *Espèces qui n'ont que 3 scutelles frontales, une en avant, et deux en arrière,*

+ *avec un loreum*,

constituant les genres *homalopsis*, *hypsirhina*,

+ + *sans loreum*,

formant les genres *helicops* et *hydrops*.

D. *Espèces qui ont cinq scutelles frontales*,

réunies dans les genres *heterodon*.

Section II. *Dents maxillaires vénéneuses et non vénéneuses.*

Fam. VI. LES SERPENS AQUATIQUES (*HYDROPHIS*).

Corps en général assez alongé, mais du reste assez diversifórme, couvert d'écailles petites et semblables, dans les premières espèces, devenant larges et très-dissemblables dans les dernières.

Tète petite; queue courte, souvent comprimée.

Dents maxillaires vénéneuses et non vénéneuses.

Observ. Cette famille, quoiqu'en partie fort naturelle, puisqu'elle renferme tous les serpens aquatiques, ne l'est cependant pas entièrement en envisageant soit l'écaillure ou le système squameux, soit le système dentaire venimeux ou non venimeux. Aussi, concevrait-on très-bien que dans l'un ou l'autre des systèmes d'ophiologie, d'après le système squameux, ou d'après les dents, cette division fût considérablement disloquée, puisqu'en effet, les premières espèces ont une écaillure partout uniforme, tandis que chez les dernières, elle est en tout semblable à celle des couleuvres. C'est même sur cette considération que va porter notre distribution des espèces.

A. *Espèces dont le système squameux est partout semblable, aussi bien sur la tête que sur et sous le corps et la queue.*

Les Acrochordes (G. *Acrochordus*) subdivisées en acrochordes proprement dits, et en *chershydrus*, suivant que la queue est conique, ou qu'elle est subcomprimée et plus longue.

B. *Espèces dont le système squameux du corps et de la queue est graduellement plus dissemblable, et dont la queue est toujours plus ou moins comprimée.*

Les Serpens d'eau (G. *Hydrophis*) partagées en *hydrus*, *pelamys*, *hydrophis*, *disteira*, *leioselasmus*, *enhydris* et *aipysurus*, suivant que les écailles du milieu du ventre sont semblables à celles du reste du corps, ou bien à peine plus grandes, ou le deviennent de plus en plus, mais en étant toujours sur un seul rang sous la queue, en y joignant quelques autres caractères : tirés du reste de l'écaillure de la tête principalement.

C. *Espèces pourvues de grandes plaques sous le ventre, et de scutelles en double rang sous toute ou partie de la queue.*

Les Fausses vipères (*Pseudechis*) réparties en *platurus*, *pseudechis*, *alecto*, *bungarus* et *oplocephalus*, suivant que la queue est ou n'est pas comprimée, et que dans ce dernier cas les scutelles de la queue sont d'abord indivises à la base, avant d'être divisées, enfin qu'elles sont toutes indivises avec une rangée médiodorsale d'écailles plus grandes ou pas plus grandes que les autres.

Section iii. *Dents maxillaires toutes vénéneuses.*

Fam. VII. LES VIPÈRES (*VIPERINA*).

Corps en général assez peu alongé, surtout dans sa partie caudale, toujours conique, couvert d'écailles en dessus, et de plaques simples ou doubles sous la queue.

Dents maxillaires, peu nombreuses, tres-grandes, et toutes vénéneuses, portées par des mâchoires très-dilatables.

Observ. On peut faire pour cette famille la même observation que pour la précédente, c'est-à-dire, qu'à ne considérer que l'appareil dentaire, elle est très naturelle; mais qu'en admettant en première considération le système squameux, il n'en est pas de même, puisqu'il y a des espèces qui, n'ayant qu'une série de scutelles sous la queue, étaient des boas pour Linné, tandis que les autres qui en ont deux étaient pour lui des couleuvres.

Cette différence jointe à celle de la présence ou de l'absence d'une grande fossette au devant de l'œil, fournit d'excellens caractères pour la subdivision intérieure de cette famille :

+ *sans fossette au devant des yeux.*

A. *Espèces dont la queue n'est pourvue en dessous que d'une seule rangée de scutelles.*

Les Boas-Vipères subdivisées en *acanthophis* et en *echis*, suivant que la tête est semi-scutellée, ou entièrement écailleuse.

B. *Espèces dont la queue est bi-scutellée dans toute sa longueur.*

Les VIPÈRES (*Vipera*), partagées d'après la considération de l'écaillure de la tête,

1) quadriscutellée sur le front, en *elaps*, *naja*, *uraeus*, *causus*, *sepedon*, suivant que la tête est extrêmement petite, que le cou est dilatable ou non, et enfin que les écailles du corps sont carénées.

2) scutellée seulement sur le museau, comme dans les genres *pelias* et *cerastes*, l'un sans corne, et l'autre en étant pourvu.

3) entièrement écailleuse dans toutes ses parties, ce qui constitue les sous-genres *vipera* et *ecihdna*, ne différant que parce que les écailles de la tête ne sont pas ou sont carénées, comme celles du reste du corps :

++ *avec une fossette au devant des yeux.*

A. *Espèces dont la queue est uni-scutellée en dessous* (les BOAS-TRIGONOCEPHALES),

constituant le genre *cenchris*.

B. *Espèces dont la queue est bi-scutellée sans crépidules à l'extrémité.*

Les TRIGONOCEPHALES (*Trigonocephalus*) réparties dans les sous-genres *trigonocephalus*, *megaera*, *botrhops*, *atropos*, *lachesis* et *tropidolœna*, suivant que la tête est toute scutellée, à moitié scutellée, ou seulement sur ses angles, que les écailles des sourcils sont relevées en crêtes, qu'il y a des écailles sur quatre rangs à la fin de la queue, ou enfin que les écailles du corps sont fortement carénées.

C. *Espèces dont la queue est bi-scutellée en dessous, et pourvue à son extrémité d'écailles disposées en grelots.*

Les Crotales (*Crotalus* L.), renfermant un assez petit nombre d'espèces partagées dans les trois genres *caudisona*, *crotalus*, et *uropsophus*, suivant que la tête est demi-scutellée, ou entièrement écailleuse, avec une seule plaque surcilière non carénée ou carénée.

Classe V. LES ICHTHYOSAURIENS (*ICHTHYOSAURIA*).

Corps pisciforme, c'est-à-dire renflé au milieu, atténué aux deux extrémités, et prolongé en avant par une tête non distincte, pourvu d'assez longues mâchoires, en arrière par une queue puissante, fort longue, conique et aiguë.

Narines ouvertes à la racine du museau immédiatement au devant des os lacrymaux.

Yeux très-grands, soutenus dans leur sclérotique par un cercle d'écailles osseuses.

Oreilles?

Membres au nombre de 4 (en deux paires) très-distans, complets, raccourcis, aplatis en nageoires, formés de cinq doigts, non séparés et non onguiculés.

Peau très-probablement nue.

Dents coniques, fortes, droites, nombreuses, presque égales aux deux mâchoires seulement, insérées dans un sillon alvéolaire commun.

Observ. Les animaux qui composent cette classe ne nous sont malheureusement connus que par leur squelette et même assez incomplétement, parce qu'ils sont fossiles des terrains anciens.

La forme particulière des vertibres, dont le corps est également excavé en avant comme en arrière; la structure de l'épaule, qui a plus de rapports avec celle de cette partie chez les reptiles, celle même des membres, ont fait assez généralement admettre que c'était dans cette classe que les ichthyosaures devaient être placés; cependant, en considérant que la double excavation du corps des vertèbres ne se trouve que dans les amphibiens et dans les poissons, il est évident que si l'on s'en rapportait à ce seul caractère, ces animaux seraient plutôt des amphibiens qu'autre chose; toutefois l'existence des côtes vertébrales nombreuses et la disposition des pièces de la ceinture antérieure, qui rappellent assez bien ce qui a lieu chez les iguanes et même chez les ornithorhynques, a dû montrer que les ichthyosaures offraient un type classique, remplissant la lacune qui se remarque dans la série animale entre les reptiles et les amphibiens, comme l'indique assez bien le nom de *proteosaurus*, qu'Everad Home proposa de donner à ces animaux fossiles, au lieu de celui d'ichthyosaures qui a prévalu comme plus ancien.

D'après cette manière de voir, il est évident que les ichthyosaures ne doivent pas être réunis avec les plésiosaures, qui sont pour nous des animaux véritablement reptiles, intermédiaires aux tortues et aux crocodiles par tous les caractères tirés de la composition du squelette, et dont nous nous sommes bornés à former un ordre distinct.

Classe VI. LES AMPHIBIENS (*AMPHIBIA*).

Corps très-diversiforme, quelquefois très-court et déprimé, d'autrefois lacertiforme, et même serpentiforme, à queue entièrement nulle ou assez longue, à tête peu ou point distincte, pourvu de deux paires de membres ou d'une seule paire, ou entièrement nullipède, couvert d'une peau constamment nue, ou plus ou moins muqueuse et épithélifère.

Observ. L'organisation des animaux qui constituent cette classe ne peut sous aucun point être comparée à celle des reptiles, comme il est aisé d'en juger par la nature des os, par le mode d'articulation des vertèbres, et surtout par celle de la tête pourvue d'une double condyle; en un mot, par toutes les parties du squelette et par suite du système myologique.

Les appareils des sens offrent également des différences classiques, et entre autres la structure de la peau qui, quoique susceptible d'offrir constamment dans l'épaisseur du derme des espèces de granulations plus ou moins nombreuses, plus ou moins aplaties, ne présente jamais d'écailles comme dans les reptiles, résultat du pincement de toutes les parties de la peau et de l'extension d'un véritable épiderme.

Les appareils de la respiration et de la circulation ont aussi une structure particulière : le cœur étant réduit à n'être plus composé que d'un seul ventricule et d'une seule oreillette, ce qui entraîne un seul gros vaisseau sortant et rentrant.

L'appareil de la génération n'est pas moins différent, n'étant jamais en connexion directe avec des appendices extérieurs.

Le produit même de la génération est si particulier dans les différens états qu'il parcourt avant de parvenir à l'état adulte, que l'on a été obligé de dire qu'il subit des métamorphoses à peu près comme les insectes hexapodes : et ce qu'il y a de plus remarquable, ce qui se trouve aussi dans quelques-uns de ceux-ci, c'est que l'animal, à l'un de ses degrés de développement, est pourvu de branchies, et par consequent obligé de vivre dans l'eau, tandis qu'à son état adulte, il les perd et devient complétement pulmoné, et par suite obligé de vivre et de respirer l'air en nature. C'est même de cette particularité de vivre successivement dans l'eau et dans l'air, que le nom d'*amphibia* a été tiré et donné à cette classe avec juste raison; mais aucun des reptiles n'étant dans ce cas, cette dénomination ne pouvait leur être étendue.

Les mœurs et les habitudes des animaux de cette classe sont aussi fort différentes de celles des reptiles, et surtout quant au séjour qui est constamment plus ou moins aquatique pour les amphibiens, et accidentellement pour le plus petit nombre des reptiles.

La position des amphibiens dans la série animale ne pouvait réellement varier depuis que Linné a abandonné la considération du nombre des membres pour premier élément de la classification des animaux vertebres. Aussi, tous les zoologistes sont-ils d'accord pour les placer immediatement avant les poissons.

Quant à la disposition des grands genres linnéens, qui sont devenus des ordres plutôt que des familles, elle offre un peu plus de variations. La moins pour l'un d'eux, celui qui doit comprendre les cœcilies. En effet, les erpétologistes qui se sont décidés à ranger, à l'imitation d'Oppel, ces animaux parmi les amphibiens ou reptiles nus, les ont placés, les uns à la tête de la classe établissant le passage aux amphisbènes par lesquelles ils terminent les reptiles, les autres à la fin, adoptant ainsi l'opinion de Schneider, que ce sont les animaux les plus rapprochés des poissons, et ensuite pour répéter jusqu'à un certain point la disposition établie dans la classe des reptiles, d'après la dégradation des membres dans la locomotion et la prédominance du tronc.

La distribution et par suite la distinction des espèces sont au contraire assez difficiles dans cette classe d'animaux, ce qui est une conséquence nécessaire de l'absence de toute espèce d'écaillure. Aussi, les subdivisions génériques que les zoologistes ont établies dans les grands genres linnéens, sont-elles extrêmement vacillantes et d'une grande difficulté à préciser d'une manière convenable; surtout parce qu'en général elles reposent sur des nuances différentielles dans le séjour ou le mode de locomotion.

Quant à nous, notre ordre général est tracé d'après la forme du corps, la dégradation des membres, toujours en rapport inverse, et pour la disposition des espèces, nous tâchons constamment de suivre le même principe, d'après lequel, dans chaque groupe, un ordre déterminé de dégradation est, autant que cela se peut, indiqué et apprécié.

Ord. I. LES BATRACIENS (*BATRACHIA*) (G. *Rana* L.).

Corps très-court, tronqué en arrière, sans apparence de queue, et pourvu en avant d'une tête aussi large que lui, sans trace de cou.

Bouche très-grande, très-fendue, parabolique.

Anus arrondi, terminal et plus ou moins dorsal.

Membres complets, constamment au nombre de deux paires, les postérieurs toujours bien plus longs que les antérieurs, et pourvus de doigts plus ou moins inégaux, au nombre de quatre en avant et de cinq en arrière.

Narines petites, rondes, suboperculées.

Yeux latero-supérieurs, pourvus de paupières.

Oreilles visibles ou non par un tympan large et superficiel.

Peau nue, lisse ou plus ou moins verruqueuse.

Système de coloration généralement par plaques ou zébrures.

Dents nulles ou fort petites, aiguës, irrégulièrement espacées à la mâchoire supérieure et au palais.

Langue le plus souvent fort large et retournée, la pointe bifurquée en arrière.

Observ. Cet ordre est tellement distinct, tellement tranché même dans presque tous les points de l'organisation, et surtout par l'absence totale de queue à l'état parfait, qu'il n'y a jamais eu de doute sur sa circonscription ni sur sa caractéristique, ce qui est exactement en rapport avec les mœurs et les habitudes des animaux qui le constituent à toutes les époques de leur vie. L'une des plus grandes particularités de ces animaux, c'est de sortir de l'œuf sous une forme qui n'a réellement presque aucun rapport avec celle qu'ils auront quand ils seront parvenus à l'âge adulte, puisque alors ils sont pourvus d'une assez longue queue qui disparaîtra, et qu'il n'ont aucune trace de pattes qui, au contraire, se montreront alors. Aussi, est-ce sur le durée de cet état de larve avec une queue que nous avons essayé d'établir la dégradation, les salamandres qui constituent l'ordre suivant, ayant pour caractère d'avoir cet organe à tous les âges.

Malheureusement, la connaissance de la durée de cet état dans lequel le batracien est pourvu de queue et de pattes à la fois, est fort peu connue pour la plupart des espèces, dès-lors on a été obligé de faire porter leur disposition et leur distribution sur des caractères de séjour plus ou moins aquatique, indiqué par la proportion des membres postérieurs, et sur la disposition des doigts plus ou moins palmés, ou dilatés et pourvus de pelottes visqueuses à l'extrémité. En y joignant l'état de la peau plus ou moins tuberculeuse et crypteuse, quelquefois au moyen d'amas de cryptes que leur position vers les oreilles a fait nommer *parotides*, l'existence ou l'absence de dents mxillaires et du tympan, on peut arriver à une distribution des espèces peut-être assez naturelle, mais aussi à des subdivisions génériques fort peu tranchées et par conséquent très-difficiles à définir, et réellement, sans aucune importance. Nous savons cependant que c'est le pipa dont les petits restent le moins long-temps à l'état de têtard, et au contraire les jackies qui y restent le plus long-temps, et c'est ce qui nous a conduits à adopter l'ordre suivant.

Fam. I. LES B. DORSIPARES.

Corps très-large, très-déprimé, disposé de manière à donner asile aux produits de la génération, se développant dans des cellules extérieures du dos de la femelle.

Tête large, déprimée, triangulaire et fort courte.

Bouche très-fendue et triangulaire.

Anus terminal inférieur.

Narines fort petites, linéaires, et tout-à-fait terminales.

Yeux fort petits, sans paupières mobiles, et à pupille circulaire.

Tympan caché.

Membres empêtrés, les antérieurs médiocres, à quatre doigts grêles, subégaux, libres, et quadrifides à l'extrémité, les postérieurs très-grands à cinq doigts fort longs, subégaux, et très-largement palmés.

Peau tuberculeuse et granulée.

Dents entièrement nulles.

Langue complétement adhérente si ce n'est à ses bords.

Observ. Cette famille, si singulière par tous les caractères qui viennent d'être énoncés, ne contient que le genre *pipa*, qui lui-même ne renferme qu'une espèce, remarquable, il est vrai, par la manière dont les œufs déposés sur le dos de la femelle y occasionnent une sorte de phlogose par suite de laquelle se produit pour chacun d'eux une cellule dans laquelle le jeune crapaud acquiert toute la forme de l'adulte, en passant sans doute par les mêmes phases que les batraciens ordinaires.

Fam. II. LES B. AQUIPARES.

Corps assez peu déprimé, arrondi, ovale ou même quelquefois assez élancé, pourvu de membres à doigts inégaux, d'yeux avec des paupières mobiles, de dents palatines au moins, et d'une langue très-développée, retournée, adhérente seulement en avant, libre et bifide en arrière.

Observ. Les animaux de cette famille, séparée de la précédente, parce que les œufs ne peuvent se développer que dans l'eau, diffèrent réellement trop peu entre eux, pour qu'il soit possible de les répartir en familles : comme l'usage plus que la science semble le demander; tout au plus même si l'on peut en former trois genres suffisamment caractérisés et définis, tant les espèces se nuancent depuis la première la plus crapaud, jusqu'à la dernière la plus grenouille.

Toutefois, en prenant en considération la faculté de courir, de sauter et de nager, nécessairement en rapport avec la disposition et la proportion des membres et des doigts, on arrive à les subdiviser assez bien en trois genres principaux.

Crapaud (*Bufo*).

Corps plus ou moins ramassé et bombé, couvert d'une peau en général tuberculeuse, les membres postérieurs assez peu alongés; dents nulles aux maxillaires, ou du moins fort petites; des parotides plus ou moins prononcées.

Observ. Il est véritablement fort difficile de définir ce genre d'une manière un peu rigoureuse, à moins de n'avoir égard qu'à l'absence des dents à la mâchoire supérieure; mais alors plusieurs espèces, qui, du reste, ont tous les caractères des crapauds, devront en être retranchées.

Nous disposerons les espèces de crapauds d'après la considération des dents maxillaires, la présence ou l'absence du tympan, et d'après le degré de palmature des doigts.

A. *Espèces qui n'ont pas de dents maxillaires, ni tympan visible, et dont les doigts antérieurs sont grêles et subégaux, les postérieurs largement palmés.*

Les Pipa-Crapauds (G. *Xenopus* Wagl.) (*Dactyletra* Cuv.).

B. *Espèces sans dents maxillaires, ni tympan visible.*

Les Moyens crapauds, répartis suivant que le dos est couvert d'une partie de la peau comme ossifiée, que les yeux, ou la tête sont très-petits, ou enfin, que les doigts postérieurs assez longs sont palmés ou tout-à-fait libres dans les sous-genres *kaloula*, *brachycephalus* ou *ephippipher*, *microps*, *breviceps*, *pelobates* et *paludicola*.

B. *Espèces sans dents maxillaires, mais avec le tympan visible.*

Les Crapauds, partagés en *bufo* proprement dit, *otilophus*, *chaunus* et *physalaemus*, suivant que la tête n'est pas ou est angulaire vers les tempes, ou qu'elle est extrêmement petite, ou enfin que les doigts sont très-grêles.

C. *Espèces à dents maxillaires fort petites, et le tympan non visible.*

Les Petits crapauds, subdivisés suivant que le museau est un peu proéminent, ou est arrondi et obtus, en *rhinella* et en *bombinator*.

D. *Espèces à dents maxillaires et à tympan visible.*

Les C. Accoucheurs (G. *Atytes*).

Rainette (*Hyla*).

Corps en général assez svelte, à peau lisse, pourvu de membres grêles, très-dégagés à leur racine, avec des doigts élargis en disque visqueux à l'extrémité; des dents palatines et maxillaires, comme dans les grenouilles.

Observ Ce genre, qui semble au premier aspect se distinguer aisément des crapauds et même des grenouilles, contient cependant des espèces dont les doigts sont à peine dilatés. On peut donc les disposer d'après l'intensité de ce caractère et de la palmature des doigts.

Le tympan n'est pas visible et les doigts sont courts, très-dilatés, et entièrement libres dans les *phyllomedusa*. Le tympan est visible, les doigts courts et palmés à tous les pieds dans les *hypsiboas;* les antérieurs sont palmés entièrement, et les postérieurs demi-palmés dans les *phyllodytes;* les antérieurs sont libres et les postérieurs demi-palmés dans les *calamites*, *auletus*, *scinax* et *hyas*, distingués par des caractères tout-à-fait insignifians; enfin les dernières espèces sont de véritables grenouilles-rainettes par la longueur, la disproportion des doigts, aussi bien que par la petitesse de leur dilatation terminale.

Grenouille (*Rana*).

Corps en général moins élancé que dans les *rainettes*, et moins ramassé que dans les crapauds, couvert d'une peau assez lisse, sans parotides en arrière des yeux, les membres postérieurs fort longs, avec les doigts inégaux; des dents maxillaires et palatines.

Observ. Les espèces de batraciens qui constituent ce genre sont également assez difficiles à distribuer d'une manière à la fois naturelle et facile.

A. *Espèces à tympan évident,*

réparties dans les sous-genres *cystignatus*, *rana*, *houlema*, *occidozyga*, suivant que les doigts postérieurs ne sont pas palmés ou le sont fortement, et d'après quelques autres caractères plus insignifians.

B. *Espèces à tympan caché*, *avec des dents aux deux mâchoires*, *et la tête comme cuirassée*,

formant le G. *Hemiphractus.*

C. *Espèces à tympan caché avec des dents à la mâchoire supérieure seulement*,

partagées en *ceratophrys*, *megalophrys* et *pseudis*, suivant qu'il y a des espèces de cornes sur les yeux, ou que le museau est un peu prolongé, ou qu'enfin les doigts sont largement palmés.

Ord. II. LES PSEUDO-SAURIENS (*PSEUDO-SAURIA*).

Corps lacertiforme, plus ou moins alongé, prolongé en arrière en une queue plus ou moins grande, conique ou comprimée.

Tête petite, assez peu distincte, déprimée, à museau court, et plus ou moins obtus.

Bouche assez large, terminale.

Anus longitudinal.

Narines petites, arrondies ou linéaires.

Yeux médiocres ou obsolètes.

Tympan constamment caché.

Membres au nombre de deux ou d'une paire seulement, et dans le premier cas faibles, étalés, terminés par des doigts courts, un peu inégaux, libres ou palmés.

Peau muqueuse, poreuse, quelquefois un peu tuberculeuse.

Dents toujours fort petites, faibles, à peine implantées.

Langue large, molle, courte et entière.

Observ. L'organisation des pseudo-sauriens est assez différente dans un grand nombre de points de celle des batraciens pour mériter leur établissement en un ordre distinct, et toutes ces différences, sauf l'existence de côtes rudimentaires, indiquent évidemment une marche de dégradation, comme on peut s'en assurer en étudiant le mode d'articulation des vertèbres, et surtout la constitution des membres.

Quant à leur distribution systématique, il n'y a, en apparence du moins, aucune difficulté en prenant en première considération, comme l'ont fait tous les zoologistes et moi-même pendant long-temps, l'absence ou l'existence de trou ou de branchies sur les côtes du cou; mais, comme j'admets en principe que ces dispositions doivent être considérées comme transitoires et comme pouvant disparaître lorsque l'animal sera arrivé à l'état parfait, j'ai dû préférer le nombre des membres et la forme générale du corps comme principe de la disposition des espèces.

Fam. I. LES SALAMANDRES (G. *SALAMANDRA* L.).

Corps, y compris la queue, médiocrement alongé, constamment pourvu de deux paires de pattes, assez rapprochées.

Observ. Cette famille doit comprendre, à ce qu'il me semble, non-seulement les salamandres proprement dites, qui perdent entièrement la disposition branchiale, lorsqu'elles sont parvenues à l'état adulte, mais encore un certain nombre d'espèces qui conservent plus ou moins long-temps, soit cette disposition tout entière, soit une partie seulement de cette disposition.

Les Axolotls ou *siredon* sont les salamandres batrachoïdes, dont le corps est court, la queue médiocre, très-comprimée, et qui conservent pendant un long temps de la vie tout l'appareil branchial de l'état de larve.

Les Salamandres (*Salamandra*) renferment les espèces dont les corps est lacertiforme, plus ou moins déprimé, les membres médiocres, pourvus de doigts assez longs, sans trace de dispositions branchiales; elles se subdivisent ensuite, suivant que la queue est conique, en *salamandra* et *salamandrina*, et suivant qu'elle est comprimée en *triton* et *pleurodeles*.

Les *salamandrops* sont les salamandres à queue comprimée dont les membres et les doigts sont fort courts, au nombre de quatre, aussi bien en arrière qu'en avant, et qui conservent, soit seulement le trou branchal, comme dans les *menopoma*, ou bien ce trou et trois branchies de chaque côté, comme dans les *menobranchus*.

Fam. II. LES PROTÉES.

Corps cylindrique, fort alongé, terminé par une queue comprimée, pourvu de deux paires de pattes extrêmement petites, à peine digitées, et conservant plus ou moins long-temps tout ou partie de la disposition branchiale.

Observ. Quoiqu'au fond il y ait assez peu de différences entre les salamandes du dernier genre de la famille précédente et les protées, cependant la disposition du tronc et des membres indique chez les salamandropsiens une dégradation trop évidente pour n'être pas signalée.

Le petit nombre d'espèces qui constituent cette famille sont encore partagées en deux genres, suivant que le tronc est plus ou moins alongé, les pieds plus rudimentaires, et surtout que les unes conservent la disposition branchiale complète, ou seulement les trous latéraux, ce qui constitue les genres *proteus* et *amphieuma*.

Fam. III. LES SIRÈNES (*SIRENA*).

Corps assez alongé, cylindrique, terminé par une longue queue comprimée, pourvue en avant d'une seule paire de pattes, tétradactyles, et conservant plus ou moins long-temps la disposition branchiale complète.

Observ. Cette famille, ne comprenant qu'un genre qui lui-même n'est formé que d'une seule espèce, n'en présente pas moins un degré d'organisation tranché parmi les pseudo-sauriens; elle représente ici les bimanes de l'ordre des saurophiens.

Ord. III. LES PSEUDO-PHIDIENS (*PSEUDO-PHIDIA*).

Corps fort alongé, cylindrique, sans distinction de tête ni de queue, ni de membres.

Bouche fort petite, inférieure, ou dépassée par le museau.

Anus rond ou sublongitudinal, et presque terminal.

Narines ponctiformes et très-distinctes.

Yeux entièrement cachés.

Peau entièrement nue, muqueuse, à peine épithéliée, mais granulifère dans son derme comme dans les salamandres.

Dents assez fortes, peu nombreuses, inclinées en arrière, aux deux mâchoires et au palais.

Langue plate, large, immobile, à peine distincte dans son contour.

Observ. Les animaux qui constituent cet ordre ont une organisation si particulière, que l'on considère la structure de leur tête, son mode d'articulation par deux condyles, celle des vertèbres à double cavité; comme dans les poissons, l'absence de côtes, l'immobilité de l'os carré, la structure du cœur et des gros vaisseaux, la présence d'une vessie bifide, la forme et la position de l'anus, la nudité de la peau, malgré le grand nombre de corpuscules solides qui sont immergés dans le derme, l'absence de tout organe extérieur de génération, que l'on a pu depuis assez long-temps les considérer comme n'étant pas de véritables serpens, et comme devant plutôt être rangés parmi les amphibiens; seulement on a pu varier sur la position de ce genre au commencement ou à la fin de la classe. En les plaçant au commencement et en terminant les saurophiens par les amphisbènes, on établissait une sorte de passage que repousse cependant, à ce qu'il me semble, toute l'organisation, malgré une certaine ressemblance de forme générale. En les plaçant, au contraire, à la fin des amphibiens, on répète pour ainsi dire les premières subdivisions des reptiles dont le corps d'abord large et suborbiculaire devient de plus en plus alongé, à mesure que les membres tendent à disparaître, et finissent par être tout-à-fait serpent. Au reste, la découverte faite par M. Muëller de trous de chaque côté du col des cœcilies, comme dans les amphieumes, ce qui fait supposer que, dans le premier âge, ces animaux sont pourvus de branchies, ne peut guère laisser de doute sur la place à assigner aux cœcilies.

Quoique cet ordre ne contienne qu'un assez petit nombre d'espèces, vivant dans les lieux marécageux, comme tous les animaux de cette classe, les erpétologistes récens ont encore trouvé à y former trois genres : *siphonops*, *cœcilia* et *epicrium* ou *ichthyophis*, d'après la considération de la présence ou de l'absence d'une fossette au devant des yeux, ou d'espèce de barbillons à la mâchoire supérieure.

Venons maintenant à la description du petit nombre d'espèces de reptiles rapportées de la Californie par M. P. E. Botta.

Agama (Phrynosoma) coronata.

Pl. XXV, fig. 1, 1*a*, 1*b*, 1*c*.

Le corps, comme dans toutes les espèces de cette section, est court, déprimé et suborbiculaire; il est terminé par une queue brusquement conique, assez aiguë, et plus longue que dans les autres espèces, puisqu'elle est deux fois et demie dans la longueur totale, tandis que dans l'agame orbiculaire de Wiegman, elle y est au moins trois fois.

La tête, également très-ramassée, très-distincte par un étranglement du cou bien marqué, a ses carènes suroculaires et bilatérales très-prononcées; le museau, arrondi et déprimé, est extrêmement court, et comme pincé pour constituer les lèvres.

Les narines, situées en dedans de l'angle sourcilier prolongé, sont grandes et entourées d'un cercle d'écailles ou de scutelles polygonales au nombre de quatre.

Les yeux sont également assez grands, de couleur noire, et pourvus de deux paupières également mobiles.

Le tympan est large, oblique, assez enfoncé, et situé dans une fente verticale en arrière de la tête.

Les membres sont robustes, assez courts, et terminés par des doigts de médiocre longueur, assez inégaux, surtout aux pieds de derrière.

La tête est couverte de scutelles polygonales dans une disposition régulière, et fort compliquée.

Elle est hérissée d'épines comprimées, une sur chaque sourcil, cinq le long de chaque branche de la mâchoire inférieure, quatre le long de l'arcade zygomatique, et qui, en se joignant à deux grandes occipitales, constituent une sorte de couronne, entourant obliquement les deux tiers postérieurs de la tête.

Tout le reste du corps, le col excepté, ainsi que la queue et les parties externes des membres, est couvert de petites squames irrégulières, aplaties, entremêlées de tubercules épineux, polyédriques, irrégulièrement épars, si ce n'est de chaque côté du ventre, où ils forment une double série fort serrée, ainsi qu'à la queue; en dessous, les écailles sont plus larges, plus imbriquées, et entièrement lisses.

A la face interne de la cuisse, elles sont encore un peu plus grandes, surtout celles qui sont percées de pores fémoraux, et qui forment une rangée.

Celles qui recouvrent les doigts sont fortement carénées, et assez imbriquées sur les côtés, pour donner à ceux-là l'apparence d'être denticulés ou serrés.

Les ongles qui terminent ces doigts sont petits, arqués et fort aigus.

La couleur générale est d'un gris cendré en dessus, avec une grande tache brune, bordée de noir de chaque côté du cou, se prolongeant sur le dos, avec deux paires de bandes arquées qui en partent sans se réunir dans la ligne médio-dorsale blanche.

Tout le dessous est d'un blanc jaunâtre, ainsi que les rangées d'épines latérales, avec quelques taches grises entremêlées.

Les deux épines occipitales sont d'un brun plus ou moins foncé, ou noirâtre, les autres sont roses.

Observ. Des quatre espèces que les erpétologistes les plus récens ont établies dans cette division des agames, celle qui se rapproche le plus de l'animal que nous venons de décrire est bien certainement l'A. *orbicularis* de M. Weigman; mais notre A. *coronata* en diffère : 1° par le nombre des pores fémoraux, et l'étendue de la série qu'ils forment commençant presqu'à la ligne medio-abdominale et se prolongeant dans toute la longueur de la cuisse; 2° par un beaucoup plus grand développement des cornes de la tête; 3° par une longueur proportionnelle de la queue plus grande; 4° par le plus grand développement des tubercules épineux de tout le corps; 5° par une coloration toute différente, puisque le ventre n'est nullement tacheté, et que les bandes du dos sont tout autrement espacées; 6° enfin parce que l'orifice des narines est en dedans de l'angle sourcilier prolongé, et qu'il y a cinq épines à la mandibule, au lieu de quatre, caractères sur lesquels M. Wiegman a le plus insisté comme constant pour l'espèce qu'il a nommée A. *orbicularis*.

Callisaurus draconoides.

(Pl. 24, fig. 2, 2*a*.)

Corps généralement assez court.

Tête distincte assez large, un peu déprimée, anguleuse, à museau court, comme pincé et arrondi.

Cou distinct, avec un double pli cutané sous la gorge, indice d'un goître plus ou moins extensible.

Tronc court, renflé, convexe en dessus, aplati en dessous, élargi sur les côtés par un pincement cutané étroit, étendu de la racine d'un membre à l'autre.

Queue assez courte, conique, ou mieux déprimée, aplatie en dessous, et assez convexe en dessus.

Narines rondes, fort grandes, supérieures et subtubuleuses.

Yeux fort grands, à deux paupières épaisses, également mobiles, l'inférieure denticulée à son bord.

Tympan visible, ovale, assez superficiel.

Bouche large et parabolique.

Anus médiocre, subtransverse.

Membres très-grands, surtout les postérieurs, à doigts extrêmement alongés, inégaux, principalement en arrière.

Système squameux, presque uniforme ou à peine dissemblable; les écailles supérieures, beaucoup plus petites cependant que les inférieures, toutes lisses, même sur les membres, celles de la tête irrégulières et fort petites, les scutelles labiales assez larges et très-obliques.

Des pores fémoraux, serrés, nombreux, occupant toute la longueur de la cuisse.

Ongles excessivement alongés, aigus et presque droits.

Dents maxillaires simples, subcylindriques, petites, appliquées.

Dents palatines nulles.

Langue épaisse, large et entière, ou à peine échancrée, finement papilleuse.

Couleur générale d'un gris cendré, parsemé de petites taches blanches, avec des bandes transversales brunes sur le dos, la queue et les membres, d'un blanc jaunâtre en dessous, avec trois taches d'un bleu très-foncé de chaque côté du ventre, et cinq dans la longueur de la queue.

Observ. Cette jolie espèce de reptiles se rapproche assez difficilement des autres groupes nombreux que les erpétologistes modernes ont établis dans le sous ordre des sauriens. En effet, n'ayant pas de dents palatines, ce serait un agame pour M. Cuvier; mais les écailles très-petites et toutes lisses, ce qui est très-rare dans cette famille, et l'absence de dents canines empêchent de le considérer réellement comme appartenant à ce genre. Ce serait donc un ignane; mais les dents sont entières et non pectinées, et il n'y a pas de dents palatines, caractères qui distinguent essentiellement cette famille. Nous sommes donc portés à penser que ce saurien doit être rapproché de la petite famille qui contient les dragons et les sitanes, ce qui se trouve confirmé par l'existence d'un rudiment de membranes sur les flancs; toutefois, comme il n'a pas de dents canines, ni de goître, nous avons été forcés d'en former un petit genre distinct auquel nous avons donné le nom de *callisaurus* pour indiquer la gentillesse de ce petit animal, remarquable en outre par la forme de ses ongles et par plusieurs autres caractères indiqués plus haut.

Ignana (Cyclura) acanthura

(Pl. XXIV, fig. 1.)

Corps en totalité fort alongé et conique; tête quadrangulaire; cou avec deux ou trois plis cutanés transversaux; tronc assez long, cylindrique; queue fort longue et conique.

Yeux noirs, encadrés d'un cercle doré.

Les membres sont robustes, les doigts longs, inégaux, surtout en arrière.

Les écailles qui recouvrent la tête sont inégales, polygonales et fort petites; celles du dos sont également très-petites; celles de la ligne medio-dorsale, un peu plus élevées, forment une crête paléacée extrêmement basse; mais à la queue, toutes les écailles sont epineuses, et forment des verticilles épineux, sans interception jusqu'à son extrémité.

Les pores fémoraux sont au nombre de sept seulement, et n'occupent que la moitié de la longueur de la cuisse.

La couleur est d'un vert gai en dessus, avec 3-4 bandes incomplètes, noires sur le dos seulement, outre des indices de quelques-unes plus effacées. La tête est d'un gris verdâtre; la queue est grise, annelée de brun; enfin, le ventre est d'un gris rosé, pointillée de noir, avec les pattes d'un gris verdâtre, marbré de noir.

Observ. Cette espèce d'ignane a certainement un assez grand nombre de rapports avec l'*ignana cyclura* de M. Wiegman, mais elle en diffère notablement par la position de la crête dorsale, parce que la queue n'est nullement comprimée, et qu'elle est garnie de verticilles épineux sans interruption et non alternativement épineux et lisses, et enfin par la couleur qui n'offre pas de points bruns.

Cordylus (*Gerrhonotus*) *multi-carinatus.*

(Pl. XXV, fig. 2.)

Corps alongé, lacertiforme; tête pyramidale, régulièrement scutellée; cou assez étroit; tronc assez long, déprimé, avec un sillon latéral profond, depuis le tympan jusqu'à l'anus; queue fort longue, plus que la moitié du corps en totalité, et comprimée à l'extrémité.

Membres fort grêles, assez petits, assez distans, pourvus chacun de cinq doigts grêles, et assez inégaux.

Écailles dorsales sur le corps et sur la queue, toutes carénées, et disposées de manière à former quatre paires de lignes carénées dans toute l'étendue du corps.

Ecailles ventrales, larges, imbriquées, parfaitement lisses, et formant six paires de rangées longitudinales.

Couleur générale d'un gris brun verdâtre, traversé par des bandes plus foncées sur le dos et la queue.

Observ. Cette jolie espèce de scinque à sillons latéraux et à écailles carénées ressemble assez à l'espèce que M. Wiegman a nommée *gerrhonotus taeniatus* dans son *Herpetologia mexicana*, par la couleur, et au *gerrhonotus imbricatus* du même auteur par le système squameux ; mais elle diffère notablement de l'un et de l'autre; du premier, parce que les écailles sont carénées, et du second, parce que toutes celles du dos le sont également, outre la couleur qui n'est pas uniforme, mais zonée de noir sur un fond gris-verdâtre. D'ailleurs, la forme ou la disposition des scutelles céphaliques diffère notablement.

Rouleau de Botta (*Tortrix Bottæ*).

(Pl. 26, fig. 1, 1*a*, 1*b*.)

Corps assez alongé, cylindrique, s'atténuant un peu aux deux extrémités, l'une et l'autre obtuses.

Tête fort petite, peu ou point distincte, à museau court et obtus, couverte de plaques scutellées.

Bouche petite; anus transverse fort reculé.

Narines latérales fort petites, entre les deux scutelles nasales, dont la supérieure semble une frontale.

Yeux médiocres.

Scutelles céphaliques : — 1 incisive grande, 4 frontales, 1 interorbitaire très-grande, 2 occipitales fort petites, 1 loréale, 3 oculaires antérieures et 3 postérieures.

Dorsales : — formées d'écailles petites, égales, lozangiques, lisses et imbriquées.

Ventrales : — formées de scutelles transverses, assez étroites, sur un seul rang, sous le ventre et sous la queue.

Système de coloration uniforme, isabelle, plus foncé sur le dos et la queue.

Longueur totale : $0^m.60$, dont $0^m.06$ pour la queue.

Observ. Ce serpent, dont les dents sont proportionnellement assez fortes, peu nombreuses et non venimeuses, est un véritable tortrix dans la définition que Oppel donne de ce genre; mais en portant une plus grande attention à la disposition des scutelles de la tête, on voit qu'il appartient à une division particulière plus rapprochée des couleuvres à un seul rang de scutelles sous la queue. En effet, l'œil n'est pas percé dans une seule plaque, comme chez les véritables rouleaux, et cependant ce ne peut être un *aspidurus*, qui est un degré de plus vers les couleuvres; en sorte que dans le système erpétologique des auteurs allemands, on devrait en former un genre particulier.

La Couleuvre porte-chaîne (*Coluber catenifer*).

(Pl. XXVI, fig. 2, 2, 2.)

Corps fort alongé, assez gros, cylindrique; tête petite, peu distincte, et colubriforme; queue proportionnellement fort courte, conique et très-aiguë.

Bouche médiocre; anus fort reculé.

Narines latérales, petites, entre les deux scutelles nasales, au point de contact et de jonction avec la scutelle frontale antérieure.

Yeux médiocres.

Scutelles céphaliques : — 4 frontales, 1-4 oculaires, 2 loréales.

Scutelles abdominales fort larges, et uniseriées sous le corps, biseriées sous la queue.

Écailles ovales, alongées, foliacées, carénées, assez grandes, et s'accroissant de la ligne médio-dorsale aux flancs.

Couleur générale d'un gris jaunâtre, avec une suite de taches noires, formant une chaîne serrée dans toute la longueur du dos, et une autre série plus petite sur chaque flanc.

Longueur totale : 1^{m}.08, dont 0^{m}.15 pour la queue.

Observ. Cette belle espèce de couleuvre, remarquable par l'ordre dans lequel les taches noires du dos se disposent de manière à ressembler à une chaîne à anneaux serrés, et par une série de taches séparées sur chaque flanc, offre aussi une disposition toute particulière dans les scutelles oculaires.

Son système de coloration a été assez mal rendu dans la figure.

La C. infernale (C. infernalis).

(Pl. XXVI, pl. 3, 3*a*.)

Corps assez grêle, fortement atténué aux deux extrémités; tête petite, colubrine, à museau arrondi; queue assez courte, conique et très-aiguë.

Bouche très-fendue; anus très-reculé.

Narines latérales, échancrées également dans les deux scutelles nasales.

Yeux médiocres.

Scutelles céphaliques : — 4 frontales, 1 loréale, 1-3 oculaires.

Ecailles du corps longues, ovales, foliacées, carénées, imbriquées, croissant de la ligne médio-dorsale à la ligne latérale.

Couleur d'un bleu foncé, presque noir, offrant sur le milieu du dos une bande longitudinale jaunâtre, et une série de points rouges le long de chaque flanc; le ventre d'un bleu d'ardoise clair.

Longueur totale : $0^m.84$, dont $0^m.16$ pour la queue.

OBSERV. Cette grande espèce de couleuvre (puisque l'individu que nous avons observé avait deux pieds et demi de long) appartient à une division particulière, d'après la considération des plaques oculaires que nous n'avons encore observées que dans un assez petit nombre d'espèces. Du reste, c'est une véritable couleuvre sous tous les autres rapports.

Coluber (Ophis) Californiæ.

(Pl. XXVII, fig. 1, *1a*, *1b*.)

Corps de forme ordinaire et de médiocre longueur; tête assez grosse, déprimée, à museau court et obtus; queue assez courte, conique et aiguë.

Narines latérales, ovales, échancrant également les deux scutelles nasales, et touchant la frontale extérieure.

Yeux médiocres, latéraux.

Scutelles céphaliques : — 4 frontales, 1 loréale, 1-2 oculaires.

Ecailles assez petites, lozangiques, imbriquées et parfaitement lisses.

Couleur générale jaune, plus claire en dessous, plus foncée en dessus, avec six bandes longitudinales plus ou moins effacées, plus ou moins anastomosées, de couleur noire, les supérieures plus larges et souvent confondues.

Longueur totale : $0^m.66$, dont $0^m.095$ pour la queue.

OBSERV. Cette belle espèce de couleuvre véritable, appartenant à la division à laquelle Wagler a donné le nom d'ophis, caractérisée par la forme en couteau, et la plus grande longueur des dents maxillaires postérieures, me semble ne pouvoir être rapprochée d'aucune des six espèces que les erpétologistes ont rapportées à cette section, ce qui nous a déterminé à la considérer comme nouvelle.

C. vertébrale (C. vertebralis).

(Pl. XXVII, fig. 2, 2*a*, 2*b*.)

Corps alongé et grêle dans ses parties antérieures; tête petite, assez distincte, à museau atténué; queue courte, conique, et aiguë.

Narines latérales, petites, rondes, échancrant également le milieu des deux scutelles nasales.

Yeux grands, latéraux, presque entièrement entourés de scutelles oculaires.

Scutelles céphaliques : — frontales 4, loréales 2 superposées, 2-3 oculaires.

Scutelles venrales 245, caud. 64.

Ecailles assez petites, lozangiques, lisses, imbriquées.

Couleur d'un jaune roussâtre ou fauve, marbré de brun foncé, formant des taches enchaînées, en forme de vertèbres de poissons coupées, se détachant et s'éloignant entre elles de plus en plus, à mesure qu'elles deviennent plus postérieures.

Longueur totale : o.^{m}53o^{r}, dont o^{m}.o75 pour la queue.

Observ. Cette couleuvre, des mêmes pays que les précédentes, offre une particularité assez remarquable dans le système de scutelles qui entourent l'œil, et que je n'ai encore rencontré que dans une espèce que je crois nouvelle, et qui nous a été rapportée du mont Liban par M. P. E. Botta.

C. zonée (Zacholus) zonatus.

Corps cylindrique, à dos surbaisse, subcaréné, atténué brusquement en arrière, et assez peu en avant; tête petite, tétragone, à museau obtus et épais; queue courte, petite et fort aiguë.

Narines latérales, grandes, infundibuliformes, au milieu des deux scutelles nasales.

Yeux grands et latéraux.

Bouche très-fendue; anus très-reculé.

Scutelles céphaliques : — 4 frontales, 1 loréale très-petite, 1-2 oculaires fort petites.

Ecailles assez grandes, rhomboidales, subimbriquées, croissant du dos aux flancs, subconvexes et parfaitement lisses.

Couleur générale d'un blanc fauve, entièrement zoné de noir foncé, avec deux demi-anneaux de la même couleur sur la tête.

Longueur totale : $0^m.415$, dont $0^m.055$ pour la queue, ce qui fait un peu plus d'un septième de la longueur totale.

Observ. Cette jolie espèce est remarquable par son système de coloration qui se trouve fréquemment chez les serpens aquatiques et plus ou moins vénimeux. C'est cependant une véritable couleuvre tout-à-fait innocente.

C. tête-plate (*C. planiceps*).

(Pl. XXVII, fig. 3, 3*a*, 3*b*.)

Corps grêle, assez alongé, cylindrique; tête petite, très-déprimée, peu distincte, à museau court et elliptique; queue peu longue, grêle et très-aiguë.

Narines latérales, fort petites, échancrant profondément le milieu de la plaque nasale, alongée et unique.

Yeux médiocres et très-écartés.

Bouche large, assez fendue; anus reculé.

Scutelles céphaliques : — 4 frontales, 0 loréale, 1-1 oculaires.

Scutelles abdom. 134, commençant à quelque distance de la tête, sous-caudales, 56.

Ecailles larges, bombées, fort lisses, luisantes, imbriquées obliquement.

Couleur uniforme, roussâtre en dessus, d'un blanc sale en dessous, avec une plaque noire sur l'occiput, et le commencement du cou.

Longueur totale : $0^m.240$; $0^m.060$ pour la queue, dont la longueur est ainsi un quart de la longueur totale.

Observ. Cette petite espèce de couleuvre dont nous avons observé un individu mâle adulte, puisqu'il avait les appendices générateurs entièrement sortis, nous a paru devoir être distinguée, non-seulement comme espèce, mais comme type d'une division particulière, à cause de la disposition de ses scutelles céphaliques. En effet, la combinaison de l'absence de loreum avec une seule scutelle oculaire en avant et en arrière, ne s'était pas encore présentée à notre observation, et nous n'avons encore trouvé dans aucun erpétologiste moderne rien de semblable.

La distribution des espèces de couleuvres des erpétologistes modernes (c'est-à-dire des serpens sans aucune dent maxillaire vénéneuse, à système squameux très-dissemblable, les scutelles inférieures unisériées avant, et bisériées après l'anus), est encore loin d'être naturelle; peut-être même faudrait-il, pour y parvenir, ne considérer la vénénosité que secondairement à l'écaillure. Mais dans les idées généralement admises, on peut, à ce qu'il me semble, distribuer les espèces de ce genre autrement que je ne l'ai fait plus haut, en prenant en première considération les scutelles frontales, puis les oculaires et enfin les loréales, dont le nombre et la forme peuvent très-bien conduire jusqu'aux espèces, tant il y a de fixité dans la disposition et la proportion des scutelles céphaliques.

I. Espèces qui ont des crochets à l'anus (G. *python* et *constrictor*).
II. Espèces sans crochets à l'anus.

A. *Scutelles frontales paires*,

a] au nombre de 2 (G. *calamaria*, *xenopeltis*),
b] au nombre de 4.

1) — oculaires 0-1 (G. *ophites*, *brachyorrhos*).
2) — oculaires 1-1 (*C. planiceps*, *elapoides*, *catostoma*).
3) — oculaires 1-2.

+ Avec un loreum (G. *homolosoma*, etc., comme dans la distribution précédente.

+ + Sans loreum (G. *clœlia*, etc.).

4) — oculaires 1-3.

+ Avec loreum (G. *tropidonotus*, *leionotus*).

+ + Sans loreum (*C. pondecerianus*).

5) — oculaires 1-4 (*C. catenifer*).

6) — oculaires 2-2.

+ Avec un loreum (G. *coronella*, *dryophylax*, etc.).

7) — oculaires 2-3.

+ Avec un loreum (*C. montis Libani*, *vertebralis*).

8) — oculaires 2-4 (G. *langaha* (1) ou *rhinurus*).

9) — oculaires 3-2 (G. *tragops* ou *drynus*).

10) — oculaires 3-4 (G. *periops*).

B. *Scutelles frontales impaires*,

a] au nombre de 3.

+ Avec loreum (G. *homalopsis*, *hypsirhina*).

+ + Sans loreum (G. *helicops*, *hydrops*).

b] au nombre de 5 (G. *heterodon*).

---

(1) J'ai observé, dans les collections du Muséum, trois individus et le squelette du serpent qui constitue ce genre, et j'ai pu aisément m'assurer d'abord qu'il n'est nullement vénimeux, comme l'avait dit Brugnières, les dents maxillaires étant égales, peu nombreuses, distantes et coniques, et ensuite qu'il ne diffère des couleuvres du genre *tragops* que par un système scutellaire de la tête un peu différent, car sa queue est également biscutellée en dessous. Du reste il n'en est pas moins remarquable par l'espèce de queue squameuse qu'il porte à l'extrémité du museau, et qui semble avoir poussé entre la scutelle incivise, la première paire frontale alors un peu dérangée, et les nasales, très-latérales, en sorte que si l'on persiste à en former un genre distinct, on pourrait le nommer *rhinurus*, la dénomination de *xiphorhynchus* proposée par Wagler pour remplacer celle d'*amphistrate* donnée par Schneider étant un peu longue.

PARIS — IMPRIMERIE ET FONDERIE DE FAIN,
Rue Racine, n° 4.

Prêtre pinx.t Impr: Lith. de Thierry Frères Franke del.t

1. Cyclura acanthura.
2. Callysaurus draconoïdes.

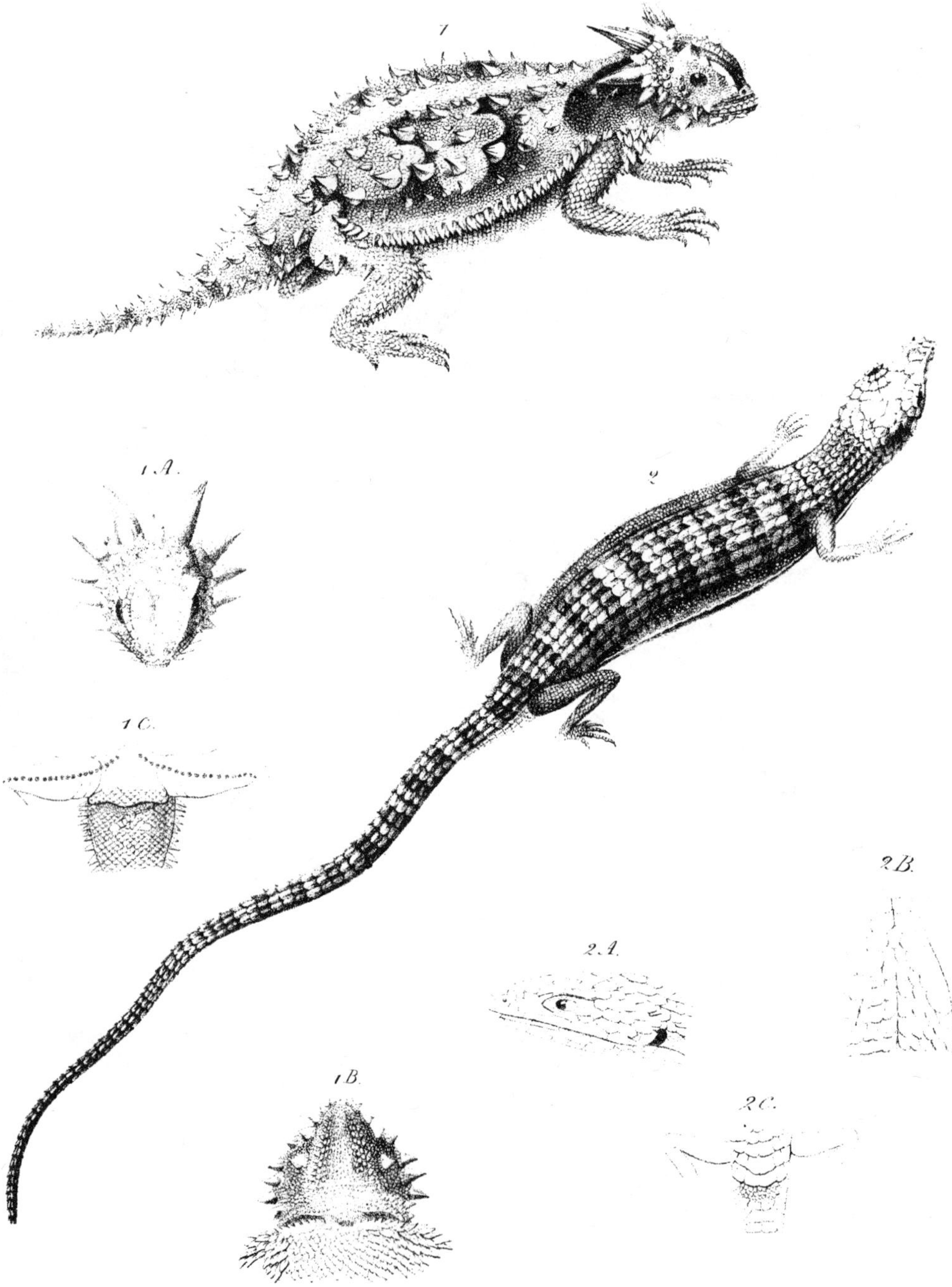

Prêtre pinx^t. Impr: Lith. de Thierry Frères. Franke del.

1. Phrynosoma Coronatum
2. Gerrhonotus multicarinatus.

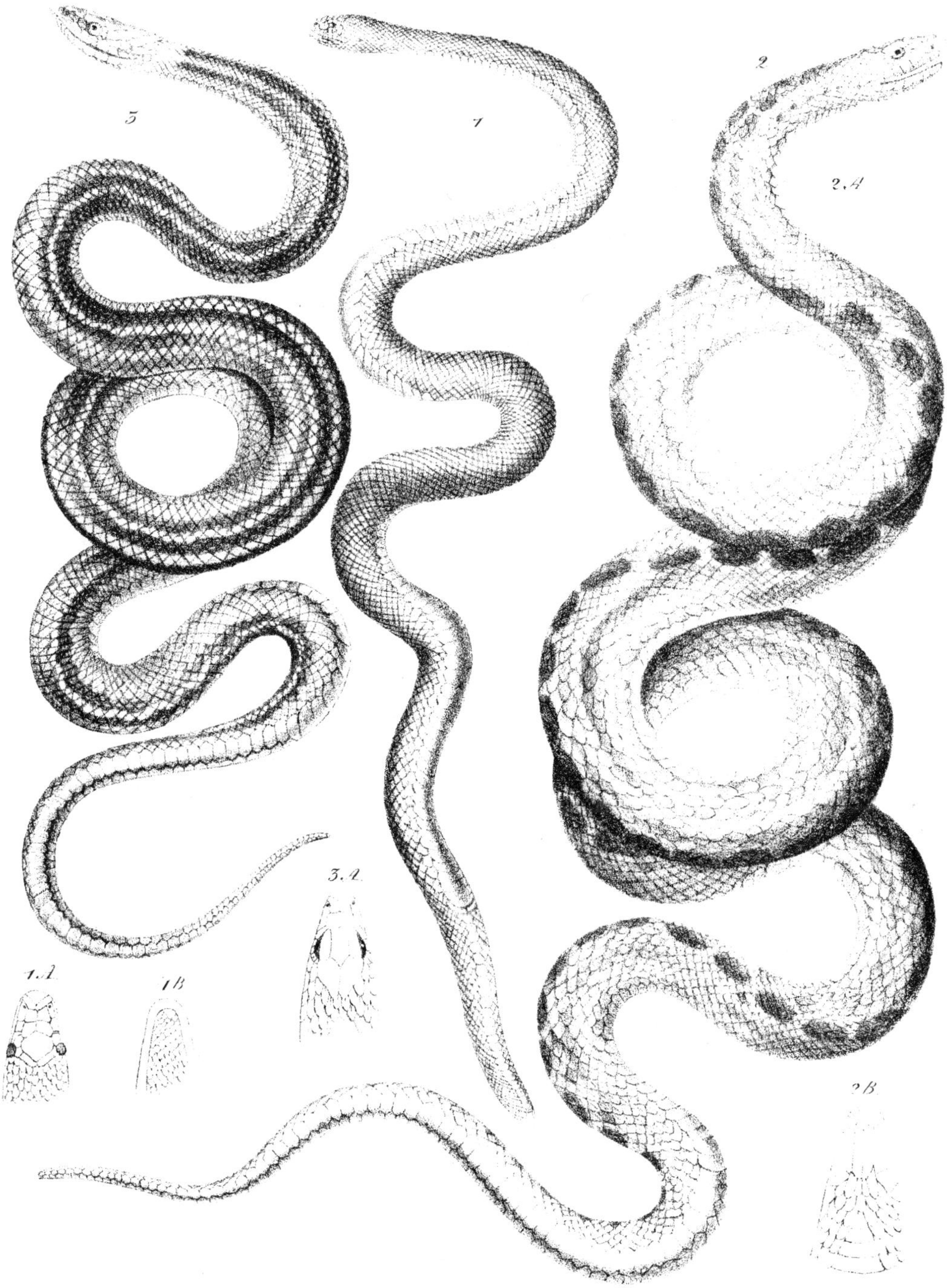

Prêtre pinx.t Impr. Lith. de Thierry Frères Franke del.t

1. Tortrix Bottæ.
2. Coluber Catenifer.
3. Coluber infernalis.

Prêtre pinx.t Impr. Lith. de Thierry Frères. Franke del.t

1. Coluber (Ophis) Californiæ.
2. Coluber Vertebralis.
3. Coluber planiceps.

REPORT UPON THE REPTILES COLLECTED ON THE SURVEY

J[ames] G[raham] Cooper

No. 4.

REPORT UPON THE REPTILES COLLECTED ON THE SURVEY.

BY J. G. COOPER, M. D.

CLASS **REPTILIA**. Reptiles.

Order I. CHELONIA. The Turtles.

ACTINEMYS MARMORATA, Agass.

The Western Pond Turtle.

Emis marmorata, B. & G. Proc. Acad. Nat. Sc. Phil. VI, 1852, 177.

Emys nigra, HALLOWELL, Proc. Acad. Sc. VII, 1854, 91.—IBID. P. R. R. Surv. Rep. vol, X, 1858, p. 3, pl. I.

Actinemys marmorata, AGASS. Cont. to Nat. Hist. of N. A. I, 1857, 444; II, pl. 3, fig. 5-8.—GIRARD, U. S. Expl. Exped. Herpetology, 466, pl. XXXII

SP. CH.—Carapax elliptical, convex, with an obtuse vertical ridge. Plastron sub-ellipsoid, broadest anteriorly. Limbs scaly, scales in front large, not imbricated; behind, moderate sub-tubercular. Nails rather stout. Tail above keeled sub-conical or tapering. Color black above, or olive with small black mottlings; beneath yellowish, with a black blotch in the young.

This, the only turtle yet known from the west of the Rocky mountains, is common in fresh-water ponds and rivers west of the Cascade mountains, though less so in the Columbia than in the warmer ponds. Mr. Gibbs also saw turtles at the mouth of the Yakima which were probably of this species. It is rather difficult to catch, being very watchful, but will sometimes bite at a hook. The specimen I preserved was taken when about to deposit its eggs, on the 9th of June. I tried to hatch some of them, but without success. Found in the ponds about Fort Steilacoom.—C.

It attains, when full grown, a length of about eight inches, and a corresponding width. The black of the back has a brownish or rusty tinge. It is called by the Nisquallies *El-la-chick*—S.

Order II. SAURIA. The Lizards.

ELGARIA PRINCIPIS, Baird & Girard.

The Spotted Elgaria

Elgaria principis, B. & G. Proc. Acad. Sc. Phil. VI, 1852, 175.—GIRARD, U. S. Expl. Exped. Herpetology, 214, pl. XXII, figs. 9-16.

SP. CH.—Dorsal scales in 48 transverse, and 14 longitudinal series. Three unequal post-nasals. Preanal shields larger than the abdominal. Tail longer than body and head together.

Several specimens of this graceful and harmless little animal were obtained about the end of July, on the "Yakolt," and another prairie on the banks of the Cathlapoot'l river. They were found under stones, and would seem to be more in the habit of hunting their insect prey on the ground than on trees or fences. Their movements were so rapid as to make it very difficult to catch them, and in struggling to escape nearly all broke off part of their tails, which, as in other reptiles, is probably a slight loss to them, a cartilaginous growth soon supplying its place. This species seems to approach nearer the cold regions than any other, as the near vicinity of the perpetual snows of the Cascade mountains and the considerable elevation of the valley causes a very moderate temperature, even in summer, in the place where it was found.

I have seen this lizard but once since, near the mouth of the Columbia among the warm sand and drift-logs on its northern bank.

When alive the colors were light brown, with black spots; belly pinkish white.

ELGARIA GRANDIS, Baird & Girard.

The Banded Elgaria.

Elgaria grandis, B. & G. U. S. Expl. Exped. Herp. 212, pl. XXII, fig. 1.9.

SP. CH.—Dorsal scales disposed upon fifty transverse and fourteen longitudinal series. Four unequal post-nasal plates. Preanal shields smaller than the abdominal. Tail nearly twice the length of the body and head together. Above brownish olive, with eleven brown bands across the neck and back, covering one row of scales which are tipped with white. Beneath unicolor.

Found at Fort Steilacoom by Dr. Suckley.

SCELOPORUS OCCIDENTALIS, Baird & Girard.

The Western Fence Lizard.

Sceloporus occidentalis, B. & G. Proc. Acad. Nat. Sc. Philad. VI, 1852, 175.—GIRARD, U. S. Expl. Exped. Herpetology, 383, pl. xix, fig. 8—14.
S. frontalis ? B. & G. in the same works.

SP. CH.—Cephalic plates rugose; anterior nearly equal to the vortex plate. One or two unequal lorals; temporal scales small keeled and acuminated. Preanals moderate, rounded, sub-emarginated; post-anal small, smooth, subtruncated, sub-emarginated. Posterior surface of thighs covered with minute lanceolated and keeled scales.

Differs from *S. undulatus*, *Wiegm*, of the eastern States, chiefly in its larger and rougher dorsal scales, while the abdominal are smoother. The blue abdominal patches are also wider apart, and of different shape.

I first met with this little lizard on the western border of the Great Columbia plains, about 2,000 feet above the sea, on the eastern slope of the Cascade mountains. It frequented the pine trees, and did not occur on the open plain. I again met with them at the Pisquouse river, about latitude 48°, and at about the same elevation. In agility and grace, as well as in colors, it is precisely like its eastern congener. I obtained one also on the edge of the plain near Steilacoom west of the Cascade mountains, where it is not uncommon; but I never saw it west of the Coast range, in the damper and cooler climate.—C.

Color.—Back, wood brown, tinged with greenish and ash gray, and spotted with triangular waved markings, on each side of the median line, of *dark* brown. At the dorsal median line these approach so nearly to each other on the opposite sides as to give, at a little distance, the appearance of waved transverse bars, extending entirely across the back. Each of these bars is bordered by a narrow margin of a greenish gray tint, resembling in color the pale mildew of new cheese.

Belly *white*, bordered on each side by a *steel* blue streak one inch long, one quarter of an inch wide; its *inner* border one line in width, black. Outer border of said streak *greenish* blue. A patch of steel blue, divided on the throat by a narrow white line, on the median line.

Found on the edges of the prairies and in the more open portions of the forests. It appears to choose the neighborhood of rotten logs. It is quite active, running and dodging about as nimbly as a squirrel.—S.

SCELOPORUS GRACIOSUS, Baird & Girard.

The Slender Fence Lizard.

S. graciosus, B. & G. Proc. Acad. Nat. Sc. Phil. VI, Apr. 1852, p. 69.—Ib. Ex. to Great Salt Lake, 1852, p. 346; pl. v, fig. 1.

S. gracilis, B. & G. Proc. Acad. Sc. Phil. 1852, p. 175.—Girard, Herpet. U. S. Expl. Exp. 1858, p. 386; pl. xx, fig. 1-9.

Sp. Ch.—Cephalic plates smooth; anterior parietal larger than vertex plate; middle occipital very large, pentagonal. Post-anal scales large, sub-lanceolate, smooth, posteriorly notched. Posterior surface of thighs covered with small, rounded, smooth scales. Olivaceous above, with a double series of crescent-shaped black spots on the back and two lateral light streaks, between which is a row of black spots. Beneath yellowish; under surface of head clouded with bluish. Male with an elongated blue patch on each side of the abdomen.

Found at the Dalles, Oregon Territory, or at Steilacoom. At Fort Dalles they live in the cracks of the basaltic rocks, which their colors resemble in tint, being darker than those at Steilacoom. They usually emerge from their hybernation about the middle of March.—S.

CROTAPHYTUS WISLIZENII, Baird & Girard.

C. wislizenii, B. & G. Pr. Acad. Nat. Sc. Phila. VI, Apr. 1852, p. 69.—Ib. Ex. to Great Salt Lake, 1852, p 340; pl. iii.

C. gambelii, B. & G. Pr. Acad. Nat. Sc. Phila. VI, 1852, p. 126.

C. fasciatus, Hallowell, Pr. Acad. Nat. Sc. VI, 1852, p 207.—Ib. Sitgreaves Ex. to Zuñi, 1853, p. 115; pl. V.

Sp. Ch.—Head proportionally narrow and elongated. Cephalic plates and scales on the back very small. Yellowish brown, spotted all over with small patches of deep brown or black, becoming partial or complete rings on the tail.

Found at the Dalles, Oregon Territory, and up Snake river, near Fort Boisé.—S.

TAPAYA DOUGLASSII, Girard.

The Oregon Horned Toad.

Agama douglassii, Bell, Trans. Linn. Soc. London, XVI, 1833, 105; pl. x.—Harlan, Med. & Phys. Researches, 1835, 141, fig. 3.

Phrynosoma douglassii, Wagl. Natural Syst. Amph. 1820, 146.—Holbrook, N. A. Herpet. II, 1842, 101; pl. xiv.—Girard, Stansbury's Ex. to Salt Lake, 1852, 361 and 362; pl. vii, figs. 6-10.

Tapaya douglassii, Girard, U. S. Expl. Exp. Herpet, 397; pl. xxi, figs. 1-5.

Sp. Ch.—Head large, depressed; vertex slightly declivous; snout sub-convex or rounded, sub-depressed at the nostrils. Cephalic plates moderate, very rugose, occipital and temporal spines reduced to small acerated cones. Sub-maxillar shields moderate and ridged. Auricular aperture granular, sub-tubercular, or sub-denticulated in front. Labial plates moderate. Mental scales small and unequal; gular folds minutely scaly. Abdominal scales moderate, sub-rhomboid, posteriorly obtuse. Femoral pores distant; the series from either side approximating upon the inter-femoral region without being continuous.

I obtained numerous specimens of this singular animal in the vicinity of the Yakima river, in August, and saw them as far as latitude 48° 30′ north, on the open plains, usually among rocks and sand. They all had the gray color which they retain in alcohol, excepting one, which was of a brick red on the back, but beneath white like the rest. These colors resemble those of the stones among which they live, and it is supposed by some that they have the power of changing their hue like the chameleon, and like the better known tree toad of the Atlantic States. Though ferocious in appearance they are perfectly harmless; yet the Indians believe them to have the power of producing a poisonous wound with their blunt spines, and it is possible that such a wound, *if made*, might sometimes be troublesome. They do not attempt to bite, and are so slow in motion as scarcely to get out of the way. They are, no doubt, like the ugly but useful toads, more useful than ornamental, as their food consists chiefly of insects.

As might be supposed from the locality, the specimens found in Washington Territory are smaller than those from Oregon and Utah.

I never saw nor heard of its occurrence west of the Cascade mountains.—C.

This animal was found by me at Fort Benton, on the Upper Missouri, and again on the plains west of the Rocky mountains, as far as the Cascade range.

One was caught in September near the Snake river, Oregon Territory. At Christmas, although having been shut up since its capture in an empty match box, it was still quite lively. When irritated it would spring in a most threatening manner at anything pointed at it, at the same time opening its mouth widely, and *audibly hissing*, after which it would inflate its body and show other evident marks of anger. It died about February 1, probably from starvation, as the heat of the house prevented torpor, and there were no insects upon which to feed it.—S.

Order III. OPHIDIA. Serpents.

CROTALUS LUCIFER, Baird and Girard.

The Western Rattlesnake.

Crotalus lucifer, Baird & Girard, Proc. Acad. Sc. Phil. VI, 1852, 177.—Ibid. Cat. N. A. Reptiles, 1853, 6.—Girard, U. S. Expl. Exped. Herpetology, 187, pl. XV, figures 1 to 6.

Sp. Ch.—Dorsal rows of scales, twenty-five ; exterior one smooth ; second and third obsoletely carinated. Tail and posterior part of body with fifteen to twenty half rings. A series of dorsal hexagons or octagons, also two small irregular series on each side. A light stripe from the supra-ocular crosses the angle of the mouth on the third and fourth series of supra-labials.

A specimen of this rattlesnake shot in the Yakima valley agreed very well with the description given by Baird and Girard in their Catalogue of Reptiles. Another, however, killed about September 25, in latitude 48°, on the Columbia river, was of a pure white ground color with beautiful bright sea-green blotches on the back. It had probably just changed its skin and had not acquired its permanent brown or yellow ground color with dark brown blotches.

Rattlesnakes are much less numerous north of the Columbia than south, these two being all I met with in two months' travelling through the interior. None are found west of the Cascade range, except an occasional straggler carried down the Columbia river.—C.

Specimens vary in the ground color from white to yellowish brown. The spots vary from greenish to chestnut brown. In the young a light stripe crosses the vertex between the supra-orbital scales, and another connects it with the ash color of the back of the head.

One specimen was killed, from the *mouth* of which three young ones were said to have escaped.

Found from the Dalles up the Columbia and Snake river. At the Dalles they are so numerous as to be very annoying, having been known to enter dwelling houses. Since the introduction of hogs in the vicinity they seem to have diminished. The Indians use the tail of the rattlesnake as a medicine to produce abortion.—S.

CROTALUS CONFLUENTUS, Say.

Prairie Rattlesnake.

Plate XII.

Crotalus confluentus, Say in Long's Exped. Rocky mountains, II, 1823, 28.
Crotalus confluentus, B. & G. Cat. N. A. Serpents, p. 8.

SP. CH.—Head sub-triangular. Plates on top of head squamiform, irregular, angulated, and imbricated ; scales between superciliaries small, numerous, uniform. Four rows of scales between the sub-orbital series (which only extends to the centre of the orbit) and the labials. Labials 15 or 18, nearly uniform. Dorsal series 27–29. Dorsal blotches quadrate, concave before and behind ; intervals greater behind. Spots transversely quadrate posteriorly, ultimately becoming 10 or 12 half rings. Two transverse lines on superciliaries, enclosing about one-third. Stripe from superciliary to angle of jaws crosses angle of the mouth on the second row above labial. Rostral mangined with lighter.

Milk river, Nebraska.

This species is very numerous on the Missouri river and its tributaries, between Fort Union, Nebraska, and the Rocky mountains. In July and August they are found very common in the dry cañons, and among the willow brush, and cottonwood forests along the banks of the rivers. They are then sluggish and stupid, being, according to popular belief, "blind," and are said to be at that season exceedingly venomous. This stupid condition during the drought of summer is not uncommon to many species of snakes, the torpidity being analogous to that of hybernation, and may therefore be called *aestivation*. Hunters have told me that the serpents are "blind," because they are at that time about shedding the cuticle, and that as evidence of loss of vision the snake, when provoked, will "strike wildly."—S.

EUTAINIA ATRATA, Kennicott.

SP. CH.—Body compact, cylindrical, moderately stout. Head small and narrow, eye very small ; eight upper labials sixth largest. Dorsal rows of scales seventeen, exterior row largest, higher than long, and very slightly carinated ; the next row smaller, but considerably larger than the third, distinctly carinated. Scales of the central dorsal rows proportionately shorter than in *E. concinna* and *E. pickeringii*. A very broad, deep lemon yellow dorsal stripe, covering nearly three rows, and distinct from head to tip of tail. The rest of the upper parts entirely deep blue black, without a trace of the lateral stripe or of light spots. Abdomen entirely uniform greenish slate, yellowish green under the head.—KENNICOTT.

California.—C.

EUTAINIA COOPERI, Kennicott.

PLATE XV, REPTILES, FIG. 1.

The Red Striped Garter Snake.

Eutainia cooperi, KENN. Proc. Acad. Nat. Sc. Philad. 1859, p. —.

SP. CH.—Body stout, compact, and cylindrical, as in *E. radix*, dorsal rows of scales only seventeen. Head short, depressed anteriorly. Labials seen above; the 5th twice as large as the 7th, being the largest of all, and greatly developed. Colors (in alcohol,) above, uniform blackish brown, without spots, or olivaceous brown with two rows of black spots, as in *E. vagrans*, but which do not encroach upon the stripes. Dorsal stripes yellowish, distinct on one and two half rows ; lateral stripe usually distinct, covering the second and part of the third rows ; the row below being of the same color as the back. Abdomen usually slate color, sometimes lighter. Body frequently suffused with red, especially the dorsal stripe ; abdomen sometimes tinged with red.—KENNICOTT.

The most highly colored specimens of this snake were caught in one spot, on the 2d of August, in a small prairie in the Cathlapoot'l valley. They had the colors represented in the plate, but which fade after long keeping in alcohol. Others obtained in the Willopah valley, in 1854, had dark stripes, and *young* ones were without spots.—C.

EUTAINIA PICKERINGII, Baird & Girard.

Pickering's Garter Snake.

E. pickeringii, B. & G. Cat. N. Amer. Serpents, 1853, p. 27.—GIRARD, Expl. Exped. Reptiles, p. 150, pl. XIII, fig. 14–20.

SP. CH.—Body rather slender ; dorsal rows of scales nineteen, the first large and moderately carinated. Head large and high, with the upper labials well developed, seven in number, the fifth largest. Color, black above ; the stripes three, greenish or brownish yellow, (fading in alcohol.) A series of irregular vertical bars on the sides from the third to sixth rows, confluent with the lateral stripe, which covers the second and third rows. Abdomen and first dorsal rows, slate color, bluish, or greenish.

Var. *α*. Without the lateral spots, the space being black like the back.

β. Entirely black, the stripes being so narrow as to be almost obsolete ; specimens showing every degree of these variations. Sides of head usually more or less mottled with brown and green, the most so in the largest specimens.

Obtained in all the western portions of the Territory, usually in or near the forests. The commonest species, but more rare, east of the Cascade range.—C.

The variety with red spots was not found by me near Puget Sound, though common at the Dalles and Vancouver.

The colors of my specimen, *when living*, were as follows: In the spring, dorsal scales of a *very dark* invisible green, (almost black.) Dorsal and lateral stripes *pea green*. Chin *white*. Belly greenish white anteriorly, becoming posteriorly bluish white and bluish slate, and the under surface of tail "blue black."

In midsummer the colors remain the same, except that they are darker, (deeper.) This species exists in great abundance near Steilacoom and Nisqually. They are found on the gravelly prairies and in the vicinity of the numerous small lakes of this section of country. About the first of April they begin to come out of winter quarters, and can be seen at mid-day sunning themselves near small clumps of scrub-oak bushes, to which they retreat when alarmed. A little later they are found in couples or in small companies. Although they are rarely ever found more than one-eighth of a mile from water, they are, nevertheless, still more anxious to be close to it as the season advances. They will then (in May and June) be found lying close to the water, on the lake shores in the grass, and among the sedge of the marshes, and even upon small bog islands, as much as fifty yards from the shore.

In summer, like other members of this genus, they are found lying in small pools, and in the water at the edges of the lakes, during the heat of the day. They appear to be a harmless, lazy species, and, as above stated, exceedingly fond of the water.

They rarely exceed twenty-five inches.—S.

EUTAINIA LEPTOCEPHALA, Baird & Girard.

The Small-headed Striped Snake.

Eutainia leptocephala, B. & G. Cat. N. Amer. Serpents, 1853, p. 29.—Girard, Expl. Exped. Herpetology, p. 151, pl. XIII, fig. 7-13.

Sp. Ch.—Body rather long and slender. Head small, narrow, and flattened; not much larger than neck. Dorsal rows 19, the two exterior rows larger, the outer not carinated. Scales on most of the tail not carinated. Labials narrow ; 7 above, fifth and sixth largest. Above, light olive brown, or darker, with about 130 small brown spots in two series on each side the vertebral line, occuring on alternate scales, sometimes wanting. Vertebral stripe faint, on a single row of scale ; lateral stripe in young specimens partially apparent anteriorly. A blackish stripe from the post-orbital back to the angle of the mouth, sometimes touching the labials.

Found in all the western portion of the Territory.—C.

Found apparently in all parts of the Territory and in Oregon, specimens having been obtained in the St. Mary's valley by Lieutenant Mullan, and on the Snake river, near Fort Boisé, by myself.—S.

EUTAINIA VAGRANS, Baird & Girard.

The Large-headed Striped Snake.

Eutainia vagrans, B. & G. Cat. N. Amer. Herpetol. I, 1853, p. 35.—Girard, Expl. Exped. Herpetol. p. 154, plate xiv. figs. 5-10. (See plate XVII, Reptiles, Beckwith's P. R. R. Report, vol. X.)

Sp. Ch.—Body long and slender; head short, broad, and high, thicker than neck. Dorsal rows 20 or 21; stripes faint or partly wanting, with two rows of brown or black spots, about one hundred in number, on each side, confluent with the stripes. Scales of tail and outer dorsal row distinctly carinated. Labials, 8 above, the sixth and seventh largest, higher than wide. Ground color above light olive brown, the stripes paler; beneath, slate color.

Found in the Yakima valley in August, but is known to extend also west of the Cascade range.—C.

Found at Puget Sound and at Fort Boisé, on Snake river. While on Boisé river I saw a snake apparently of this species, which, upon being pursued, retreated to a small hole in the ground, the calibre of which was just sufficient to admit its occupant. Whether the hole had been excavated by the serpent, or was simply the burrow of some small *rodent*, I had no means of ascertaining.—S.

EUTAINIA CONCINNA, Baird & Girard.

Plate XV, Reptiles, Fig. 2.

The one-striped Garter Snake.

Tropidonotus concinnus, Hallow. Proc. Acad. Nat. Sc. VI, 1852, p. 182.

Eutainia concinna, B. & G. Cat. N. Amer. Serpents, 1853, p. 146.

Ch.—Body moderately lengthened. Head small. Dorsal rows, 21, carinated. Vertebral stripes, yellow: *no* lateral stripes, but in their place a series of 75 salmon red spots or bars extending from the second to sixth row of scales, and covering om half to two scales width. Ground color black, belly slate or greenish black, whitish anteriorly. Head reddish yellow, inged with brown.

Found at Vancouver, and not observed along the coast.—S.

Were it not for the distinction in the scales of the head and back, which in this genus are relied upon as specific characters, and even divide the species into groups, these five species might be combined into two by their colors. *E. cooperii*, *leptocephala*, and *vagrans*, approach each other very closely, and *pickeringii*, especially the variety with one stripe, much resembles *concinna*; in fact, some specimens have much the appearance of being hybrids. More observations and specimens will be required to decide what characters are truly specific.

All these garter snakes have very similar habits, and are equally harmless. It is generally believed that those with carinated scales, like the three last species, are more aquatic in their habits, and I did notice this to be the case with *vagrans* and *leptocephala;* but all of them live not far from water, and in the dry season, at least, are found almost always near it. Another favorite resort is the Indian burying-grounds, where the corpses, placed above ground, attract numerous insects; and it is these, not human flesh, which form the food of the beautiful little garter snakes, notwithstanding the association of graves and serpents so common in the minds of poets and people.

The Indians have something of the same natural (?) horror of even harmless snakes that pervades more civilized races; and it is due to the same ignorance, for they have frequently told me that snakes were poisonous which I handled with perfect impunity. Perhaps, also, superstition is connected with this antipathy.—C.

EUTAINIA HAYDENII, Kennicott. N. S.

Plate XIV.

Sp. Ch.—Head broader and more depressed in front than in *E. radix*. Form stout, compact, and cylindrical, most so of the genus, except *E. radix*. Ground color light olive green, with three longitudinal yellow stripes, and six series of distinct black spots. In life some red coloring visible on the sides. Lateral stripe on the third and fourth rows less sharply defined than in *E. radix*. Dorsal rows, 21.—Kennicott.

Fort Pierre, Nebraska.—Dr. Evans.

EUTAINIA FAIREYI, B. & G.

PLATE XIII, adult; XVI, young.

Eutainia faireyi, B. & G. Cat. N. A. Serpents, p. 25.

SP. CH.—Body above deep uniform glossy black. A dorsal stripe of yellowish green one and less than two half scales wide; and one lateral stripe on each side on the third and fourth rows, usually of the same color. Sometimes the dorsal stripe is more deeply yellow than the lateral. Dorsal scales below the lateral stripe as black as above it. Abdomen greenish white. Form a little stouter than *E. saurita*. Head large. Tail rather less than one-third of the total length. Dorsal scales in nineteen rows.—KENNICOTT.

Illinois.

EUTAINIA RADIX, B. & G.

E. radix, B. & G. Cat. N. A. Serpents, p. 34.

SP. CH.—Body stout, compact, and cylindrical; tail short. Head short, narrow; nose pointed. Ground color above very dark olive brown, sometimes black, with three sharply defined, rather narrow, yellow longitudinal stripes, and six series of very indistinct black blotches, which are not visible in the darker specimens. Dorsal rows, 21. Lateral stripe on the third and fourth rows.—KENNICOTT.

Fort Snelling, Minn.—S.

REGINA KIRTLANDII, Kennicott.

PLATE XX, FIG. 2.

Regina kirtlandii, KENNICOTT, Proc. Phil. Acad. Nat. Sci. 185.

SP. CH.—Body stouter than in any other species of Regina or Nerodia; head smaller and tail shorter. Head depressed, very small, short, proportionately broad behind, continuous with the body, the neck being without any visible contraction. Crown very convex, sloping to the snout. A single nasal, with the nostril in its centre. One large ante-orbital, two large post-orbitals. Vertical, broad, sub-hexagonal. The body enlarges rapidly from the head to its full size, continuing of about the same size to near the anus, where it contracts suddenly; the tail being very small. Ground color light reddish brown, with four dorsal series of circular black spots, the two central series smallest. Abdomen uniform reddish, with a row of small black spots on each side. Dorsal rows 19, all strongly carinated.—KENNICOTT.

Illinois.

REGINA GRAHAMII, Baird & Girard.

The Prairie Water Snake.

PLATE XIX, FIG. 1.

Regina grahamii, B. & G. Catal. N. Amer. Serpents, 1853, p. 47.

SP. CH.—Dull dark brown, with a dorsal light brown line, margined on each side by a narrow indistinct black line. A broad yellowish stripe on the first, second, and third rows, margined above by an indistinct black line on the fourth and fifth rows, and below by a distinct narrow black line on the lower fourth of the first lateral row and extreme end of the abdominal scutellae. Abdomen yellowish, tinged posteriorly in the adult with olive, with a single central row of small subtriangular black spots posteriorly, which disappear on the anterior third of the body, and are sometimes obsolete in young specimens. Dorsal rows of scales, 19.—KENNICOTT.

Illinois.

SCOTOPHIS VULPINUS, Baird & Girard.

The Fox Snake.

PLATE XXII.

Scotophis vulpinus, B. & G. Catal. N. Am. Serpents, 1853, p. 75.

SP. CH.—Body stoutest of the genus; tail thick. Head large, very stout, broad, and rounded throughout. Snout short, broad; vertical plate as broad as long; superciliaries broad; loral elevated as high as long. Eye smallest of the genus.

Dorsal rows, 25, the central distinctly carinated. Abdominal scutellae about 203. Light yellowish brown, with a dorsal and two lateral series of chocolate brown blotches. No suffusion of black in old specimens as in *S. alleghaniensis*. A frontal brown bar and a post-ocular vitta ; obsolete in old specimens.—KENNICOTT.

Fort Snelling, Minnesota.

A full grown specimen of this serpent was brought to me alive, at Lake Amelia, near Fort Snelling, Minnesota. When provoked it expressed its irritation by vibrating the tip of its slender tail, which, when striking a crumbled dead leaf or any other small object, would produce a well-marked rattling noise, very similar to that made by the rattlesnake under the same circumstances.—S.

PITUOPHIS WILKESII, Baird & Girard.

The Oregon Bull Snake.

Pituophis wilkesii, B. & G. Cat. N. A. Reptiles, I, 1853, 71.—GIRARD, U. S. Expl. Exped. Herpetology, 137 ; pl. ix, figs. 1-7.

Pitnuphis catenifer, B. & G. ? *P. aunectens*, B. & G. ? op. cit.

SP. CH.—Two pairs of post-frontal plates. Dorsal scales in 29 to 31 series ; the three outer series smooth. Tail about a sixth of the total length. Post-ocular vitta running over the last labial to the angle of the mouth. Ground color (yellowish,) with a dorsal series of sub-quadrate, and two lateral series of sub-circular blotches.

From a large number of specimens in the Smithsonian Institution, lately examined, it appears probable that the two latter names quoted as supposed species will become synonyms, in which case *P. catenifer* will be the name of the species. The Californian specimens, on which they were founded, differ chiefly in the *proportions* of the cephalic scales, and the size of their spots, which are largest in *catenifer*.

A specimen caught in the Yakima valley, on August 28, differed from the description above given only in the ground color, which was pale gray instead of yellowish—a tint doubtless caused by the alcohol.

This large snake, sometimes three or four feet long, is, like the others of the genus commonly called "Bull" and "Pine snake," quite harmless, and must destroy a great many mice, gophers, and other vermin. Not found west of the Coast range.—C.

The ground color gray, sometimes tinted with brownish or yellowish. Spots brown, most dark posteriorly, sometimes margined with black. Ground color of belly yellow; spots black. Found sparingly at Puget Sound, but common at the Dalles and up Snake river to Fort Boisé. It is, in fact, mostly confined to the open country. The length of the adult rarely exceeds three feet.—S.

PITUOPHIS SAYI, Baird & Girard.

Prairie Bull Snake.

PLATE XXII.

Coluber sayi, SCHLEGEL, (non Holbr.)—Essai Phys. Serp. Part. descr. 1837, 157.

" *Coluber mellanoleucus*, var. SAY."—HARL. Jour. Acad. Sc. Philad. V, 1827, 360.—IB. Med. and Phys. Researches, 1835, 123.

Pituophis sayi, B. & G. Catal. N. Am. Serpents, 1853, p. 152, (under *Coluber sayi.*)

SP. CH.—Head proportionately small, crown rounded throughout, snout very pointed. Rostral very narrow, projecting, the apex elongated and pointed, extending far back between the prefrontals. Verticals short, broad, nearly as wide anteriorly as long. Nasals and loral large. One ante-orbital, rarely two ; three post-orbitals, rarely four. Dorsal scales shorter and more rounded than in *P. bellona*. Ground color whitish or reddish yellow, a dorsal series of sub-quandrangular blotches, with 3 or 4 small series on each side, the blotches brown on the middle of the body, black towards the head and tail. Abdomen yellowish white, with an external series of rather large black spots on each side, inside of which are two irregular series of wavy black blotches. Head much spotted with black and labials heavily margined with the same. A pitch black post-ocular with a frontal band, and sub-orbital vertical bar.—KENNICOTT.

The prairie bull snake is very numerous in eastern Minnesota, where I obtained many specimens. They are said to be occasionally ploughed up by the settlers when breaking the prairie sod.

During the rutting season they seem to follow each other by the scent, as I noticed upon one occasion that having captured a fine living individual, and placed it with others in a barrel near my tent, a very large snake of the same species was found a short time after but a short distance from my reptile prison, being on its way in a direct line from the locality whence the other had just been brought.

A nearly similar action was observed with another species. The facts might have been simply accidental, although if so, seemed strangely to confirm the popular belief on the subject. Whether the mates of these and other species do follow each other by the scent is well worthy of future investigation. The fact seemed so impressed on the minds of my companions that the naturalist's tent was for a time *tabooed.*

A very large *Pituophis* was collected by me in Nebraska. Its colors were considerably darker than those of the Minnesota specimens. The length of the skin, stretched somewhat in taking off, was eighty-seven inches.—S.

BASCANION VETUSTUS, Baird & Girard.

The Green Racer.

PLATE XX, FIG. 1.

Bascanion vetustus, B. & G. Catal. N. A. Reptiles, I, 1853, 97.—GIRARD, U. S. Expl. Exped. Herpet. 127. Pl. VIII, figs. 12—19.

SP. CH.—Width of head equal to half its length, taking the greatest dimensions. Inferior post-orbital in a notch between the fourth and fifth labials. Colors, when fresh, pale green above, straw yellow beneath, sometimes varying to brownish above.

Three specimens of this handsome snake were caught in or near the Yakima valley, in August, 1853. Like its relative, the black snake of the Atlantic States, it is perfectly harmless, and does not even bite usually when handled. It runs with great rapidity, and the popular name given both to this and to the allied, if not identical, *B. flaviventris*, of Nebraska and Kansas, is very appropriate. Not found west of the Coast range.—C.

A specimen caught August 8, at Fort Steilacoom, had the following colors when fresh: *Upper parts of a uniform brownish slate color.* Edges of scales blackish. Skin colored the same as the central portions of the scales. Belly bluish white, with a tinge of yellow anteriorly and under the chin. Caudal scutella much darker than those of the middle.

Specimens from the Dalles partake of the characters of *B. flaviventris*, and are of a pale olive, becoming brown posteriorly, sides bluish. Young specimens are ash colored above, with chestnut spots; head olive and spotted. Beneath pale greenish yellow, with small reddish brown spots anteriorly. In older ones the blotches of the back become confluent and finally unicolor.

It has the same habit of climbing in bushes common to the black snake of the eastern States. Found sparingly at Puget Sound.—S.

MASTICOPHIS TÆNIATUS, Baird & Girard.

The Californian Whip Snake.

Leptophis tæniata, HALLOWELL, Proc. Acad. Nat. Sc. Phil. VI, 1852, p. 181.
Masticophis tæniatus, B. & G. Catal. N. Am. Rep. 1853, p. 103.—BAIRD, P. R. R. Rep. Vol. X, (Beckwith's Exp.) Pl. XXIII.

SP. CH.—Dorsal rows 15; tail about one-third of the total length. A broad brown dorsal stripe margined by a darker line. The four outer rows of scales on each side yellow, with a darker line through the centre of each. A dark line along the edge of the abdomen, making six dark lines on each side. Beneath yellowish.

A young specimen of this species was obtained on Snake river, Oregon Territory. Nothing regarding its habits was observed.—S.

DIADOPHIS PULCHELLUS, Baird & Girard.

The Western Ring Snake.

Diadophis pulchellus, B. & G. Cat. N. Am. Serpents, 1855, p. 115.

SP. CH.—Dorsal scales in fifteen rows; occipital ring broad; colors, (in alc.,) above uniform dark greenish brown, beneath orange with small black spots irregularly scattered both above and below from head to tail, sometimes wanting. Color of abdomen extending on to lateral dorsal scales.

A living specimen had the following colors: Top of head glossy, of a dark brownish olive with a bluish tinge, several of the plates being indistinctly spotted with black. Occipital ring deep reddish orange, having a diagonal width of three scales. A band along the back eleven scales wide of a slaty brown, rather pale, and with a slight greenish tinge anteriorly; posteriorly it becomes more bluish, losing its greenish cast and narrowing on the tail from 9 to 2 scales in width. Chin, belly, and first two rows of dorsal scales bright, fiery orange. This color deepens posteriorly, becoming more red, the sub-caudal scutella being almost the color of red sealingwax. A row of black spots on each side, each spot being on the posterior edge of the first row of dorsal scales. The abdominal scutellæ have two well-marked rows of small black spots besides several other partial or irregular rows. All these spots are found on the posterior edges of the scales.

The occipital ring is margined posteriorly by a series of black spots, showing an apparent attempt in nature to form a black linear margin.

This is the only specimen of the genus or the species that I have seen west of the Rocky mountains. The natives do not say that it is abundant, and the white settlers have only occasionally observed it. Fort Dalles, Oregon Territory, May, 1855.—S.

CELUTA AMŒNA, B. & G.

PLATE XIX, FIG. 2.

Ground Snake; Worm Snake.

Celuta amœna, B. & G. Cat. N. A. Serpents, 1853, p. 129.
Coluber amœnus, SAY, Journ. Acad. Nat. Sc. Philad. IV, 1825, 237.—HARL. Journ. Acad. Nat. Sc. Philad. V. 1827, 355; and Med. & Phys. Res. 1835, 118.—STORER, Rep. Rept. Mass. 1839, 226.
Calamaria amœna, SCHL. Ess. Phys. Serp. Part. descr. 1837, 31. Pl. i, figs. 19 and 20.
Brachyorrhos amœus, HOLBR. N. Amer. Herp. III, 1842, 115. Pl. xxvii.

SP. CH.—Above uniform chestnut brown, opalescent; light yellow (bright salmon color in life) beneath. Dorsal scales in 13 rows.

Western Missouri.

WENONA PLUMBEA, Baird & Girard.

The Brown Wood Snake.

Wenona plumbea, B. & G. *Wenona isabella?* B. & G. Proc. Acad. Sci. Philad. VI, 1852, 176; Ibid. Catal. N. A. Reptiles, I, 1853, 139, 140.—Girard, U. S. Expl. Exp. Herpet. 1858, 112, 113, pl. VII, f. 1–14.

The distinctness of the above supposed two species is uncertain from the small number of specimens examined. They differ in the frontal plates and in the orbitals, but their colors are described as nearly the same, *plumbea* being lead color above, and *isabella* light brown; both yellow beneath, the latter more dull.

Not very uncommon at Puget Sound. Specimens appear to unite the characters of the species described by Baird and Girard.—S.

The short, thick, and clumsy form of these snakes, with their small pointed head covered with scales like those of the back; very small eye and mouth; numerous small smooth scales, and blunt tail, at once distinguish them, though in color they much resemble the "green racer," heretofore described, when in its old skin, before changing it in July.

About June 1, 1855, I obtained a large specimen of the above snake under the loose bark of a log, where it seemed to have crawled to escape from the light. It appeared dazzled, and did not attempt either to escape or to bite. This snake, though unpleasant in appearance, is perfectly harmless, and, like most others, useful in destroying insects and mice. It is the nearest representative of the *boa constrictor* in the United States, but little is known of its habits.—C.

Order IV. BATRACHIA. Soft Skinned Reptiles.

Tribe I.—**Anoura**.—Tailless Batrachians.

BUFO BOREAS, Baird & Girard.

Bufo boreas, B. & G. Expl. Exped. Herpet. p. 74, pl. VI, fig. 4–9.

Sp. Ch.—Upper surface of head plain; skin not adhering to the skull, and smooth; green, with a dorsal yellow vitta; pustules reddish; beneath of a soiled white, maculated with black.

Distinguished from *B. columbiensis* by having the fore arm longer than the hand or arm, both pairs of legs longer, and hind foot proportionately longer, and the soles nearly smooth. The bright colors above described are those of the young. When alive the colors vary in shade, but are similar in pattern in all examined. That best marked has the back olive brown, darker posteriorly; tubercles dark brown; spots on legs nearly black, both surrounded by a narrow edging of straw yellow, becoming greenish, and mixed with the brown on the sides. The dorsal stripe is bright yellow, (wanting in some specimens;) angles of mouth and inner sides of thighs yellow also; belly pale brownish white, with black spots posteriorly; iris rich gold color, veined with black; canthi black. Another specimen had a brown circle around iris, and green lids.

This toad is abundant west of the Cascade mountains, and has habits similar to those of most other true toads, hopping about at dusk after its insect prey; and, although disagreeable to many persons, is perfectly harmless and very useful.—C.

BUFO COLUMBIENSIS, Baird & Girard.

The Columbia Toad.

Bufo columbiensis, B. & G. Proc. Acad. Nat. Sci. Philad. VI, 1853, 378.—GIRARD, Proc. Acad. Sci. VII, 1854, 87.—IBID. U. S. Expl. Exped. Herpet. 77, pl. V, figs. 4—9.

SP. CH.—Upper surface of head plain; skin adhering to the skull and granulated. Parotids and tympanum small. A membranous tarsal fold; toes palmated. A dorsal white vitta and an oblique dark patch beneath the eye.

Found in Washington Territory west of the Cascade mountains. Nothing distinctive was noticed in its habits.—C.

RANA PRETIOSA, Baird & Girard.

The Salmon Frog.

Rana pretiosa, B. & G. Proc. Acad. Sci. Phil. VI, 1853, 378 —IB. U. S. Expl. Exped. Herpetology, 21, 1858; pl. II, figs. 13—18.

SP. CH.—Tympanum rather small. Hands smooth. Feet granulated underneath; fingers slender and tapering. Toes webbed; terminal joint of longest free; a small flattened horny process at the base of the inner toe and a minute conical tubercle between the fourth and fifth Two depressed dorsal series of pores, one on each side; a glandular ridge along the upper jaw.

When living the colors in full-grown specimens are: nose and cheek pea green, tinged with gold; a stripe of the same hue down each side of the back. Remainder of back from eyes to tail reddish brown, with a black streak. Legs paler, with tranverse black bars. Flanks and inner sides of thighs salmon red, growing darker towards feet. Belly white, sometimes with brick-red spots; thighs posteriorly and sides spotted with white. A wood brown stripe on each side of head; lips tinged with gold color. Size of body from two to three inches long.

Younger specimens pale green above; thighs pale brown; no reddish tints beneath, white spots larger and more numerous.

I found frogs nowhere common in the Territory. The specimens described above were found about springs near Shoalwater bay, and when alive are very beautiful and interesting little animals.

RANA HALECINA.

The Spotted Frog.

PLATE XXIX, FIG. 7.

Although the distance of locality, which gives the species a wider range than any other reptile of the Atlantic coast, would *a priori* indicate distinctness of species, a careful comparison by Professor Baird does not show any tangible points of difference. Specimens of this frog were collected at Fort Dalles.—S.

HYLA REGILLA, Baird & Girard.

The Oregon Wood Frog.

Hyla regilla, B. & G. Proc. Acad. Nat. Sci. Phil. VI, 1852, 174; and 1853, 301.—IBID. U. S. Expl. Exp. Herpet. 60, pl. III, figs. 13—18.

Hyla scapularis, HALLOWELL, Proc. Acad. Sci. VI, 1852, 183.

SP. CH.—Vomerine teeth disposed upon two circular groups. Fingers free, toes semi-palmated. Skin above tuberculous, beneath glandulous. Green, with orange reflections, sometimes maculated or banded with blackish brown.

This brilliant little wood-frog abounds in some localities and at some seasons, but at others is rarely met with, as it seems to seek the high trees. In September I found vast numbers of them around a spring on the high mountain spur between the Yakima and Columbia rivers, over 4,000 feet above the sea, and also obtained a few at Chequass, in the Klickatat Pass, at the same elevation. I was inclined to think that the great dryness of the summer east of the Cascade range had driven them from the trees to seek for water. I found them also, but rarely, at Shoalwater bay, where they seem to leave the trees less often. Those which have brown markings are much larger than the others, and not so frequently found.

The shrill "peep" of this frog is often heard, but it is difficult to see the performer in its shelter among the green leaves.—C.

Specimens of this beautiful little tree toad, caught April 3, 1855, had the ground color of upper parts pale green, mottled by small irregular patches of a slightly deeper shade, which are bordered by streaks of black. Inside of thighs and legs of a yellowish flesh color. Black streak from the shoulder to the end of the nose, passing through the centre of the eyes. Caught in a dry situation on the upland near Fort Dalles. It is found at a considerable distance from water or trees. It is fond of clinging to rocks, but otherwise had the habits of a tree toad.—S.

Tribe II.—**Uradela.**—Tailed Batrachians.

TARICHA TOROSA, Gray.

The Warty Salamander.

Triton torosus, Esch. Zool. Atl. V, 1833, 12 Taf. xxi, fig. 15.

Triton ermani, Wiegm. in Erman's Reise um die Erde, 1835, and Arch. fur Naturg, 1836, II, 250.

Salamandra beecheyi, Gray; Beechey's voyage, Zool. 1839, 99.

Triton granulosus, Skilton, Amer. Journ. Sc. VII, 1849, 202; pl. figs. 4 and 5.

Notophthalmus torosus, Baird, Journ. Acad. Nat. Sc. Philad., I, 1850, p. 284.

Pleurodeles californiae, Bibr. Mus. Zool. Soc. (Gray.)

Taricha torosa, Gray, Catal. Amph. Brit. Mus. II, 1850, p. 25.—Girard, U. S. Expl. Exped. Herpetology, 1858, p. 5, figs. 1–8.

Sp. Ch.—Tail longer than the head and body together, compressed, provided with a slightly elevated membranous keel upon its upper and lower edges. Tip of toes callous; inner toes in both pair of limbs very small. Skin granular. Color, rich dark brown above, with minute pale dots; sides and beneath, orange or yellow; iris, gold color.

This salamander is usually found during the day crawling slowly about the woods, often at a considerable distance from the water. They are so slow and stupid as never to attempt to get out of the way, and seem, therefore, to be soon exterminated where hogs are allowed to range in the woods. I found them far most abundant in the damp spruce forests near the coast, but have also found them everywhere as far east as the summits of the Cascade mountains, where, at a height of over 4,000 feet above the sea, I obtained the first full grown specimen met with, on August 5, 1853. This one, as if to verify the popular fable of the salamander, was caught while leisurely crawling through the ashes of our camp fire, no doubt driven out of the ground by the heat. I have met with this animal crawling about in January and February, but in severe whether it hybernates in the mud at the bottom of ponds, where, if shallow, it sometimes is killed by freezing, as I have found a large number dead in this way. In March and April they seek the water to spawn, and their loud piping may then be heard at night, everywhere, though they are silent at other seasons. In May their spawn is found attached to twigs in the water, after which time they seek the land again. I found young ones about two inches long in July, which were entirely aquatic, and had external branchiae or gills, but in color were exactly like the adult animal. When in the water this is a very active animal, and has a greater expansion of the thin membrane of its tail than during its terrestrial life.

The spawn mentioned as deposited by this species in the water may have belonged to some other animal, as most salamanders are believed to deposit it on land, under stones, &c.; but I never found any in such situations.

NOTE.—Both Mr. Gibbs and myself found near Steilacoom a small kind of salamander under logs and stones, which seems to be undescribed. The specimens were, in some way, lost or destroyed. When alive, they were pale purplish, with yellow longitudinal stripes.—C.

Among the lost collections from the Dalles was a small salamander, found in January, 1855, which had the sides deep blue, with light markings; belly spotted with black. No such species is known to naturalists.—S.

SIREDON LICHENOIDES,? Baird.

"Ground Puppy;" "Four-legged Fish."

Siredon lichenoides,? BAIRD, Stansb. Salt Lake Exp. 1852, p. 336, pl. 1.

A specimen of siredon collected by Dr. Suckley, near Fort Union, Nebraska, differs from the typical *S. lichenoides*, in fuller form, and in the presence of distinct rounded black spots on a grayish brown ground, the spots larger and more distinct than in *S. mexicanus;* the eyes are more anterior. In the absence of tangible, distinctive characters, it is difficult to say if this animal be really distinct from *lichenoides;* but it may, at any rate, be considered as a well marked variety, *S. melanosticta*. The total length of the specimen, No. 4073, is about nine inches.—BAIRD.

This siredon was obtained on the route between Fort Union and Fort Benton, Nebraska. Vast numbers of a species apparently identical are found in the small brackish lakes of central Minnesota, where I procured some thirty or forty specimens, which were, unfortunately, afterwards lost on their way to Washington city. The species is, apparently, permanently aquatic; the only occasion upon which I noticed them out of their native ponds was when, during a terrific thunder shower, I found many in the wet grass within a few feet of a small prairie lake, to which they had retreated from the water, apparently from fright produced by the loud claps of thunder. I generally obtained them by dragging a seine through the small, shallow, muddy ponds, so numerous in that region. The contents of my net were generally confined to those animals, and vast numbers of aquatic insects.—S.

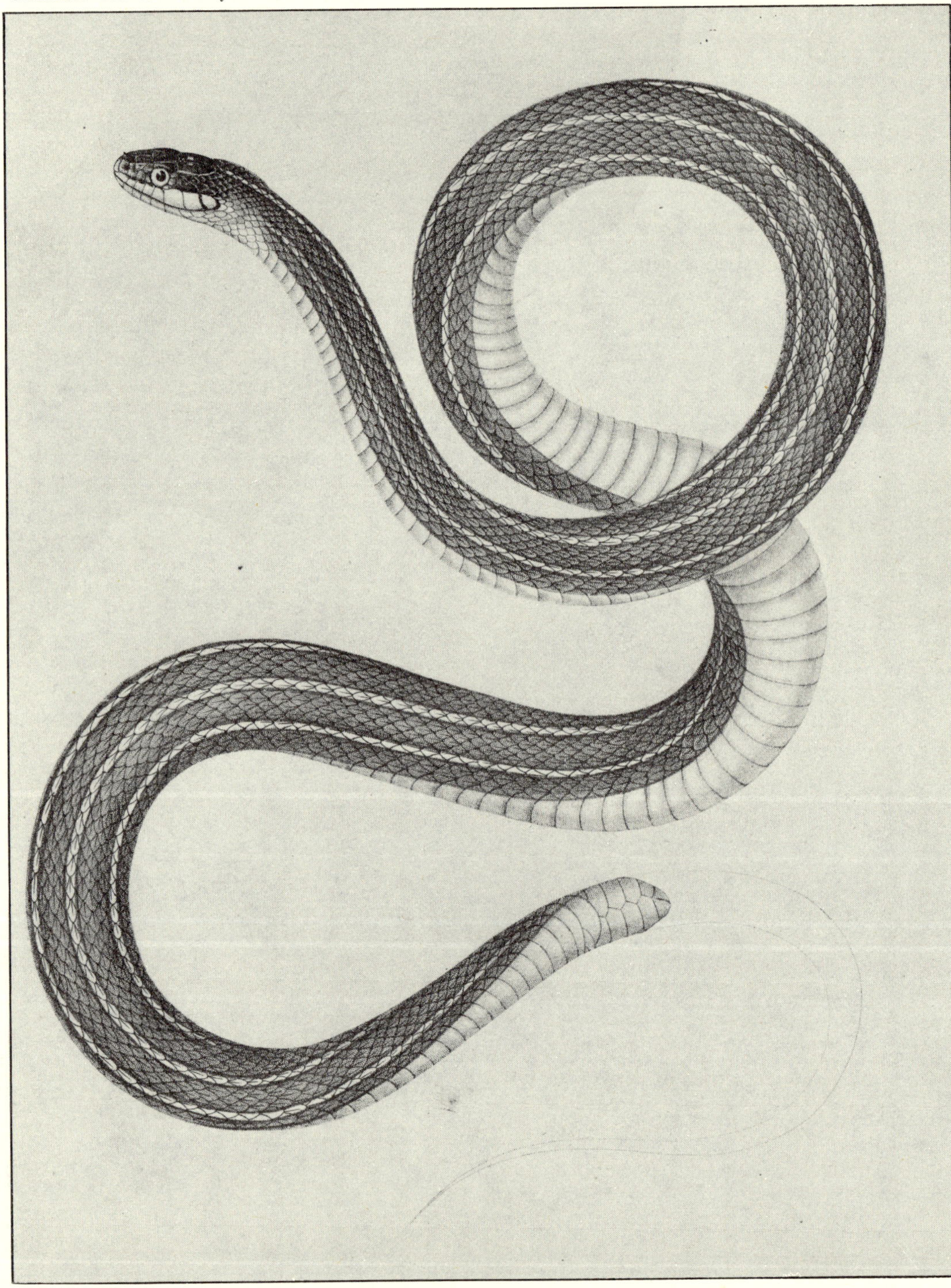

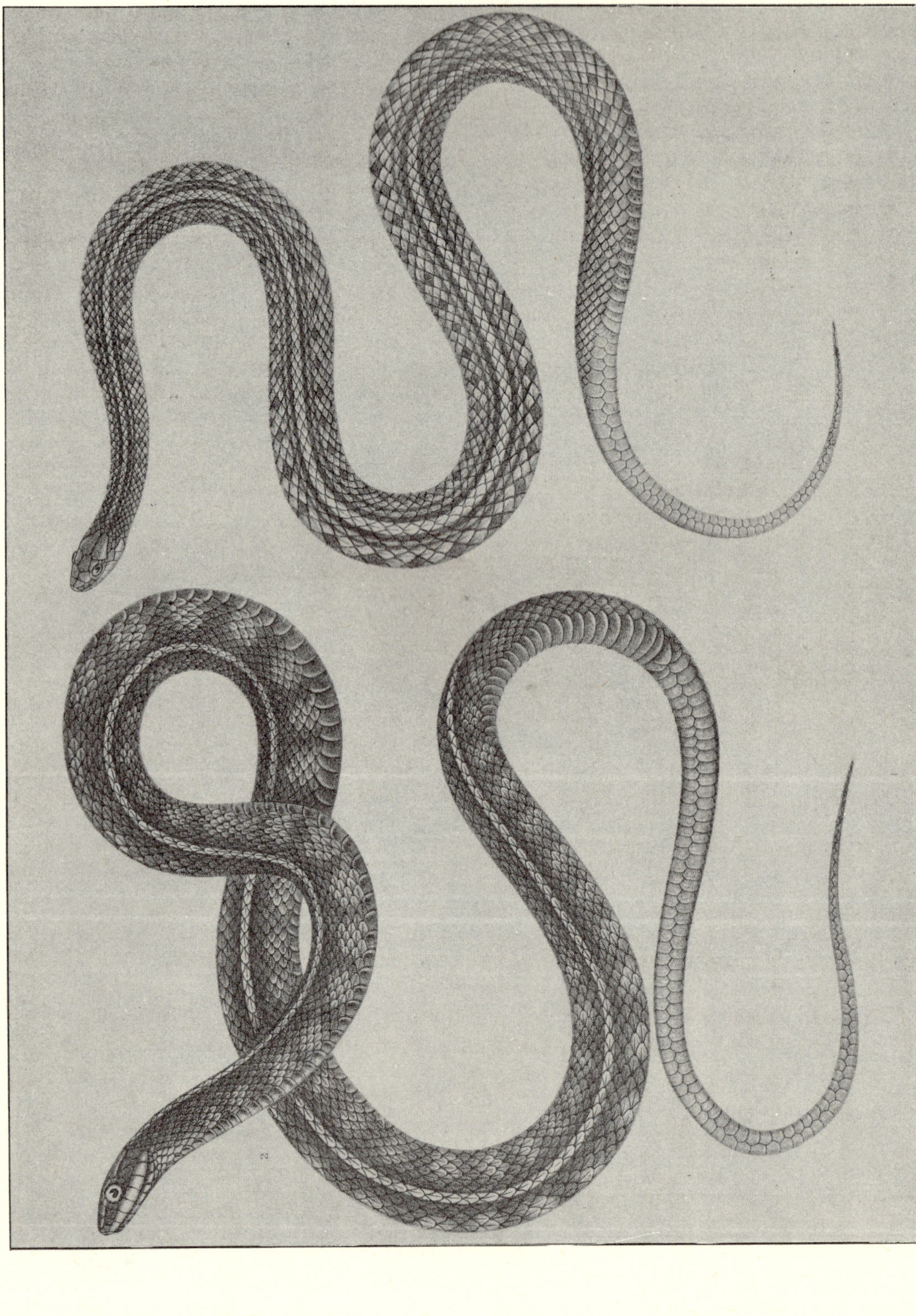

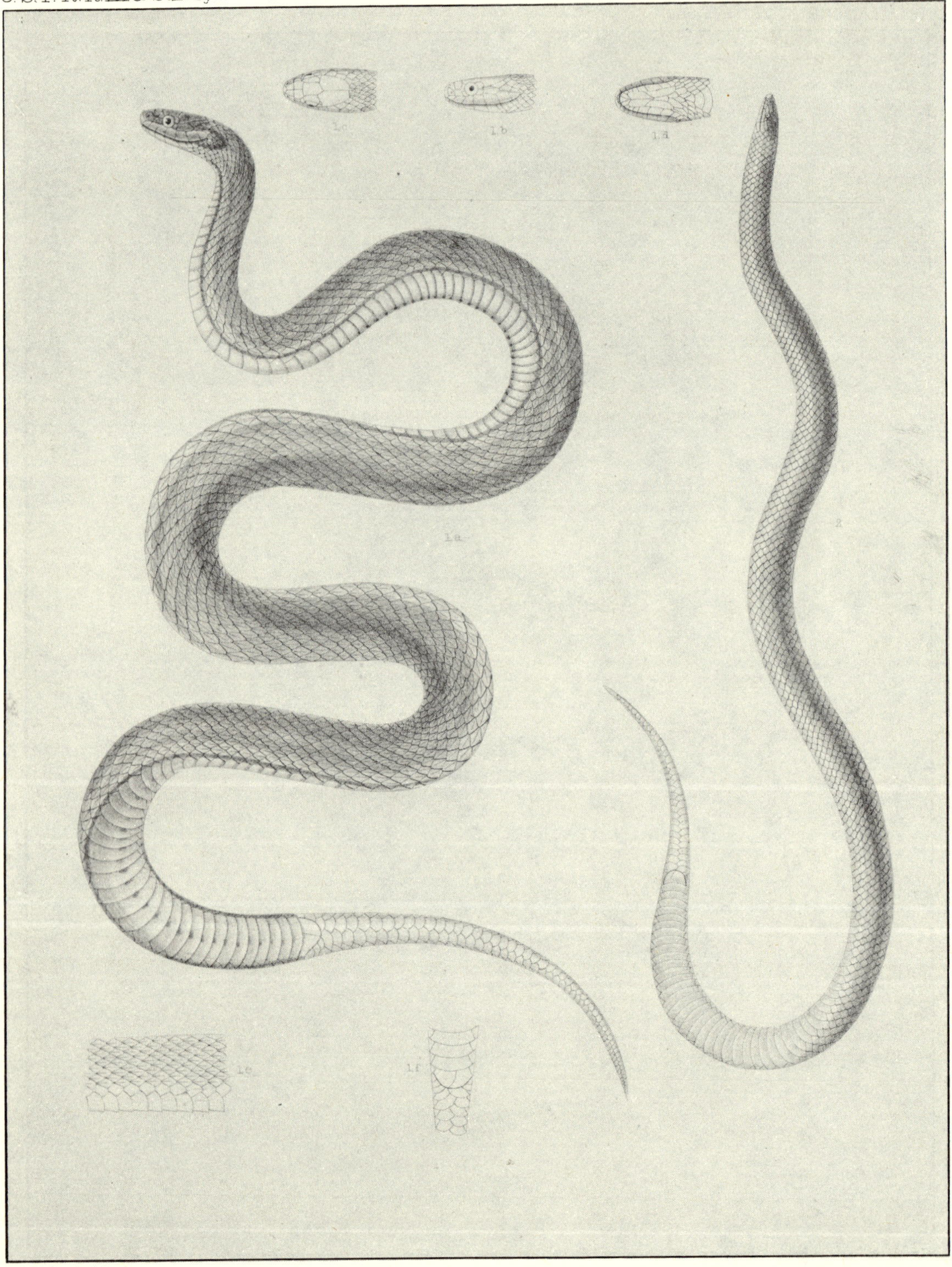

J. H. Richard. del.

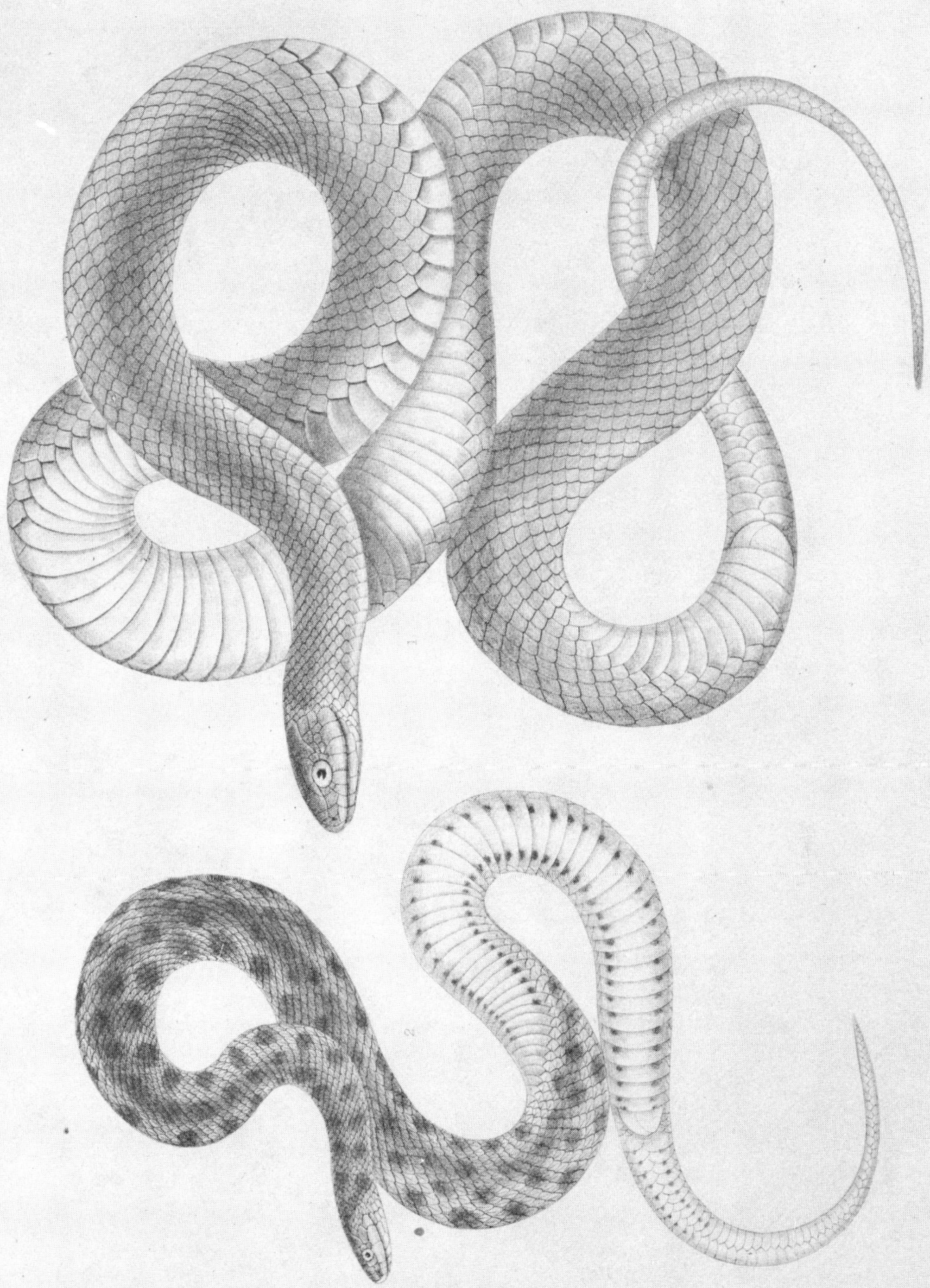

J. H. Richard del.

J. H. Richard del. C. Metzeroth sc.

J. H. Richard del. C. Metzeroth sc.

J. H. Richard. del. C. Metzeroth. sc.

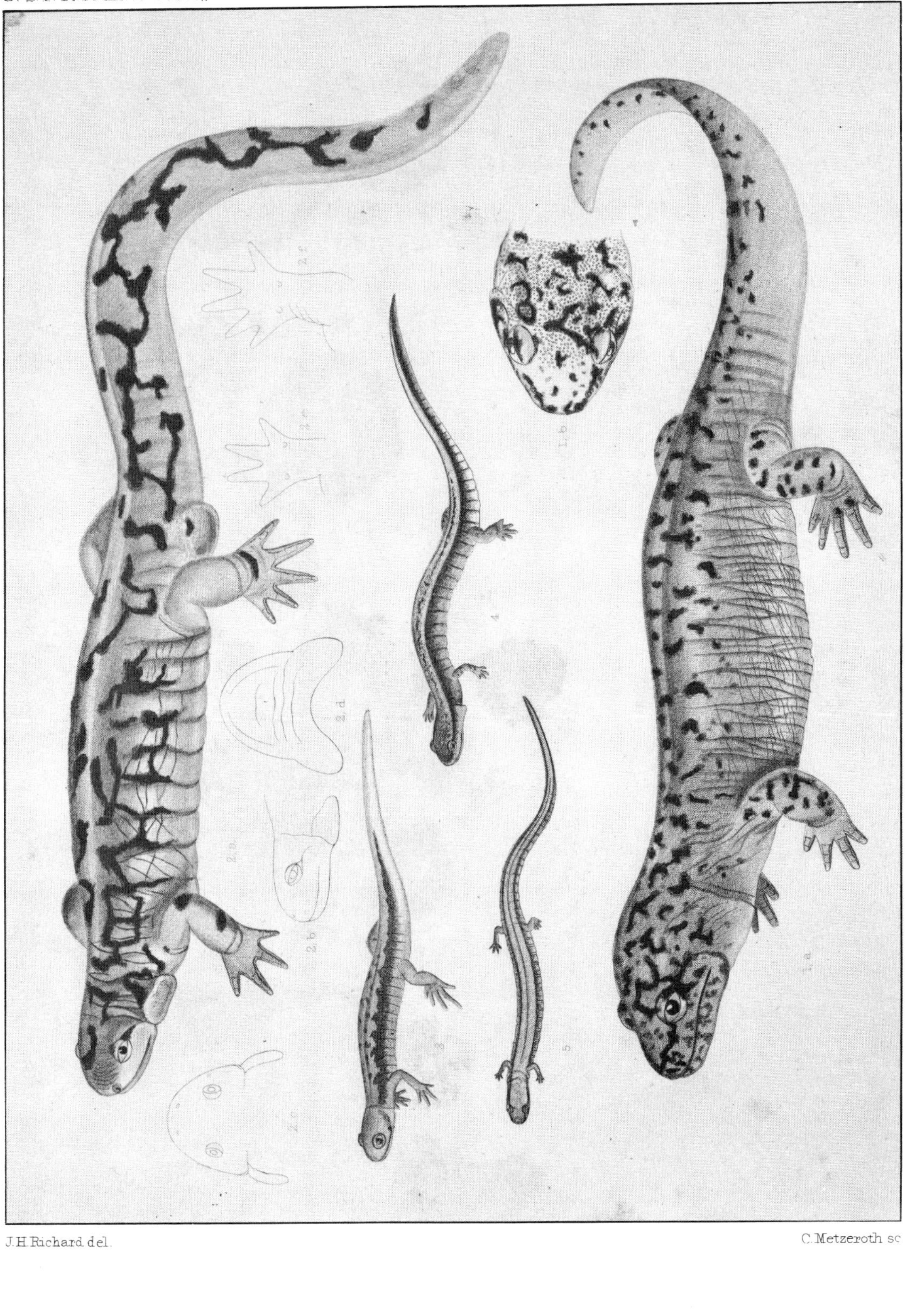

J. H. Richard del. C. Metzeroth sc.

REPORT ON THE RECENT REPTILES AND FISHES OF THE SURVEY COLLECTED BY CAMPBELL CARRINGTON AND C. M. DAWES

Edward D[rinker] Cope

REPORT ON THE RECENT REPTILES AND FISHES OF THE SURVEY, COLLECTED BY CAMPBELL CARRINGTON AND C. M. DAWES.

BY EDWARD D. COPE, A. M.

REPTILIA.

OPHIDIA.

CAUDISONA CONFLUENTA, Say; *var.* with transverse spots narrowed. Ogden, Utah.

EUTÆNIA VAGRANS, B. and G., (Catalogue, p. 35.)

Fish Creek, Montana; Yellowstone Basin; between Copenhagen, Utah, and Fort Hall, Idaho; Fort Hall, Idaho; Salt Lake City; Ogden, Utah.

Var. a. With colors like *E. sirtalis;* the sides olive, with about fifty pairs of black spots, the vertebral band yellow, black-bordered. From Camp Carling.

EUTÆNIA PARIETALIS, Say.

Salt Lake City; lake ten miles east of Logan, Utah, (salt;) Fish Creek, Montana.

PITYOPHIS BELLONA, B. and G.

One specimen without the anterior frontal (vertical) shield; from Ogden, Utah.

BASCANIUM FLAVIVENTRIS, Say, (B. and G., Catalogue, p. 96.)

Ogden, Utah.

LACERTILIA.

PHRYNOSOMA DOUGLASSII, Bell.

Var. α.—The usual form: Salt Lake City.

Var. β, exilis.—A small form not more than one-half or two-thirds the usual size, but nearly identical in details of structure and coloration.

The differences observable are: the rather shorter muzzle, which is entirely vertical in profile; the smaller scale above the canthus of the mouth, and the temporal horns; the less prominence of the posterior superciliary angle, and the much reduced size. A geographical variety. Carrington's Lake, Montana; Fort Hall, Idaho.

SCELOPORUS CONSOBRINUS, B. and G.

This species is very abundant and variable. Its varieties are four, as follows:

Var. 1.—Typical; scales large, especially on the sides; crural cross-series, 10–11; rows between interscapular and crural points, 33; 2 prefrontals on each side; lateral and dorsal spots distinct. Localities, Yellowstone Basin; Blackfoot Fork.

Var. 2.—Like the last, but the scales smaller on sides and back; 13 rows on rump. Salt Lake City, Utah; south of Fort Hall, Idaho.

Var. 3.—Scales still smaller; 16–17 interscapular, 14 crural cross-rows; 3 prefrontal plates on each side. Dorsal spots large, their borders touching the lateral spots; both pale-edged behind, forming an angular border in ♀. This form grades into the last. A male has the border color of variety 2. South of Fort Hall, Idaho.

Var. 4.—Like variety 2, but only half the size. South of Fort Hall, Idaho, and Salt Lake City.

SCELOPORUS GRACIOSUS, B. and G.; *S. gracilis*, B. and G.

This species is very near the last, but the scales are still smaller. There are 38 transverse dorsal rows and 20 interscapular. The lateral scales are twice emarginate. From Salt Lake, Utah, to Oregon.

CNEMIDOPHORUS TESSELLATUS, Baird; *Amiva tessellata*, Say, (*vide* Pac. R. R. Surv., vol. X, Beckwith's Report.)

Salt Lake, Utah.

TESTUDINATA.

CHRYSEMYS OREGONENSIS, Harlan; Agass.

The Yellowstone Lake.

BATRACHIA.

ANURA.

BUFO COLUMBIENSIS, B. and G., (United States Exploring Expedition, Herpetology, by Girard, p. 77.)

Pleasant Valley and Yellowstone Basin. Specimen from latter locality like types; that from Montana different in coloration. It is a bright green, with numerous blackish speckles on upper surfaces, and brown interscapular spots; below uniform.

CHOROPHILUS TRISERIATUS, Wied; *Helœcetes*, Baird.

Carrington's Lake, Yellowstone Basin.

SPEA BOMBIFRONS, Cope.

Blackfoot Fork.

RANA HALECINA, Bosc.

Fort Hall, Idaho.

RANA PRETIOSA, B. and G., (United States Exploring Expedition; Herpetology, p. 20.)

Pleasant Valley, Montana.

This frog is a near ally of the European *R. temporaria*, and is, as Girard remarks, distinguished from the other west-coast species (*R. aurora*, B. and G.) by its much shorter limbs. The present species was originally found at Puget's Sound, Washington Territory.

RANA SEPTENTRIONALIS, Baird, (Proc. Acad. Nat. Sci., Phila., 1854, 61.)

Abundant; Carrington's Lake, Yellowstone Basin, and Fish Creek, Montana.

PISCES.

ISOSPONDYLI.

COREGONUS WILLIAMSONII, Girard; the Rocky Mountain white fish.

THYMALLUS TRICOLOR, Cope, (Proc. Acad. Nat. Sci., Phila., 1865, p. 80; Gunther, Catalogue Brit. Mus., VI, 201.)

Specimens from Yellow Creek and the Gallatin Fork of the Missouri in Montana. This species was originally discovered in the Grand River, Michigan. It seems to be a rare fish east of the Mississippi; but my friend, J. Dickinson Sergeant, informs me that it has been found abundantly in a stream in the northern part of the peninsula of Michigan. The number of specimens brought by Dr. Hayden from the head-waters of the Yellowstone indicates that this region is its home. They maintain well the characters by which it was originally distinguished from the *T. vulgaris* and *T. signifer*. The muzzle is shorter and the gape of the mouth larger than in the former; the maxillary bone is narrower and longer, reaching to below the middle of the pupil instead of to near the front of the orbit. The length of the head equals the depth of the body and enters the length without caudal fin, 4.5 times. There is some variation in the radial formula as follows: D. 20–22; A. 13–14. Scales, 8–9—86–90—10–12. In the younger specimens the small blue spots tend to form short longitudinal bars.

SALMO, Linn.

The species of this genus, found in the streams rising in the Rocky Mountains, are numerous, and, as elsewhere, nearly allied. Those I have observed in Dr. Hayden's collections number three, while a fourth is described by Dr. C. Girard, which I have not met with. The allied species differ as follows. They all belong to the group *Salar:*

Depth, 5.75 in length; eye, 4.5 times in head; snout obtuse; caudal fin scarcely emarginate; Br. IX................*S. virginalis*.

Depth, 4.75 in total, (to point caudal;) eye, 5 times in head; muzzle acute; scales larger, 26 below dorsal fin; cranium not keeled above; head one-fourth length; dorsal fin nearer muzzle than end caudal scales; caudal fin scarcely emarginate; Br. X .... *S. spilurus.*

Head large, broad, flat, not keeled, 4.25 in total, equal depth of body; muzzle obtuse; eye nearly 5 times in head; scales, 42 below dorsal first ray; dorsal fin equidistant; caudal fin not notched .... *S. stomias.*

Head smaller, 4 times in length to notch of caudal, (which is well emarginate;) upper surface keeled; muzzle obtuse; eye 4 times in length; depth, 4.5 in length, to end caudal scales; dorsal midway between latter and end of muzzle; scales small, 40–43 below dorsal first ray; Br. XI .... *S. pleuriticus.*

Head acuminate, keeled above, 4.66 times in length to notch of caudal fin, which is well marked; eye, one-fifth head; depth, 5.23 to caudal notch; dorsal nearer muzzle than end of caudal scales; scales large, 33 below dorsal first ray; Br. XII; spots large, distant .... *S. carinatus.*

Head one-fourth total length; eye, 5 times in head; dorsal fin equidistant between insertion of caudal and end of muzzle; muzzle rather pointed; Br. X, XI .... *S. irideus.*

Of the above species *Salmo spilurus* and *S. carinatus* are distinguishable by their large scales and smaller orbits; while in *S. stomias* and *S. pleuriticus* the scales are very small. On the other hand, *S. carinatus* and *S. pleuriticus* agree in the strong median carina on the superior aspect of the cranium. *S. stomias* is readily separated by the large head and mouth. Its habitat, so far as known, is the Kansas River, far to the eastward of the Rocky Mountains.*

SALMO SPILURUS, Cope, *sp. nov.*

This species is represented by six specimens from the Sangre de Christo Pass, in Colorado, from one of the sources of the Rio Grande.

It is rather a fusiform fish, with small head and acuminate muzzle, which is very little decurved at the end to the lip margin. The eye enters the muzzle 1.33 times, the inner border of the adipose eyelid being regarded as the dividing line. The top of the head is slightly convex, but entirely without keel. The maxillary bone extends to a little beyond the posterior margin of the orbit, and is flat and considerably wider distally than it is proximally. In this it differs from the *S. stomias*, Cope, where that bone maintains an almost equal width throughout. All the teeth are well developed, including both rows of vomerines. Scales in about 33 rows between the base of the first dorsal ray and the lateral line, or 26 rows between the middle of the dorsal fin and the same.

Radii, Br. X; D. II. 11, (10 in one;) A. II. 10, (in two, 11.)

Ground color, pale in spirits; the caudal peduncle from the middle of the anal fin, with the caudal and dorsal fins, thickly spotted with large, irregularly disposed black spots. Those on the caudal peduncle are darkest between the scales; each one having, therefore, a reticulated appearance. Above the lateral line they extend to the dorsal fin, continually contracting their distribution from the lateral line upward.

*In Hayden's Report, Geology of Wyoming, 1871, p. 433, this is erroneously stated to be the Platte, a very different river.

A few scattered spots are found all the way to the head, and four or five mark the side of the latter. Adipose fin spotted; others unicolor.

The largest specimen obtained measures 13 inches in length.

The affinities of this species to the *S. virginalis* appeared to be close, and Girard cites specimens from the locality from which it was procured as belonging to the latter. Its description cannot, however, be reconciled with the *S. spilurus*, especially in the relations of the depth to the length, by which it would appear that *S. virginalis* is a much more slender fish. The figure also agrees with the description.

SALMO PLEURITICUS, Cope, *sp. nov.*; *Salmo (Salar) virginalis*, Cope, not Girard, (Hayden's Survey Wyoming, 1871, 433.)

This is the abundant mountain trout of the head-waters of the Green and Platte Rivers, and even of the Yellowstone. It is rather a stout species, with obtusely descending muzzle, and large eye entering the head only four times. The cranial keel is a marked character; its elevation is greater between the orbits than on the posterior part of the frontal bones. The interorbital width is 1.33 times the long diameter of the interpalpebral opening of the eye The dorsal fin is nearer the origin of the marginal rays of the caudal fin than to the end of the muzzle, but is midway between the latter and the termination of the scales on the sides of the fin. Radii, D. II. 11–12 and 13; A. II. 11. Br. XI. The scales range from 40 to 45 below the first dorsal ray to the lateral line. The maxillary bone extends to a little beyond the orbit, and is not expanded.

This is a spotted species, and the spots are chiefly found above the lateral line and on the whole caudal peduncle, and on the dorsal and caudal fins. They are usually rather scattered, less numerous on the peduncle than in *S. spilurus*, and more so anteriorly; those on the fins are smaller and less numerous. There is, however, variation in the size and number of the spots. The sides are ornamented with short, broad longitudinal bars of crimson; a band of the same color occupies the fissure within each ramus of the mandible and skin on the median side of it. The fins are all more or less crimson; but none of these are black-bordered. The largest specimens are 10–12 inches long.

Seven specimens of this species are in the collections from the heads of Green River; from Medicine Lodge Creek, Idaho, (two specimens;) four from the Junction, Montana. A specimen each from Yellow Creek and the Gallatin Fork of the Missouri, Montana, represent at least a color variety of this fish. The spots are much smaller and much more numerous, though distributed over the same regions; they are less numerous on the caudal fin. In the Gallatin specimen there are 51 scales above the lateral line; in the other 44. Another variety from the Yellowstone Basin is only represented by young specimens. They have no spots on the caudal fin.

A number of dried specimens from the Yellowstone Lake, of larger size than the specimens above described, probably belong to this species. They are rather more closely spotted on the caudal peduncle and fin, but are similar in all important respects. The only discrepancy which I find is the relatively smaller eye, (not orbit,) which enters the head five times, and the greater prolongation of the maxillary bone. These characters are, perhaps, due to the larger size attained by the individuals. They are from a foot to eighteen inches in length.

SALMO CARINATUS, Cope, *sp. nov.*

The characters of this species, pointed out in the table, show its marked distinction. It is a more slender fish than any here described, approaching more nearly the proportions assigned by Girard to the *S. virginalis*. From this species the numerous branchiostegals, more distinctly forked tail, &c., distinguish it.

The head is elongate, but not wide, and the muzzle descends regularly, but not abruptly, to the lip margin. The eye is contained in it 1.5 times and enters the head five times in a specimen of the size of those of *S. pleuriticus*, where it enters four times, as well as in larger animals. The interorbital region contracts at the front part of the orbits so as to be little wider than the long diameter of the latter. The frontal keel extends the whole length of the vertex, and is very conspicuous; it is most elevated posteriorly. The maxillary extends to beyond the orbit. The dorsal fin is equidistant between the end of the muzzle and the base of the marginal radii of the caudal fin, therefore nearer the former than to the termination of the lateral caudal scales.

Radii, Br. XII, XI; D. II–11; A. II–11.

The ground is light, perhaps rosy in life, and is marked with round, black spots, sparsely but equidistantly distributed over the whole body. The dorsal and caudal fins are spotted, but rather sparsely; those of the former being arranged in two or more longitudinal series. There are indications that the fins and sides of the head were crimson, and that there were large spots of the same color on the middle of the sides.

Two specimens of uncertain locality; fragments of, perhaps, a third from the Yellowstone Geyser Basin.

Another species of trout was obtained from Carrington's Lake, Montana, but the specimens are in too bad a state for determination.

PLECTOSPONDYLI.

SEMOTILUS CORPORALIS, Mitch.

Crow Creek, at Camp Carling.

CERATICHTHYS NUBILUS, G.; *Argyreus nubilus*, G., (U. S. Pac. R. R. Surv., X, p. 244.)

This species has the physiognomy of the genus to which Girard referred it, but the premaxillary is projectile and the upper lip separated from the muzzle by a fold. This is the only point of distinction yet stated by which the two genera may be separated. Teeth, 4.2–2.4; barbels inconspicuous. Axils of the fins crimson.

Grass Creek, Idaho. Collection No. 4.

APOCOPE, Cope, *gen. nov.*

Teeth of the raptorial type, 4.1–1.4, without grinding surface. Barbels present; upper lip separated from muzzle by a fold. Anterior part of the lateral line only present.

APOCOPE CARRINGTONII, Cope, *sp. nov.*

This is a small species allied to the last, but the muzzle is broader and less prominent, and the mouth larger. The muzzle is quite obtuse in profile and overhangs the mouth very little, and the end of the maxillary bone does not quite reach the line of the margin of the orbit. Barbels minute; teeth, 4.1–1.4; isthmus wide; eye a little smaller than one-fourth the length of the head, and 1.5 times in interorbital width. Scales, 10–60–11. Dorsal fin originating behind the point above the ventrals, and markedly nearer the basis of the caudal than the end of the muzzle. Caudal well forked. Radii, D. 8; A. 7; length of head a little more than four times in length to basis of caudal; depth five times in the same; length, 20 lines.

Color olivaceous, with a dark lateral band from end of muzzle, and dark shades on the back.

Four specimens from the Warm Springs, Utah. The species is named in honor of Campbell Carrington, zoologist of Dr. Hayden's expedition, to whose zeal in the cause of science we are indebted for the materials analyzed in this report, and that on the same subject in the Survey of Wyoming, 1871. Collection No. 9.

APOCOPE VULNERATA, Cope, *sp. nov.*

The head is broad and the muzzle wide and obtuse, not projecting over the mouth. Barbels minute. The end of the maxillary bone does not quite reach the marginal line of the orbit; orbit five times in head, one and a half times in the interorbital space. Length of head one-fourth, depth of body one-fifth length without caudal fin. Dorsal a little behind line of origin of ventrals, nearer caudal fin than muzzle. Radii, D. 8; A. 7. Length, 24 lines. Scales small, 15–72–10.

Color olive, with a broad dark band from end of muzzle to caudal fin, paler above and below it; belly yellow; a crimson spot on the chin.

This species differs from *A. carringtonii* in the wider muzzle, smaller scales, and greater development of the lateral line. In this species it is continued to the end of the anal fin; in the type of the genus it scarcely extends to the dorsal. Collection No. 1.

From Logan, Utah; discovered by Cam. Carrington.

RHINICHTHYS MAXILLOSUS, Cope, (Proc. Acad. Nat. Sci., Phila., 1864, 278.)

Crow Creek; Grass Creek, Idaho.

TIGOMA RHINICHTHYOIDES, Cope, *sp. nov.*

Size small; form elongate. Head 4.3 times in length without caudal fin; depth, 5.2 times in same. Muzzle obtuse, not projecting; mouth inferior horizontal, maxillary not reaching line of orbit. Head nearly four times longer than diameter of orbit, which equals length of muzzle, and is 1.5 times interorbital space. Teeth, 4.2–2.4. Scales, 12–67–12, lateral line with occasional interruptions at the posterior part. Dorsal fin behind line of ventrals, nearer basis of caudal fin than end of muzzle. Length, 24 lines.

Olive-brown above, yellow below; a broad brown lateral band and longitudinal blackish line on the thoracic region on each side. The median band is darker spotted, and there are blackish spots on the dorsal region. Head black above; chin red.

The upper lip of this species is separated by a fold, but the species has a general resemblance to those of *Rhinichthys*.

From Logan, Utah. Collection No. 2.

PROTOPORUS, Cope, *gen. nov.*

Teeth raptorial; outer row, 4–4; no grinding surface; no barbels; upper lip separated by a fold. Lateral line represented only by a short anterior series of pores.

This genus is related to *Tigoma*, as *Apocope* is to *Ceratichthys*.

PROTOPORUS DOMNINUS, Cope, *sp. nov.*

Form rather stout, front convex, muzzle obtuse, mouth horizontal, end of maxillary not quite attaining orbit. Scales, 9–56–6. Lateral line not attaining the point opposite to the origin of the ventral fins. Length of head, 4.33; depth of body five times into length without caudal fin; diameter of eye four times in head; once in muzzle, 1.3 times in interorbital width. Fin radii, D. I. 8; A. I. 7. Rays of pectoral thickened, not reaching ventral; ventral reaching anal. Length, 24 lines: teeth, 4.2–1.4, somewhat hooked.

Color silver-gray, with a broad dark lateral band, which is darker spotted; back and sides of thorax dark shaded; top of head dark.

Numerous specimens from Fort Hall, Idaho, collection No. 3.

HYBOPSIS BIVITTATUS, Cope, *sp. nov.*

Muzzle narrow, very obtusely descending, not projecting; mouth horizontal, end of maxilla reaching the line of the orbit. Orbit four times in length of head, 1.5 times in the rather flat interorbital width. Length of head, 4.66 times; depth of body four times in length minus caudal fin. Dorsal region arched; basis of dorsal fin descending posteriorly, and originating very little behind origin of ventrals. Pectorals only reaching half-way to ventrals; latter about half-way to anal. Radii, D. I. 8; A. I. 7. Teeth, 4.2–2.4, with grinding surface. Scales, 12–53–11. Lateral line complete. Dorsal fin nearer muzzle than origin; caudal fin midway between former and notch of latter. Length, 3 inches.

Color silvery; the scales blackish punctulate; dorsal region blackish; a dark band from epiclavicular region to basis of caudal fin; another from end muzzle, across operculum, to basis of anal.

From Warm Springs, Utah; Campbell Carrington, collector. Collection No. 10.

It is interesting to note that in the distribution of color, especially in the two lateral bands, this species is identical with *Protoporus domninus*, *Tigoma rhinichthyoides*, and *Apocope carringtonii*, a well-marked case of mimetic analogy. Another case of this kind is exhibited by two species of eels of different genera from Costa Rica. The species are undescribed.

MURÆNA AQUÆDULCIS, Cope, *sp. nov.*

Branchial fissure small; posterior nostril not tubular, situated halfway between eye and anterior nostril. Eye contained nearly twice in the muzzle. Maxillary teeth in two rows, ethmoids in a single one. Former, 4 long in the inner, 17 in the outer row, counting from the front margin of the orbit. Vomerine teeth well developed; dentaries in two rows.

Form stout; dorsal fin extending two-thirds the total length, the anal 2.33 times in the length. Color brownish-black, with rather distant yellow spots, which are accompanied by a darker shade on same side, and which become confluent into yellow marblings on the gular and pectoral regions.

From the Rio Grande, near the city of San José, Costa Rica, Central America.

PŒCILOPHIS NOCTURNUS, Cope, *sp. nov.*

Branchial fissure small; posterior nostril just above the orbit at the anterior margin. Muzzle rather obtuse, twice as long as the small eye. Maxillary teeth in a single row; the vomerines obtuse, continued in line to the ethmoids. Latter in three rows, with interspaces. Dentaries one-rowed. Form moderately stout; dorsal fin extending nearly to nape. Anal fin a little over one-third total length.

Color deep purplish-brown, with rather scattered, irregular, yellow spots with black margins; rather paler below.

From the Rio Grande, at San José, Costa Rica. Dr. Van Patten.

This species, in general proportions, is a little more slender than the last, but in general appearance, as size and coloration, can scarcely be distinguished from it. It is a curious case of mimetic analogy.

SIBOMA ATRARIA, G., (U. S. Pac. R. R., Rep. X, 297.)

Abundant in Grass Creek, Idaho. *Siboma* differs from *Clinostomus* in the anterior position of the dorsal fin. Collection No. 7.

MYLOLEUCUS, Cope, *gen. nov.*

Teeth raptorial, but with well-developed masticatory surface. 5–4 in outer row. No barbels; lateral line well developed. Dorsal fin above or in front of line of ventrals.

This genus is *Siboma*, with developed grinding surfaces of the teeth.

MYLOLEUCUS PULVERULENTUS, Cope, *sp. nov.*

Form stout; head short, muzzle not decurved; mouth terminal, slightly descending, the maxillary bone nearly attaining the anterior line of the orbit. Head 3.75 times in length, exclusive of caudal; depth three and a half times. Eye 4.2 times in head, once in muzzle. Preorbital bone deeper than long. Scales, 13–58–9. Radii, D. I. 9; A. II. 7. Caudal well forked. Length, 3½ inches.

A dark plumbeous band extends from the origin of the lateral line above to the caudal peduncle, and on it to the caudal fin. Below this the color is silvery, thickly dusted with black dots; above it is an olive-yellow band; then a dark dorsal region, all dusted. Sides of head silvery, dusted. Fins unspotted.

Numerous specimens from the Warm Springs, Utah; type No. 6.

CLINOSTOMUS PANDORA Cope, *sp. nov.*

This species is nearly a *Siboma*, in consequence of the position of the dorsal fin being so little removed from the vertical above the ventrals. Dentition refers it to *Clinostomus*, while its appearance is that of a *Ceratichthys* or *Semotilus*. I adhere to the technical characters until others are found which will give us the clew to the truer affinities.

Shape fusiform, head small, one-fifth of total length, (or fourth without caudal fin,) broad, and rather flat at the muzzle. The latter does not project, and the lips are equal. Maxilla not attaining the line of the orbit by some distance. Eye small, 6.25 times in length of head, twice interorbital width. Scales, 17–61–9, small in front of dorsal fin. Radii, D. I. 8; A. II. 8. Isthmus moderately wide; depth of body 4.2 times in length without caudal. Dorsal fin nearer basis caudal than end of muzzle.

General color silvery, above (in spirits) brownish. A broad, ill-defined, lateral band, from epiclavicular region to basis of caudal fin, above the lateral line. Length of type specimen, 8 inches. Collection No. 11.

From Sangre de Christo Pass, from a tributary of the Rio Grande.

CLINOSTOMUS HYDROPHLOX, Cope, *sp. nov.*

This species and the following are typical forms of the genus, and interesting as the first that have been detected west of the Mississippi River. Length of head 4.75 times in total, exclusive of caudal fin; depth of body 4.5 times in same. Eye 5 times in head, one and a half times in interorbital width. Front straight; lower jaw projecting beyond upper; mouth descending; end of maxillary just reaching line of orbit. Isthmus narrow. Teeth, 5.2–2.4. Scales, 15–58–7. Radii, D. I. 8; A. I. 11. Ventrals not reaching anal. Length, 6 inches.

Color above olive, with a blackish inferior border, extending from the superior margin of the orbit. Below this, a crimson band, and still lower, a blackish band, passing from the epiclavicular region above the lateral line to the basis of the caudal fin. Below this, crimson in front, silvery behind. Fins unspotted. Suborbital bones crimson; cheek golden.

Blackfoot Creek, Idaho. Collection No. 13.

CLINOSTOMUS MONTANUS, Cope, *sp. nov.*

Muzzle decurved, obtuse; jaws equal; end of maxillary extending beyond margin of orbit. Orbit large, entering the head 3.5 times and the interorbital region once. Length of head, one-fourth length to caudal fin; depth nearly equal. Scales, 11–12—56—6. Radii, D. I. 9; A. II–12. Length, 3.5 inches.

Olive above, a dark band extending from epiclavicular region above caudal line to caudal fin. Sides crimson as high as the lateral line.

Numerous specimens (No. 8) from Grass Creek, Idaho. This brightly colored species differs from the last in the obtuse muzzle, large eye, and smaller scales above the lateral line.

CATOSTOMUS GENEROSUS, G., (Proc. Acad. Nat. Sci. Phila., 1856, 174;)
Acomus generosus, G., (U. S. Pac. R. R. Surv., X, p. 221.)

This species is closely allied to the eastern *C. teres*, Mitch.: from Utah Lake.

PERCOMORPHI.

URANIDEA PUNCTULATA, Gill, (Proc. Boston Soc. N. H., 1862, p. 41.)
Gallatin Fork of the Missouri and Warm Springs, Utah.

COTTOPSIS SEMISCABER, Cope, *sp. nov.*

Radii, D. VII–18; A. 13; V. I–4; first ray of anal below third of second dorsal. Skin prickly above the lateral line, smooth below it posteriorly. Body compressed, profile rising rather steeply to the basis of first dorsal fin. Eye 4.5 times in head, .75 time in interorbital space. Muzzle contracted, maxillary bone reaching to below middle of pupil. Two spines on preoperculum. On an inferior anterior angle of operculum. Lateral line discontinued on last fourth of caudal peduncle. Head one-third length without caudal fin.

Below yellow; dorsal line with a series of dark spots; sides with large, dark clouds.

Three specimens from Fort Hall, Idaho.

NOTES
ON THE GEOGRAPHICAL DISTRIBUTION OF BATRACHIA AND REPTILIA IN WESTERN NORTH AMERICA

E[dward] D[rinker] Cope

NOTES ON THE GEOGRAPHICAL DISTRIBUTION OF BATRACHIA AND REPTILIA IN WESTERN NORTH AMERICA.

BY E. D. COPE.

The following notes are based on collections made by myself and my assistants at various points in the Rocky Mountain and Pacific regions during the last ten years. They describe the range of various species of our terrestrial cold-blooded vertebrata, and contribute to the final definition of the zoölogical provinces and districts of the continent.

1. Lake Valley, New Mexico.

This locality is at the western border of Doñana County, twenty miles N. E. of Fort Cummings. It is in the foot-hills of the Mimbres or Negretta range. The region is rather arid, springs not being numerous; but during July and August there are frequent rains. Vegetation is abundant in the form of grass and herbaceous plants and shrubs.

Scaphiopus sp. Young.

Rana halecina Kalm.

Phrynosoma cornutum Harl.

Full of eggs in June.

Phrynosoma modestum Gird.

Very abundant in August; not seen during two days' visit in June.

Crotaphytus collaris Say. Abundant.

Uta schotti Baird.

There is but one median smaller row of dorsal scales, so that the single specimen approaches the *U. nigricauda*. Specimens of this genus are abundant.

Sceloporus

A large species seen.

Sceloporus.

A small species seen.

Holbrookia maculata B. and G. Abundant.

Var. *flavilenta*, differs from the typical form in having larger prenasal scales separated by only two flat scales in front instead of four tubercular ones, and in having only four flat scales between the nostrils above instead of six tubercular ones, and in having the scales of the front flatter. The spots are obscure or entirely wanting; when present they are more numerous than in the var. *maculata*, there being eight between axilla and groin instead of six. The sides and dorso-lateral regions are thickly marked with small yellow spots. Two specimens.

Cnemidophorus sexlineatus L.

Very abundant.

Stenostoma dulce Bd. and Gird.

One specimen.

Bascanium testaceum Say.

Eutænia cyrtopsis Kenn.

Eutænia ornata B. and G.

Crotalus scutulatus Kenn. Not rare.

Crotalus confluentus Say, var. **pulverulentus** Cope.

I propose this name for a well-marked variety of rattlesnake, which is abundant in the region of Lake Valley, especially on the grassy plains. In order to determine its relations to the species to which I refer it, I instituted a comparison with the allied forms represented in my collection. These are: Two specimens from Fort Benton, Montana; two from Central Oregon; two from Eastern California; one from Socorro, New Mexico; one from Fort Wingate, New Mexico; two from Lake Valley, New Mexico, and one from Haskell County, Texas. These represent a wide range in latitude, and are likely to give the greatest range of variation. The comparison indicates three varieties, defined as follows:—

Cephalic scales larger; four rows between superciliary plates; four rows below orbit; dorsal spots and cephalic bands light-edged; few posterior cross-bands; *confluentus*.

Cephalic scales intermediate; six rows between superciliaries; three rows below orbit (probably sometimes four); dorsal spots square, with the head-bands, not light-edged; posterior cross-bands more numerous; colors dotted with brown specks; *pulverulentus*.

Cephalic scales smallest; eight rows between superciliaries; four rows below orbit; dorsal spots and head-bands light-edged or not; numerous posterior cross-bands; *lucifer*.

The var. *pulverulentus*, at first sight, resembles the *Crotalus mitchilli*, having much the same coloration, but the head-scales and plates are quite different. It gives out a powerful musky odor when excited, which I have not noticed in the typical form of the species. It is quite probable that it is to this variety that the specimens from Arizona should be referred, which I have heretofore placed under *C. lucifer*.[1] Not having access to the specimens at this time, I cannot determine this point positively. Of those above enumerated, the specimens from Fort Benton, Fort Wingate, from Socorro, and from Texas, belong to the typical *C. confluentus*. The others are the *C. c. lucifer*.

Crotalus molossus B. and G.

I killed a fine specimen of this species, which I discovered in the act of springing through a bush. When I struck it, it was suspended over a branch, looking at me. It was heavy in its movements, except at the moment of leaping.

2. Socorro, New Mexico.

The collection from this region was made by Prof. Frank Snow, of the University of Kansas, at Lawrence. I here express my indebtedness to Prof. Snow for the opportunity of studying it.

Phrynosoma modestum Girard.

Phrynosoma cornutum Harl.

Phrynosoma douglassi Bell.

Crotaphytus collaris Say.

Holbrookia texana Trosch.

Holbrookia maculata B. and G.

Sceloporus poinsettii B. and G.

Uta stansburiana B. and G.

Cnemidophorus sexlineatus L.

Diadophis regalis Bd. and Gird.

The first time this rare species has been found within the limits of the United States. The single specimen obtained differs from the typical one in having eight superior labials, with the eye above the fourth and fifth. As the preorbital labials are very short, variation to seven in all may be anticipated, as is found in the type. This specimen is smaller than the one from Sonora originally described.

Bascanium constrictor L.

Eutænia marciana B. and G.

The common species of the Rio Grande valley.

---

[1] Proceedings Philada. Academy, 1866, 307.

Crotalus confluentus Say.

Typical variety from near the southern end of the Socorro Mountain, five miles from Socorro.

Crotalus lepidus Kennicott.

Prof. Snow was fortunate enough to obtain the first entire specimen of this species, it having been described by Kennicott from two heads. We are thus made acquainted with the most peculiar of the North American rattlesnakes. I proposed for it the genus *Haploaspis* on account of the undivided nasal plates of the typical specimens. In the present specimen, that plate is divided below the nostril. It is therefore probable that this generic name should be abandoned.

Mr. Kennicott has well described the scutellation of the head. It may be summarized here by saying that the top of the muzzle is covered by eight smooth scuta; that the rostral plate is rather low, and is in contact with the prenasal; that there are two preoculars and two loreals; and that but two scales separate the orbit from the superior labial scuta. Of the latter there are twelve. Occipital scales smooth. Scales of body in twenty-three rows, the two external on each side smooth. Urosteges, 153; gastrosteges, 27. The rattle consists of seven segments and a button, and narrows gradually towards the extremity.

The color above is a greenish gray, which is crossed by nineteen jet-black rings on the body, which do not extend on the abdomen. These rings are two and a half scales wide on the middle line, and narrow downwards on each side so as to cover but one scale in width. The scales which border the annuli are half black and half green, the effect of which is to give the edge of the ring a turreted outline. The edges of the ground-color are paler than any other part of the scales, thus throwing the black into greater relief. A large black spot, shaped like two hearts side by side, with the apices posterior, marks the nape; and there is an irregular small black spot on each side of the occiput. Some black specks between the orbits. No other marks on the head. Near the middle of the gray spaces of the body, some of the scales of many of the rows have black tips. The tail is light brown above, and has a basal broad black, and two other narrow brown annuli. Below dirty white, with closely placed shades of brown.

Total length, m. ·555; to constriction of neck, ·027; length of tail, ·074; do. without rattle, ·026.

This is one of the smallest species of *Crotalus*, and is one of the most handsomely colored. Its coloration is entirely unique in the genus. The scutellation of the muzzle places it between the two sections of the genus, typified respectively by *C. horridus* and *C. durissus*.

The specimen was captured on the summit of the Magdalena Mountains, which are northwest from Socorro twenty miles.

3. St. Thomas, Nevada.

This locality is on the Virgen River, in southeastern Nevada, nearly in the latitude of the southern boundary of Utah. The collection now referred to was made by Dr. Edward Palmer and sent by him to the Smithsonian Institution. Through Professor Baird, the distinguished Secretary, it was referred to me for identification.

Bufo lentiginosus frontosus Cope.

This is the toad of the Great Basin, representing the *B. columbiensis* of more northern regions.

Crotaphytus wislizeni B. and G.

Cnemidophorus tessellatus Say.

Ophibolus getulus boyli B. and G.

The most northern locality for this species in the Great Basin. It has been previously obtained by Palmer and Coues, near Prescott, Arizona.

Phimothyra grahamiæ B. and G.

A variety with the dorsal bands nearly obsolete, and separated by three rows of dorsal scales on all parts of the body. Two preoculars on one side and three on the other. The most northern locality for this species.

4. Santa Fé, New Mexico.

Amblystoma mavortium Baird.

Not uncommon.

Spea hammondii Baird.

Abundant in July and August, when it deposits its eggs in the pools of rain-water. It is very noisy at such times, and the open lots in the city of Santa Fé resound with its cries. They are much like those of the *Scaphiopus holbrookii*.

The range of this species is extensive. It was originally obtained near Redding in Northern California. My friend, James S. Lippincott, has sent it to me from the extreme south of California, San Diego. The Smithsonian Institution has a slightly differentiated variety from Chihuahua; and a specimen from my friend Dr. Duges, from Guanajuato, Mexico, though rather young, is apparently the same. I suspect that the *Scaphiopus dugesi* Brocchi from that locality is the same species.

This species is much like the *Scaphiopus intermontanus* described further on. It is always smaller, and the middle pair of light dorsal bands is nearly always wanting. It is still more different from the *S. varius*, which has the vomerine teeth entirely posterior to the nares, banded upper lip and marbled back.

5. San Francisco Mountains, Utah.

Lizards are very abundant in this region, especially in the Wah Wah Valley, on the west side of the range.

Phrynosoma platyrhinum Gird.

Very abundant.

Crotaphytus wislizeni Bd. and Gird.

Very abundant.

Crotaphytus collaris Say.

Very common.

Uta stansburiana Bd. and Gird.

Abundant.

Secloporus smaragdinus Cope.

Not rare.

Sceloporus consobrinus B. and G.

Cnemidophorus sp.

Many seen but not caught.

6. Provo and Salt Lake City, Utah.

Bufo lentiginosus frontosus Cope.

Abundant near Salt Lake City.

Scaphiopus intermontanus sp. nov.

I took a specimen of this species within the limits of Salt Lake City, and subsequently obtained three or four specimens from Pyramid Lake, Nevada. It resembles the *Spea hammondi* more than it does any other species. The frontoparietal bones, though ossified, are not roughened as in the other species of *Scaphiopus*. It is nearest the *S. varius* (from near San Antonio, Texas). In that species the vomerine teeth are entirely posterior to the internal nares; in this one they are between the posterior borders of the same. The lips are not cross-barred as in the *S. varius*; and the superior region has two pale lines on each side. In *S. varius* these lines are replaced by a coarse marbling. As compared with *Spea hammondi*, this frog differs in its larger size, lighter colors, and the presence of the superior pair of light lines.

It represents the *S. hammondi* in more northern regions, and the complete cranial ossification and larger size mark it as a more fully developed form.

Rana halecina Kalm. From Provo.

Rana pretiosa B. & G.

A variety without a trace of dorso-lateral folds, and of a uniform dusky color above and on the sides. Lip not striped. The posterior part of the abdomen and the inferior face of the thighs are salmon-red. Skin smooth; diameter of membranum tympani three-fifths that of the eye. Salt Lake City. This is the most southern locality of this species known.

Sceloporus consobrinus B. & G. Provo.

7. Atlanta, Idaho.

Atlanta is a small town situated on the headwaters of the South Boise River, on the southern drainage of the Sawtooth Mountain Range. The valley is quite elevated, and is shut in by granitic mountains; water and vegetation are abundant; and the snow lies on the ground late in the spring. During a short visit there in 1882, I obtained the following species:

Amblystoma epixanthum sp. nov.

Nearly related to *Amblystoma macrodactylum* Baird, and to be placed next to that species in any synopsis of the genus.[1] Costal folds twelve. No canthus rostralis. Upper jaw overlapping lower. Tail strongly compressed, as long as head and body to groin. Head wide-oval; its greatest width one-fourth its total length to the groin. Digits all rather short; four phalanges in

[1] Proc. Acad. Phila., 1867, p. 171.

fourth posterior digit. Internal nares as widely separated as the external. Eye-fissure one-half width between the anterior canthus. Median dental series presenting an angle forwards. Tongue large, deeply plicate. Length, m. ·083; length to axilla, ·017; to groin, ·040; length of anterior limb, ·012; of anterior foot, ·004; of hind limb, ·014; of posterior foot, ·0065.

Sides of body and tail, and superior surfaces of limbs, shining black. Dorsal region to end of tail and muzzle, gamboge yellow. The yellow expands on the head, and forms two cross-bands on the upper surfaces of each of the limbs. The black of the sides is occasionally interrupted by the yellow spots irregularly placed. Below, dilute black, dusted with minute white speckles. The structural differences between this and the *A. macrodactylum* are not many, but are well marked. They are: 1. The greater width of the head, which enters the length (without the tail) five times in the latter, and four times in the *A. epixanthum*; and is also seen in the greater interorbital width. 2. In the short toes, which are very much longer in the *A. macrodactylum*. In color, this species is the more brilliant; the coast species being described as brown, with gray dorsal stripes, instead of black, with yellow dorsal stripes. In it the limbs are not banded, and the belly is uniformly pale, contrary to what holds in the present species, which is the most handsome of the genus. I obtained four specimens of this salamander, under logs, in a swamp near the head of the South Boise River, on the south side of the Sawtooth Mountain Range, Idaho.

Bufo columbiensis Bd. and Gird.

Abundant. I also obtained it at Bellevue on the Wood River, about one hundred miles southeast of Atlanta.

Bascanium vetustum Bd. and Gird.

Eutænia sirtalis Linn.

These are all, except the last, species characteristic of the northern fauna of Washington Territory. The *Bufo columbiensis* ranges to the headwaters of the Missouri.

8. Mouth of Bruneau River, Idaho.

This locality is on Snake River, which cuts through the great lava outflow of southern Idaho and Oregon. The reptiles are different from those of Atlanta, and are those of the great basin of Utah. I am indebted to Mr. J. L. Wortman for these specimens.

Phrynosoma platyrhinum Gird.

Crotaphytus collaris Say.

Crotaphytus wislizeni B. and G.

Uta stansburiana B. and G.

Sceloporus smaragdinus Cope.

Pityophis catenifer Blainv.

Bascanium vetustum B. and G.

The head is a little longer than in a specimen from central Oregon, and the muzzle is less conical. The fifth superior labial just reaches the inferior postorbital.

9. From Reno to Pyramid Lake, Nevada.

The road from Reno to the southern extremity of Surprise Valley, California, passes through an arid and forbidding country. The rocks are entirely basaltic, and frequently present a rugged foundation for the road. The vegetation consists of *Artemisia*, and where alkali abounds, of *Sarcobatus*. North of Pyramid Lake, the dry alkaline flats once covered by the Alkali Lake, have a wide extent. During the hot weather of July, 1882, the region swarmed with lizards, and rattlesnakes were numerous. The greatest number of both was met with from Pyramid Lake northwards for twenty miles.

Bufo columbiensis Bd. and Gird. Pyramid Lake.

Scaphiopus intermontanus Cope.

With the preceding species in a pond near the shore of Pyramid Lake. Like other allied species, it was very noisy, almost obscuring the voice of the less vociferous *Bufo*.

Phrynosoma platyrhinum Gird.

Very abundant.

Crotaphytus collaris Say.

Crotaphytus wislizeni B. and G.

More abundant than the *C. collaris*.

Holbrookia sp.

A fine species was seen north of Pyramid Lake, but it was so swift that I did not succeed in catching a specimen. It resembles the *H. texana*, and may be an undescribed species.

Sceloporus smaragdinus Cope.

A variety with one additional row of small supraorbital scales, making six rows in all.

Cnemidophorus tessellatus Say. Abundant.

Bascanium sp. Young.

Crotalus confluentus lucifer B. and G. Cope emend. supra, p. 11.

Two specimens from Buffalo Canyon, north of Pyramid Lake. In one of the specimens the dorsal spots are first darker, then lighter-bordered, and there are twenty-three rows of scales on the body. In the other there are twenty-five rows of scales, and the spots have neither dark nor pale borders, but have pale scales scattered through them, and they have a more transverse form.

10. The Lakes of South and West Central Oregon.

This region possesses much zoölogical interest from the position which it occupies as the border-land between the faunæ of the Pacific slope and that of the great interior basin. It is here that we find the transition between the sage-brush (*Artemisia*) desert and the forest-covered Sierra Nevada Mountains and valleys. Here also we have the transition between the almost fresh lakes near the mountains, to the intensely alkaline ones east of them. An especial interest attaches to the lake faunæ; since we find in them the means of determining the characters of the fossils found in the remains of pliocene and post-pliocene lakes of the Oregon desert. This part of the subject will be more fully considered in an essay on the fishes of these lakes, now in course of preparation.

The routes on which the species of the list below given, were collected, are as follows: Along the east shore of the Great Klamath Lake to its northern extremity. From the eastern side of the lake northeastward to Silver Lake. This was part of my expedition of 1879. In 1882, I passed along the three southern Warner Lakes, and then crossed southwest to Goose Lake. Thence I traveled north to Summer Lake, crossing the Chewaucan River, which flows into Abert's Lake. Then north to Silver Lake, connecting with my route of 1879. After that, south to Goose Lake, passing along its entire eastern shore.

Bufo columbiensis Bd. and Gird.

Abundant throughout the entire region. It is especially numerous at Klamath Lake, where it covers the basaltic blocks which lie partially in the water, concealed by the *Typhæ*, which grow from the bottom. They accumulate there in large piles, sometimes as large as a bushel-measure, and afford abundant food for the *Eutæniæ*, which are scarcely less abundant. One specimen of this toad was as large as the average *Bufo marinus* of Brazil, and a specimen seen at Warner's Lake was but little smaller.

Hyla regilla B. and G.

Abundant at Silver Lake, at Warner's Lake, Goose Lake and at Fort Bidwell, twenty miles east of Goose Lake, in California, I found numbers of what I suppose to be a variety of this species. It is little over half as large in linear dimensions, and the skin is more distinctly tubercular above. Some of those from Goose Lake are more spotted; those from Fort Bidwell are nearly uniform golden-yellow and green. This species lives in swamps and on the edge of water, representing in this region the *Acris* and *Chorophilus* of the east.

Rana pretiosa Bd. and Gird.

This is the characteristic *Rana* of the northwestern interior, being accompanied by *Bufo columbiensis* and *Bascanium vetustum*. In life the posterior part of the abdomen, with the inferior faces of the thighs, are of a bright salmon-red. I obtained it the entire length of the valley of the Warner Lakes, but not at Fort Bidwell. I have found it to range as far as the eastern foot of the Rocky Mountains in Montana;[1] and the specimens assigned by me[2] to *Rana septentrionalis*, from the Yellowstone Basin, may be the variety described above from Salt Lake City. I do not now have them before me for decision. Specimens of this species are in the National Museum, from Puget's Sound (Dr. Kennerly, No. 5975 a) and from "Camp Moryie" (Dr. Kennerly, No. 5973). The first-named specimens are accompanied by the *R. temporaria aurora*. It habits are aquatic.

Phrynosoma douglassi Bell. Var.

On the elevated land which represents the Sierra Nevada Range, between Warner's Lake and Goose Lake, in the basaltic region, near the former, I found a peculiar variety of this species. The horns are even more rudimental than in the usual form, but are all represented. The prominent scales of the back are smaller and less prominent. In some of the specimens the head is shorter relatively to the body. The color is an ironrust-brown, with darker lateral spots, each with a small posterior yellow border.

---

[1] American Naturalist, 1879, p. 435.

[2] Annual Report U. S. Geol. Survey Terrs., 1871, p. 469.

Individuals are abundant; some of those taken are full of eggs. All are much smaller than the true *P. douglassi.*

Uta stansburiana B. and G.

Abundant on the crags of basalt on the sides of Warner's Valley. It is also common at Summer Lake, which is the most northern locality for the species and genus.

Sceloporus graciosus B. and G.

This very pretty species extends as far north as Summer Lake, and is quite abundant.

Sceloporus smaragdinus Cope.

Common as far north as Summer Lake. A specimen taken there has large torquoise-blue spots behind each brown cross-bar, on each side of the dorsal region.

Charina plumbea Bd. and Gird.

I found a single specimen of this curious snake in the road along the west side of Summer Lake. Although living, its muscles were alternately contracted in such a way as to give it the appearance of a knotted root. It was very tame, allowing itself to be handled without offering resistance. In life the inferior surfaces are of a rich yellow.

Pityophis mexicanus bellona B. and G.

From Summer Lake.

Bascanium vetustum B. and G.

Common in Warner's Valley, at Summer Lake and at Klamath Lake.

Eutænia pickeringii B. and G.

Very common everywhere near water, in all parts of the Lake country.

Eutænia sirtalis sirtalis Linn.

This species accompanies the preceding at Warner's third Lake, at Summer Lake and at Goose Lake, and retains its distinctive features. The specimens seen at Goose Lake have the bands brighter yellow than usual, and are very pugnacious. They preferred fighting to escaping, and bit furiously.

Eutænia sirtalis elegans B. and G.

Abundant. In young specimens the dorsal spots are distinct.

Eutænia biscutata sp. nov.

This is one of the best defined species of the genus. I have only two specimens, which agree in the following characters.

They differ in the number of rows of scales, however, one having twenty-three and the other twenty-two. All the rows of scales keeled, the median ones very strongly. Labials eight, the eye resting on the fourth and fifth. Two preoculars; three postoculars. The muzzle is rather short, the frontal plate exceeding in length the region anterior to it, and equaling the common suture of the parietal scuta. Nasals rather short; loreal as long as high; inferior preocular nearly square; superior preocular not reaching frontal. Superior labials all truncate above and none of them elevated, the sixth touching the inferior postorbital. Temporals. 1 · 2 · 3; the anterior are rather large. Pairs of genecials subequal. Gastrosteges, 156; urosteges, 79.

Color everywhere black, except on the chin and throat, and on the inferior side of the tail. The former was reddish in life. There are very faint traces of stripes on the third and fourth, and on the median dorsal rows of scales. No traces of spots on the parietal scuta.

Total length, m. 0·265; length to canthus of mouth (axial), ·012; length of tail, ·062.

This species is one of the best characterized of the genus. Its leading peculiarities are: first, the two preocular scuta; second, its twenty-three rows of scales. In both respects it is unique in the genus. Its color is characteristic. Its place is nearest the *E. radix* B. and G., with which it agrees in its rather robust proportions, and the position of the lateral stripe.

This species is not uncommon in the swamp vegetation on the borders of the lake. The specimens I took displayed little activity.

Crotalus confluentus lucifer B. and G.

This species is abundant at Warner's second Lake, and I took one at Silver Lake. The specimens are identical with those from near Pyramid Lake, Nevada.

11. The Willamet Valley, Oregon.

The fauna of this valley is that of western Oregon, and may be expected to differ from that of central and eastern Oregon. The climate of the Willamet Valley is very wet, and the soil is densely covered with forests. This is a state of things almost exactly the reverse of what obtains in central Oregon. Appropriately we have numerous species of salamanders and fewer lizards than in the latter region. This collection was made by my friend, Professor O. B. Johnson, at that time residing at Salem. The specimens were obtained at various points between that city and Portland, north of it.

Amblystoma tenebrosum Bd.

Amblystoma macrodactylum Bd.

Plethodon intermedius Bd.

Cynops torosus Esch.

Bufo halophilus B. and G.

Hyla regilla B. and G.

Eumeces skiltonianus Bd.

Gerrhonotus multicarinatus Blv.

Sceloporus undulatus thayeri B. and G.

Phrynosoma douglassi Bell.

Charina plumbea B. and G.

Diadophis punctatus pulchellus B. and G.

Bascanium vetustum B. and G.

Eutænia leptocephala B. and G.

Of three specimens, two exhibit only seventeen rows of scales. These probably represent the supposed species *E. cooperi*, which is therefore not distinct.

Eutænia concinna Hallow.

I took a specimen of this beautiful snake at Eugene City, south of Salem. Not only the lateral vertical bars, but the muzzle, lips and gular region are a brilliant red.

12. Northern California.

The species referred to in this list were found near the United States fish-hatching establishment on the McCloud River, in Shasta County. I desire here to express my indebtedness to Mr. Livingston Stone, superintendent of the hatching station, for the hospitality which he extended to me at the time of my visit there.

Amblystoma (?) **tenebrosum** B. and G.

A large siredon from a small tributary of the McCloud is probably this species. It has peculiarities of the branchial structure, and I describe it by comparison with those found in other genera of American salamanders. These are mostly derived from specimens placed in my hands by the Smithsonian Institution, to which my acknowledgments are due. The coloration which appears in the larger larvæ of the present collection, approaches nearest that of the *Amblystoma tenebrosum*. These animals were abundant in the small stream I examined, and swam with great rapidity, darting about and hiding themselves among the fallen leaves that covered the bottom.

I. Processes with two rows of rami:

Rami with many thread-like fimbriæ; *Siren.*

II. Processes with one—an outer—row of rami; processes horizontal.

A rudimental inner row of rami; fimbriæ thread-like; *Proteus.*

III. No principal rami;

A. Processes compressed; fimbriæ dependent from lower edge;

Fimbriæ thread-like, extending on both outer and inner face of process; *Necturus.*

Fimbriæ flat, long, chiefly confined to the lower margin of process; Larvæ of *Spelerpes ruber; S. bilineatus*,. and *Gyrinophilus porphyriticus.*

Fimbriæ few, subclavate; *Plethodon cinereus.*

AA. Processes long, narrow; bearing fimbriæ only on the side next the body;

Fimbriæ simple, flat, sub-equal; *Amblystoma.*

AAA. No processes nor rami; fimbriæ on the vertical septa.

Fimbriæ in numerous rows on the edge of the septa; slender, unbranched; *Larva of Amblystoma tenebrosum.*

AAAA. Processes vertical septa, with rami on the anterior edge;

Rami bearing flat, thread-like fimbriæ, which arise from the processes posteriorly and are often divided. *Larva from Simiahmoo, Washington Terr.*

Plethodon iëcanus sp. nov.

This salamander resembles the *Plethodon glutinosus* in various respects, especially in coloration. It has, however, a compressed tail like the *P. intermedius*, and short series of vomerine teeth.

The vomerine series are straight, and do not quite meet on the middle line. They are entirely behind the nares, and do not extend exterior to them. The parasphenoid patches are united into one, and are well separated from the vomerines.

Form rather stout, and the tail short, equaling (from vent) the length of the body (with vent) to the gular fold. Costal folds 13. Head a longitudinal oval, with rather narrowed, and not truncate muzzle; its length (to occiput) contained three and two-third times in length from muzzle to groin.

Limbs short; when pressed along the side they are separated by three intercostal spaces. The digits are short, and the internal ones are rudimental.

The color is black everywhere, and the superior surfaces are dusted over with minute light specks.

Measurements.	M.
Total length,	·053
Length from muzzle to axilla,	·0105
Length from muzzle to groin,	·0275
Width of head at canthus oris,	·006
Length of anterior limb,	·006
Length of anterior foot,	·002
Length of posterior limb,	·0075
Length of posterior foot,	·0032

This species is to be compared with the *Plethodon intermedius* of western Oregon. It is shorter and more robust in form, having only thirteen costal plicæ instead of fifteen.[1] The color is very different.

This species is named from the aboriginal name Iëka, of the grand peak of northern California, Mount Shasta. From the same name the town of Yreka derives its name. So I am informed by Judge Roseborough of that place, to whom I am under great obligations for many facilities and much information.

Cynops torosus Esch. *Diemyctylus torosus* Cope, Check List, Batr. Rept., N. Amer.

Bufo halophilus Bd. and Gird.

Hyla regilla B. and G.

The typical form.

Rana pachyderma sp. nov.

Represented by five specimens of different ages and sizes from

[1] On page 99, Proc. Phila. Academy, 1869, in my monograph of the *Plethodontidæ* the number of plicæ is given at 13. This is a misprint for 15. On p. 209, Proceedings for 1867, the number is correctly given as 15.

the McCloud River, and by two specimens from Ashland at the northern base of the Siskiyou Mountains, Oregon.

This species belongs to the *Rana temporaria* group, and must be compared with *Rana temporaria aurora* B. and G., and *R. pretiosa* B. and G. The vomerine teeth are opposite the posterior border of the choanæ, and form two short, transverse series. The toes are webbed to the base of the terminal phalange of the fourth digit. The hind-leg extended reaches the extremity of the muzzle with the heel. There are two plantar tubercles. The internal is narrow, rather prominent and with obtuse extremity; the other is at the base of the fourth metatarsal bone, and is rounded.

The muzzle is obtuse and the head rather wide. Its greatest width at the position of the membranum tympani, equals the length from the end of the muzzle to the line connecting the axillæ in some specimens; in others to that connecting the middle of the humeri. The skin is on all the superior surfaces thick and glandular. This condition is especially marked in the dorso-lateral fold of each side, which is so thickened in front as to resemble a parotoid gland. This becomes less visible in alcohol. The tympanic membrane is either entirely concealed, or is represented by a depression only. The skin covering it is roughened. A groove extends downwards and backwards from it. Between this and the canthus oris is a glandular thickening, and behind it are two others, one above the other. Posterior to these on the sides is a succession of rounded, roughened warts, similar to those of the toads. Similar warts, but less prominent, are scattered over the dorsal region, and are numerous near the extremity of the coccyx. The skin of the superior surfaces of the head, body and limbs is minutely but very distinctly roughened by small warts, each of which gives exit to a pore. Inferior surfaces smooth. Length of fingers beginning with the shortest, 2 · 1 · 4 · 3.

The color is dark brown or nearly black, with indistinct darker spots on the back; sides brown. Axilla and groin yellow, marbled with black. Thighs above light or dark brown, with three darker crossbars. Tibiæ similar, with three crossbars. Thighs behind, black, coarsely vermiculated with yellow, or yellow closely spotted with black. Below light yellow, spotted with brown on the gular region and on front of femora.

Measurements.

	M.
Length of head and body to vent,	·066
Length from muzzle to axilla (axial),	·030
Length from muzzle to groin (axial),	·054
Length of anterior leg,	·043
Length of manus,	·019
Length of posterior leg,	·117
Length of femur,	·033
Length of tibia,	·038
Length of tarsus,	·018

The specimens from Ashland agree with those from the McCloud, except that they are nearly black above and do not exhibit the dorsal spots.

I compare this species with the *Rana temporaria aurora* from the Russian River near the coast of California. That species has but one palmar tubercle, the internal, which is of similar proportions to that of the *R. pachyderma.* The skin is not thickened, and is much less glandular everywhere. The membranum tympani is entirely distinct. The posterior face of the femur is not vermiculated with yellow, but is covered with large black masses. The whole of the under surfaces are brown-spotted. There are four brown crossbars on the tibia: traces of the fourth sometimes appear in the *R. pachyderma.* From *Rana pretiosa* it differs in all these characters; besides those that belong to the latter, *i. e.*, the posteriorly-placed vomerine teeth and the short hind-legs.

Eumeces skiltonianus B. and G.

Gerrhonotus multicarinatus Blv.

The movements of this species are not nearly so active as are those of the *Iguanidæ* and especially of the *Lacertidæ.*

Sceloporus undulatus thayeri B. and G.

Diadophis punctatus pulchellus B. and G.

Different from the typical form of the subspecies in having no spots on the inferior surfaces. I did not admit this form as distinct in my check list, but it had best be retained. It differs from the subspecies *amabilis* in having the inferior two rows of scales unicolor with the abdomen. In life this is a brilliant orange.[1]

---

[1] At this locality I found, under bark of logs, numerous specimens of *Brachycybe lecontei* Wood. This beautiful myriapod was originally described

13. Mouth of Russian River, California.

This locality is one hundred miles north of San Francisco. The collection was made by myself, in and on the border of the great redwood forest which there covers the hills and mountains of the coast range.

Batrachoseps attenuatus Esch. Abundant.

Plethodon oregonensis Gird.

Abundant, and especially pleasing from its liquid, prominent eyes. Always under the redwoods.

Cynops torosus Esch.

Abundant. This species is entirely aquatic.

Rana temporaria aurora B. and G. *Rana draytoni* B. and G. *Rana longipes* Hallow.

Not distinguishable as a species, in my opinion, from the *Rana temporaria* of the palæarctic realm.

Gerrhonotus multicarinatus Blv.

Eutænia sirtalis elegans B. and G.

14. Los Angeles, California.

Two collections from this locality are before me. One of these was made by Mr. DeCorse, Hospital Steward, at Drum Barracks, and was sent to the Smithsonian Institution. Prof. Baird submitted it to me for determination. The second collection was given me by Mr. Horatio N. Rust, the archæologist, who made it at Passadena, a short distance from the city.

Cynops torosus Esch. Rust.

Batrachoseps attenuatus Esch. Rust.

Phrynosoma blainvillii Gray. DeCorse.

Sceloporus undulatus thayeri B. and G. Rust.

Uta stansburiana B. and G.

---

as from California, where it was supposed to have been collected by Dr. J. L. Leconte. I, however, subsequently obtained it from East Tennessee, and as Dr. Leconte had collected it in Georgia, it was supposed by Dr. Wood that the locality California was an error. Its rediscovery on the McCloud River shows that this species is found on the Pacific coast, as originally stated by Wood, and that it ranges over the width of the continent. In like manner a myriapod which I sent Mr. Ryder from the Russian River, is stated by him to be much like *Andrognathus* Cope, a genus heretofore known from the Alleghenies of Virginia.

Specimens remarkably large, and with the postinguinal black spot unusually large and distinct. DeCorse.

Gerrhonotus multicarinatus Blv. Rust.

Eumeces skiltonianus Bd. Rust, DeCorse.

Ophibolus getulus boylii B. and G. Rust.

Pityophis catenifer Blv. Rust, DeCorse.

Bascanium testaceum Say. Rust, DeCorse.

Eutænia hammondi Kenn. DeCorse.

Note on a Species of Xantusia.

The species described below was found by Dr. J. G. Cooper, Zoölogist of the State Geological Survey of California, and was placed in the collections of the University of California, where I saw it. It was kindly lent me for examination by the authorities of the University. The locality from which the specimen was derived is unknown, beyond that it is Californian.

Xantusia riversiana Cope. American Naturalist, 1879, p. 801.

The position of this genus in the system has been discussed by M. Bocourt[1] and myself.[2] I associated it with the genera *Lepidophyma* Dum., and *Cricosaura* Peters, and stated that I was not able to distinguish them from the family *Lacertidæ*. M. Bocourt places these genera in the family "Trachydermi," which also includes *Heloderma* Wiegm. This family is divided by M. Bocourt into two subfamilies, the *Glyphodonti* for *Heloderma*, and the *Aglyphodonti* for the three genera named, together with *Xenosaurus* Pet. Previously to this[3] I had examined and compared the osteology of *Heloderma* and *Xenosaurus*. On account of the differences in the form of the mesosternum, and in some other points, I regarded *Xenosaurus* as the type of a peculiar family to be placed with the *Helodermidæ* in the tribe Diploglossa. *Xantusia*, *Lepidophyma* and *Cricosaura* are, on the other hand, not Diploglossa, but are *Leptoglossa*. They are allied to the *Lacertidæ*, and especially to the Asiatic *Ophiops*, which is, like them, without eyelids. The character of the tongue is like that of the *Ecpleopidæ*, uniformly squamous, and has no resemblance to that of the *Diploglossa*. The characters of the scapular arch are those of the *Leptoglossa*. The clavicle is loop-shaped proximally, and the

---

[1] Misiou Scientifique de Mexique, Herpetology, p. 303, 1878.

[2] Proceedings of the Academy of Philadelphia, 1864, p. 229.

[3] Loc. cit. 1866, p. 322.

mesosternum is cruciform in *Lepidophyma* and *Xantusia*. I have not been able to examine *Cricosaura* as to these points. In my paper first mentioned, I stated that these genera have distinct parietal bones. I think that they should, on this account, be distinguished from the *Lacertidæ*, where they are coössified. Whether they are distinct or united in the Ecpleopidæ, I do not know, but the absence of eyelids will separate the group from that family. I use for it the name first given by Baird.[1] *Xantusidæ*, and characterize the three genera as follows:—

I. A large interfrontonasal plate; frontoparietals meeting on the middle line.

Superciliary scales none; pupil round; *Lepidophyma*.
Superciliary scales present; pupil vertical; *Xantusia*.

II. Two interfrontonasals; frontoparietals separated by interparietal.

Superciliary scales; *Cricosaura*.

All of these genera have femoral pores, and an exposed membranum tympani.

The species which has given occasion for the above discussion is the second one of the genus. It is several times as large as the type *X. vigilis* Baird, and has a different coloration. The digits are shorter.

The scales of the dorsal and lateral regions are rather coarsely and uniformly granular. The abdominal scales are quadrate, and are in sixteen longitudinal and thirty-two transverse rows. The preanal scales are in three transverse rows, the anterior two of four scales, with the median pair in both much enlarged, and the posterior row of six scales. Scales of the gular region flat and hexagonal, one row on the gular fold a little larger, and equal to the anterior gulars. Scales of the anterior aspects of the fore-leg and femur larger than the others; those of the tibia small, and those of the posterior face of the femur still smaller. Scales of the tail in whorls of equal width. The scales of equal size, and all convex in cross-section but not keeled. None of the scales of the body or limbs keeled.

The nostril is situated in a small scute at the junction of the sutures which separate the internasal, rostral, first labial, and first

[1] Proceedings Academy Philadelphia, 1858, December.

loreal scuta. Three loreals, increasing in size posteriorly. A circle of scales surrounds the eye, of which the superior or superciliary are the largest. The latter are separated by one row of scales from the parietal, supraorbital and frontal on each side. The interfrontonasal is nearly square. The frontonasals are considerably in contact. The frontal is hexagonal, and is broader than long. The interparietal is as large as each parietal. It is longer than wide, and notches the contact of the frontoparietals. The occipitals are large and quadrate. A single large temporal bounds the parietals and occipital, and it is followed by two small scuta which are in contact with the occipital. There are eight scales on the upper lip. Of these the fifth is the largest, and is part of an annulus which begins with two small scales at the posterior loreal, and terminates at the seventh scale, opposite the middle of the pupil posteriorly. The posterior labials are small, and are separated by nine rows of still smaller scales from the large temporal. No large auricular scales. The eye is rather large and its diameter is contained in the length of muzzle in front of it 1·75 times. The vertical diameter of the auricular meatus is a little less.

The first digits of both extremities are very short. The second of the pes is very little longer than the fifth. All the ungues are acute and are moderately curved. The hind-legs are remarkably short, not exceeding the fore-legs. Extended forwards the extremity of the fourth digit reaches the elbow of the appressed fore-leg. Femoral pores twelve on each side; no anal pores. The tail is not long, and its form is compressed with a flat inferior surface. The section is a triangle, higher than wide, with the apex narrowly truncate.

The color is light brown, with dark umber-brown spots on the superior surface. These spots form, in general, one median and two lateral rows, but as their forms are very irregular this order is obscure. The median dorsal are the largest, and they send branches laterally and anteroposteriorly, so that the result is rather confused. Dark brown bands cross the muzzle on the frontonasal plates and on the frontal, and form a wide U from the frontoparietals passing around the posterior edge of the occipitals. Sides of head with rather large brown spots. Inferior surfaces with minute brown spots which are least numerous on the middle line. Tail with irregular pale spots.

Measurements.	M.
Total length,	·118
Length to posterior edge of occipital plates,	·0162
Length to axilla,	·029
Length to groin,	·055
Length to vent,	·060
Width between orbits above,	·007
Width at temples,	·0115
Length of fore-limb,	·017
Length of manus,	·008
Length of hind-limb,	·023
Length of pes,	·011
Length of tibia,	·007

15. San Diego, California.

My friend, James S. Lippincott, made a collection of reptiles and batrachians at this locality, which throws considerable light on some points of geographical distribution. A catalogue of the species is here given :—

Bufo columbiensis Bd. and Gird.

A single specimen with smoother skin than the more northern forms. Gland on the surface of the tibia very distinct.

Spea hammondi B. and G.

See *antea*, page 14. Four specimens.

Eumeces skiltonianus Baird.

A specimen with the scales of the dark bands pale centered, and with a very thick tail.

Verticaria hyperythra Cope.

Cnemidophorus tessellatus tigris B. and G.

Aniella pulchra Gray.

Gerrhonotus multicarinatus Blv.

Uta stansburiana B. and G.

Crotaphytus wislizeni Bd. and Gird.

Phrynosoma blainvillei Gray.

Rhinochilus lecontei B. and G.

Hypsiglena ochrorhynchus Cope.

Bascanium testaceum Say.

General Observations.

The results to zoölogical geography obtained by the preceding identifications are as follows :—Collection No 1. The extension northwards of the ranges of *Crotalus molossus* and *Stenostoma dulce*. No. 2. The extension northwards of the ranges of *Diadophis regalis*, *Crotalus lepidus* and *Holbrookia texana*. No. 4. The extension to the Rocky Mountains of the range of *Spea hammondi*. No. 6. The discovery of a new *Scaphiopus* in the Great Basin district; and of the southern extension of *Rana pretiosa* into the same. No. 7. The discovery that the Northern Pacific fauna extends east to the Rocky Mountains. This fauna is especially represented by *Bascanium vetustum*, *Rana pretiosa* and *Bufo columbiensis*. No. 8. The fact that the Great Basin district of the Sonoran fauna extends north to the southern slope of the Rocky Mountains in Idaho, where are found several of its species. These are *Phrynosoma platyrhinum*, *Crotaphytus wislizeni*, and *Uta stansburiana*. No. 9. The discovery that the same fauna extends north along the eastern slope of the Sierra Nevada to the beginning of Surprise Valley, California. No. 10. The determination that the Northern Pacific fauna extends from Surprise Valley, eastern California; northwards as far as my explorations have extended, viz., to Silver Lake and Klamath Lakes. No. 15. The determination of a wide southern range for *Spea hammondi* and *Bufo columbiensis*, and northern range for *Verticaria hyperythra*.

These results indicate that the Pacific region has a much greater extension eastward than it has been supposed to have, but which was foreshadowed in my paper on the Zoölogy of Montana, published in 1879.[1] They also indicate that it must be divided into three districts. These I call the Idaho, the Willamet, and the South Californian districts. The first is characterized by the absence of *Gerrhonotus* and *Cynops* and of certain species of *Amblystoma*. The South Californian is characterized by the presence of *Hypsiglena* and *Rhinochilus*, and absence of *Amblystoma*. It is allied to the Sonoran region, to which it is adjacent.

As regards the relation which the Sonoran region as a whole bears to the Nearctic and Neotropical realms, some remarks may be in place here. It is a question with some naturalists to which of the two it should be referred, and some would exclude it from the Nearctic without fully determining its relations to the Neotropical realm.

There can, however, be no doubt that it lacks all the peculiar

---

[1] American Naturalist, p. 435.

features of the Neotropical realm, and if it lacks some of those of the Nearctic also, its types are mostly representative of the latter rather of the former. I content myself here with confirming this general principle by reference to the principal families and genera of cold-blooded vertebrata.

R. Nearctica.	R. Sonoriana.	R. Neotropicalia.
	PISCES.	
Salmonidæ.	Salmonidæ.	
Cyprinidæ.	Cyprinidæ.	Characinidæ.
Catostomidæ.	Catostomidæ.	
	BATRACHIA.	
Ranidæ.	Ranidæ.	
Scaphiopidæ.	Scaphiopidæ.	
		Cystignathidæ.
Amblystomidæ.	Amblystomidæ.	
	REPTILIA.	
	Lacertilia.	
Sceloporus.	Sceloporus.	Liocephalus.
Eumeces.	Eumeces.	Mabuia.
	Ophidia.	
		Bothrops.
Bascanium.	Bascanium.	Drymobius.
Tropidonotus.	Tropidonotus.	Helicops.
Eutænia.	Eutænia.	
Pityophis.	Pityophis.	
Diadophis.	Diadophis.	Rhadinæa.
		Coniophanes.
Ophibolus.	Ophibolus.	Erythrolamprus.
		Pliocercus.
		Oxyrrhopus.
		Sibon.
		Leptognathus.
		Boa.
		Xiphosoma.
	Stenostoma.	Stenostoma.

There are a good many genera which are found in the Sonoran district, which do not occur in other parts of the Nearctic realm. These genera are frequently confined to it, but when they are not, they are to be looked for in the Mexican region of the Neotropical realm. I give a list of these genera, with a corresponding one of the Mexican region, to illustrate the extent of the similarity between the two regions.

R. Sonoriana.		R. Mexicana.
	PISCES.	
Plagopterinæ.		
	REPTILIA.	
	Lacertilia.	
Heloderma.		Heloderma.
Crotaphytus.		
Uta.		
Uma.		
Callisaurus.		
	Ophidia.	
Gyalopium.		Ficimia.
Phimothyra.		Phimothyra.
Trimorphodon.		Trimorphodon.
Hypsiglena.		Hypsiglena.

It seems then that the Neotropical relationships of the Sonoran region are not great. In this consideration I have omitted the genera which are common to the Mexican region and the Nearctic realm in general. Such are *Ranidæ*, *Cnemidophorus*, *Sceloporus*, *Bascanium*, *Tropidonotus*, *Eutænia*, *Pityophis*, *Spilotes*, *Ophibolus* and *Elaps*. These forms serve to indicate the affinity between the Nearctic realm and the Mexican region. The line between the two is, however, not yet exactly drawn. The former extends on the west coast at least as far south as Guaymas, and on the plateau as far as Guanajuato. On the east coast the Neotropical fauna reaches near to the Rio Grande. See On the Zoölogical Position of Texas, by the writer, in Bulletin U. S. National Museum, No. 20, August, 1880; and Eleventh Contribution to the Herpetology of Tropical America, by E. D. Cope, Proceedings Amer. Philosoph. Society, 1879, p. 267.

SYNOPSIS
OF THE REPTILES AND BATRACHIANS
OF ARIZONA

Elliott Coues

CHAPTER V.

SYNOPSIS

OF THE

REPTILES AND BATRACHIANS

OF

ARIZONA;

WITH

CRITICAL AND FIELD NOTES, AND AN EXTENSIVE SYNONYMY.

BY

DR. ELLIOTT COUES, U. S. ARMY.

CHAPTER V.

INTRODUCTORY.

This article rests primarily upon the collections and observations made by the writer in Arizona in 1864–65. These collections, containing many new species, were placed, together with the author's field notes, in the hands of Prof. E. D. Cope, by whom the new species were described in a paper published in the Proceedings of the Philadelphia Academy for 1866. Special reference is made to this paper, which contained, furthermore, the identification of the species, forty-four in number, contained in the collections, with some of the author's field notes, critical observations of Professor Cope's, and an enumeration of the additional species then known to inhabit the Territory, sixty-eight in number.

The present paper was drawn up some years since to form part of a General Work upon the Natural History of Arizona, which the author then contemplated, but which pressure of other engagements compelled him to abandon. During the time it has lain in manuscript, it has been retouched at intervals, with the addition of various matter, until it has assumed the present shape of a synopsis of the species now known to inhabit the Territory, eighty-three in number (exclusive of numerous subspecies).

The aim has been to include none which have not actually been found in Arizona as at present bounded. It will be remembered that the present Territory includes the western half of what was formerly New Mexico, together with a considerable part of Sonora, obtained by what is known as the "Gadsden purchase"; consequently, many species described from "New Mexico" and "Sonora" are really Arizonan.

Doubtless some species already known to inhabit this Territory have not come to the writer's knowledge as such, but it is believed that these are very few; and, no doubt, also species remain to be discovered in this region, so fertile in reptilian life. But the present article is believed to be a fair *résumé* of the subject as at present understood. To render it more generally available, and useful as a work of reference, a copious bibliography has been prepared, representing a large proportion of extant references, and a nearly complete synonymy. Descriptions of lately published or less generally known species are given in many cases, together with such critical and field notes as the author found himself in position to offer.

The classification adopted is that of the eminent herpetologist, Professor Cope, whose nomenclature is followed in the main, though a different stand is taken in some instances; and the identifications of species are mainly upon the same authority, the author's work being little more in this instance than that of a compiler.

The most striking feature of the batrachian fauna of Arizona is the poverty of this region in *Urodela* (one species), which bear small proportion, in numbers of species or of individuals, to the *Anura* (eight species), in comparison with the ratio existing between these two groups in most parts of the United States; while the *Anura* themselves are poorly represented. This is probably due in greatest part to the dryness of the region as a whole. The *Chelonia* (four or five species) are likewise poorly represented, for the same reason. On the other hand, the region is rich in *Ophidia* (thirty-four species), including a larger proportion of venomous species than any other district of the United States; and it is still richer in number of individuals and of species of *Sauria* (thirty-six species). No other portion of our country exhibits such a preponderance of these forms of animal life.

NOTE.

Since the present paper was set in metal, a few *addenda* have come to my knowledge, being indicated as Arizonan in the preceding article by Dr. Yarrow. Such are *Phrynosoma blainvillii*, *Sceloporus tristichus (sp. n.)*, *S. spinosus*, *Chilopoma rufopunctatum (g. s. n.)*, *Pityophis sayi mexicanus*, *Diadophis pulchellus*, *Ophibolus doliatus annulatus*, *Bufo lentiginosus cognatus*, *B. punctatus*. But the close connection of the two memoirs renders their omission of less consequence.

November 15, 1875.

A.—REPTILIA.

CHELONIA.

CRYPTODIRA.

CINOSTERNIDAE.

1. **Aromochelys carinatus**, *Gray.*

Aromochelys carinatus, GRAY, Cat. Shield Reptiles Brit. Mus.—COPE, Proc. Acad. Nat. Sci. Phila., 1866, 310.—*Id.*, Check-List, 1875, 52.
Ozotheca tristycha, AGASS., Contrib. Nat. Hist. U. S., i, 1857, 425, pl. 5, figs. 20–22.—BD., U. S. & Mex. Bound. Surv., ii, pt. ii, 1859, Reptiles, 3.

Specimens from Fort Whipple, summer of 1865.

I ascertained the occurrence in the Territory of two other species of turtles, specimens of which were among a small part of my collections, which unfortunately was lost. One of these was procured on the Colorado Chiquito River in July, 1864; it was about eight inches long. Another entirely different species was found to be common in the headwaters of the San Francisco River, near Fort Whipple, where several specimens were secured with hook and line in January, 1865. They were very voracious, being often drawn out of the water and taken in hand before they would relinquish their hold, although not fastened upon the hook. These were of small size, about four inches in length. (One of these species may have been *Chrysemys oregonensis.*)

The two following species are introduced in the present connection on the authority of Professor Baird's Report on the *Reptilia* of the Mexican Boundary Survey:—

2. **Cinosternum sonoriense**, *LeC.*

Kinosternon sonoriense, LEC., Proc. Acad. Nat. Sci. Phila., 1854, 184.
Thyrosternum sonoriense, AGASS., Contrib. Nat. Hist. U. S., i, 1857, 428, pl. 5, figs. 8–11.—BD., U. S. & Mex. Bound. Surv., ii, pt. ii, 1859, Reptiles, 3.
Cinosternum sonoriense, COPE, Check-List, 1875, 52.

"Tucson and Guadeloupe Cañon, *Kennerly.*"

This species was originally described from Tucson, "Sonora" (*i. e.*, Arizona).

3. **Cinosternum flavescens**, (*Agass.*).

PLATE XVII, FIGS. 1, 2, 3.

Platythyra flavescens, AGASS., Contrib. Nat. Hist. U. S., i, 1857, 430, pl. 5, figs. 12–15.—BD., U. S. & Mex. Bound. Surv., ii, pt. ii, 1859, Reptiles, 3.
Cinosternum flavescens, COPE, Check-List, 1875, 52.

"Arkansas, Texas, and valley of the Gila."

The plate affords views of profile, dorsum and abdomen of the species.

3bis. **Cinosternum henrici**, *LeC.*

PLATE XVI.

Kinosternum henrici, LEC., Proc. Acad. Nat. Sci. Phila., 1854, 182.
Cinosternum henrici, COPE, Check-List, 1875, 52.

Obtained in Arizona by the Explorations West of the One-hundredth Meridian; see Dr. Yarrow's Report, *anteà.*

SAURIA

NYCTISAURIA.

GECCONIDAE.

4. **Coleonyx variegatus**, (*Bd.*) *Cope.*

Stenodactylus variegatus, BD., Proc. Acad. Nat. Sci. Phila., Dec., 1858, 254.—*Id.*, U. S. & Mex. Bound. Surv., ii, pt. ii, 1859, Reptiles, 12, pl. 23, figs 9–27, pl. 24, figs. 11–19.
Coleonyx variegatus, COPE, Proc. Acad. Nat. Sci. Phila., 1866, 310.—*Id.*, *ib.*, 1867, 85 (Owen's Valley, Cal.).—*Id.*, Check-List, 1875, 50.

"Colorado Desert, *A. Schott.*"

Head very broad; hind foot contained six times in head and body; above brownish-yellow, with irregular, small blotches of light reddish-brown, sometimes in broad transverse bands; edges of eyelids and whole under surface opaque white.—(*Descr. orig.*)

5. **Phyllodactylus tuberculosus**, *Wiegm.*

Phyllodactylus tuberculosus, WIEGM., Nova Act. K. Leop.-Car. Acad., xvii, 241.—BD., U. S. & Mex. Bound. Surv., ii, pt. ii, 1859, Reptiles, 12, pl. 23, figs. 1–8 (Durango, Mex.).—COPE, Check-List. 1875, 52.

A species of the Sonoran region, not observed by me.

PLEURODONTA.

IGUANIA.

IGUANIDAE.

6. **Phrynosoma douglassi,** (*Bell*) *Gray*.

a. DOUGLASSI.

Agama douglassii, BELL, Trans. Linn. Soc., xvi, 1828 (1833), 105, pl. 10.—HARLAN, Med. & Phys. Res., 1835, 141, f. 3.

Phrynosoma douglassii, GRAY, Griffith's An. King., ix, 1831, 44.—WAGL., Nat. Syst. Amphib., 1830, 146.—WIEGM., Herp. Mex., 1834, 54.—HOLB., N. A. Herp., i, 1842, 101, pl. 14.—DE KAY, Zoöl. N. Y., 1842, 31.—GRAY, Cat. Brit. Mus., 1845, 227.—GIR., Stans. Rep. Exp. Great Salt Lake, 1852, 362, pl. 7, figs. 6–9. COPE, Proc. Acad. Nat. Sci. Phila., 1866, 302.—ALLEN, Proc. Bost. Soc. Nat. Hist., xvii, 1874, 69.—COPE, Check-List, 1875, 49.

Tapaya douglassii, GIR., Herp. U. S. Exp. Exped., 1858, 398, pl. 21, figs. 1–5.—BD., P. R. R. Rep., x, 1859, Gunnison's & Beckwith's Route, Reptiles, 18.—*Id.*, P. R. R. Rep., x, 1859, Williamson's & Abbott's Route, Reptiles, 9.—COOP. & SUCKL., Nat. Hist. Wash. Terr., 1860, 294.

Tapaya brevirostris, GIR., U. S. Exp. Exped., 158, 377.—BD., P. R. R. Rep., x, 1859, Gunnison's & Beckwith's Route, Reptiles, 18.

Phrynosoma brevirostris, COPE, Proc. Acad. Nat. Sci. Phila., 1866, 302.

b. ORNATISSIMUM.

Phrynosoma orbiculare, HALLOW., Sitgreave's Exp. Zuñi & Col. Riv., 125, pls. 8, 9 (not of Wiegmann).

Tapaya ornatissima, GIR., U. S. Exp. Exped. Herp., 1858, 396.—BD., P. R. R. Rep., x, 1859, Whipple's Route, Reptiles, 38.—*Id.*, U. S. & Mex. Bound. Surv., ii, pt. ii, 1859, Reptiles, 9.

Phrynosoma douglassii, subsp. *ornatissimum*, COPE, Check-List, 1875, 49.

Numerous specimens in my collection, from various parts of New Mexico and Arizona. One of them, No. 407,* from Bero Springs, N. Mex., between Fort Wingate and the Rio Grande, was formerly identified by Professor Cope with *P. brevirostre*, which, however, appears to be not a tenable species.

This round bodied and plain looking species, with the cephalic spines rudimentary, exhibits a wide range of variation in color. Some specimens are uniform brown above; others have dark cross bars, with light hinder edges, or dark oval spots, yellow-margined or not; in some the spines and

*Numbers of specimens throughout this article refer to my field-register.

sides of the head are quite red. The under parts vary from a plain dull grayish-white to close nebulation with numerous squarish dusky spots. The young are generally very light colored and more uniform than the adults, a plain reddish or grayish brown prevailing on the upper parts. The orbicular shape of the body appears constant, but there is a great difference in the length of the tail and its thickness at the base. It varies from nearly half to considerably less than half the total length (contained $1\frac{3}{4}$ to $2\frac{3}{5}$ times in the total length). Specimens from the Colorado Chiquito River are conspicuous by the yellow edging of the oval brown dorsal spots.

This is the most widely distributed species of the genus *Phrynosoma* (including *Tapaya* and *Doliosaurus*) in the United States, and one of the most abundant. I have recently ascertained its extension to British America in the region of the Milk River, where it is a common animal. In these higher latitudes, its dispersion seems strictly coincident with that of *Caudisona confluenta.* In New Mexico and Arizona, it is the characteristic species. I found it abundant at all points on my journey from Santa Fé, N. Mex., to Fort Whipple, Ariz., and constantly met with it in various other parts of the last named Territory.

Like other species of the same genus, this *Phrynosoma* is slow of foot and readily captured by hand; it makes an interesting pet to one fond of observing the traits of lower animals. It may readily be secured by a thread tied behind the "horns"; and in this state of partial liberty its habits may be studied to advantage. It is one of the most inoffensive and amiable of reptiles; though some of the largest and boldest individuals sometimes make a slight demonstration in self-defense by biting weakly, it usually submits at once without remonstrance. When handled, it has a way of making itself perfectly flat, when, closing its eyes, it will simulate death in this collapsed state. Under some circumstances, it will swell up the body prodigiously till it assumes a nearly spherical shape. It has a sly way of watching for a chance to escape by bolting away when it thinks itself unobserved, and a still more curious knack of burying itself in sand or other loose soil. This is accomplished by a gradual lateral and forward insinuating wriggling of the body, with the muzzle pointed downward and the limbs drawn close to the sides. A few moments suffice for its disappearance. A certain slight means of defense which the "horns" may sometimes afford is shown by the use they are put to when the animal is irritated by poking with a finger or bit of a stick; then the head is lowered, the horns set forward, the back arched up, and the whole attitude becomes ludicrously like that of a bull in miniature. The Horned Lizards show special aversion to dogs; on the approach of one, they raise themselves to the full length of the legs, puff out the body, open the mouth, and hiss audibly, altogether presenting quite a formidable front. Their food, in confinement and otherwise, consists chiefly of flies and other insects, which they capture by a quick thrust of the fleshy tongue, lubricated with viscid saliva. I have not observed the time of coition, nor the period of gestation; but most of the females are found pregnant in July, and the young appear in great numbers in August. The male is usually smaller than the female, and of slenderer form.

7. **Phrynosoma regale,** *Gir.*

Phrynosoma regale, GIR., Herp. U. S. Exp. Exped., 1858, 406.—BD., U. S. & Mex. Bound. Surv., ii, pt. ii, 1859, Reptiles, 9, pl. 28, figs. 1–3.—COPE, Proc. Acad. Nat. Sci. Phila., 1866, 310.—*Id.*, Check-List, 1875, 49.
"*Phrynosoma solaris*, GRAY, Cat. Sauria Brit. Mus., 229."

This large species, not noted by me, appears to inhabit more particularly the desert portions of the Territory in the valleys of the Gila and Colorado.

8. **Phrynosoma maccalli,** (*Hallow.*) *Cope.*

Anota M'Calli, HALLOW., Sitgreave's Exp. Zuñi & Col. Riv., 1853, 127, pl. 10, (type of genus).
Doliosaurus mc'callii, GIR., Herp. U. S. Exp. Exped., 1858, 400.—BD. U. S. & Mex. Bound. Surv., ii, pt. ii, 1859, Reptiles, 9, pl. 28, figs. 4–6.
Phrynosoma maccallii, COPE, Proc. Acad. Nat. Sci. Phila., 1866, 310.—*Id.*, Check-List, 1875, 49.

From the same region as the last, and, like it, not noticed by me.

9. **Phrynosoma planiceps,** *Hallow.*

PLATE XXIV, fig. 1*a*, 1*b*.

Phrynosoma planiceps, HALLOW, Proc. Acad. Nat. Sci. Phila., 1852, 178.—COPE, Check-List, 1875, 49.

This species has been added to the fauna of the Territory by Dr. H. C. Yarrow since the present paper was drawn up.

10. **Phrynosoma platyrhinum**, *Gir.*

Phrynosoma platyrhinos, GIR., Stans. Rep. Exp. Great Salt Lake, 361, 363, pl. 7, figs. 1–5.—COPE, Proc. Acad. Nat. Sci. Phila., 1866, 302.

Phrynosoma platyrhinum, COPE, Check-List, 1875, 49.

Doliosaurus platyrhinos, GIR., Herp. U. S. Exp. Exped., 1858, 407.—BD., P. R. R. Rep., x, 1859, Gunnison's & Beckwith's Route, Reptiles, 18.

Specimens in my collection from Arizona, or a contiguous portion of New Mexico, but without indication of precise locality.

11. **Phrynosoma modestum**, *Gir.*

Phrynosoma modestum, GIR., Stans. Rep. Exp. Great Salt Lake, 1853, 361, 365, pl. 6, figs. 4–8.—COPE, Proc. Acad. Nat. Sci. Phila., 1866, 302.—*Id.*, Check-List, 1875, 49.

Doliosaurus modestus, GIR., Herp. U. S. Exp. Exped., 1858, 409.—BD., U. S. & Mex, Bound. Surv., ii, pt. ii, 1859, Reptiles, 10.—*Id.*, P. R. R. Rep., x, 1859. Whipple's Route, Reptiles, 38.

Bero Springs, June, 1864, one specimen of this small and inconspicuous species.

12. **Sceloporus consobrinus**, *Bd. & Gir.*

Sceloporus consobrinus, BD. & GIR., Marcy's Rep. Exp. Red Riv., 1853, 224, pl. 10, figs. 5–12.—BD., P. R. R. Rep., x, 1859, Whipple's Route, Reptiles, 37.—*Id.*, U. S. & Mex. Bound. Surv., ii, pt. ii, 1859, Reptiles, 5.—HAYD., Trans. Am. Phil. Soc., xii, 1862, 177.—COPE, Proc. Acad. Nat. Sci. Phila., 1866, 303.—ALLEN, Proc. Bost. Soc. Nat. Hist., xvii, 1874, 69.—COPE, Check-List, 1875, 49.

A wide ranging species in the West, whose habitat, as now known, extends north to the Yellowstone, where it was lately discovered by Mr. J. A. Allen, and to Oregon. I found it in various parts of Arizona, not only in the southern desert portions where reptiles most abound, but also in the pine covered mountainous tracts, as about Fort Whipple, for instance, where it was a common inhabitant of the high dry woods, with *Cnemidophorus gularis.*

13. **Sceloporus clarki**, *Bd. & Gir.*

PLATE XXIII, FIG. 1 (var.).

a. CLARKI.

Sceloporus clarkii, BD. & GIR., Proc. Acad. Nat. Sci. Phila., 1852, 127.—BD., U. S. & Mex. Bound. Surv., ii, pt. ii, 1859, Reptiles, 5.—COPE, Proc. Acad. Nat. Sci. Phila., 1866, 310.—*Id.*, Check-List, 1875, 49.

Sceloporus magister, HALLOW., Proc. Acad. Nat. Sci. Phila., 1854, 93.

b. ZOSTEROMUS.

Sceloporus zosteromus, COPE, Proc. Acad. Nat. Sci. Phila., 1863, 105 (Lower California).
Sceloporus clarkii, subsp. *zosteromus*, COPE, Check-List, 1875, 49.

Judging from the citations of numerous specimens from various localities in the Report of the Mexican Boundary Survey, this species is common along the southern border of the United States. It was not met with by me. Var *zosteromus* is the Lower California form.

14. **Sceloporus poinsetti**, *Bd. & Gir.*

Sceloporus poinsettii, BD. & GIR., Proc. Acad. Nat. Sci. Phila., 1852, 126.—BD., U. S. & Mex. Bound. Surv., ii, pt. ii, 1859, Reptiles, 5, pl. 29, figs. 1–3.—COPE, Check-List, 1875, 48.

Quoted from "Sonora" in the Mexican Boundary Report. Obtained in Northern Arizona by Mr. H. W. Henshaw.

15. **Sceloporus scalaris**, *Wiegm.*

Sceloporus scalaris, WIEGM., Herp. Mex., 1834, 52, pl. 8, f. 2.—BD., U. S. & Mex. Bound. Surv., ii, pt. ii, 1859, 6.—COPE, Check-List, 1875, 49.

Los Nogales, "Sonora."

16. **Sceloporus marmoratus**, *Hallow.*

Sceloporus marmoratus, HALLOW., Proc. Acad. Nat. Sci. Phila., 1852, 178.—*Id.*, Sitgreave's Exp. Zuñi & Col. Riv., 1853, 110, pl. 2.—BD., U. S. & Mex. Bound. Surv., ii, pt. ii, 1859, Reptiles, 6.—COPE, Check-List, 1875, 48.
Sceloporus delicatissimus, HALLOW., Proc. Acad. Nat. Sci. Phila., 1852, 178.—*Id.*, Sitgreave's Exp. Zuñi & Col. Riv., 1853, 109, pl. 1.

Not observed by me. The known range of this species, however, extends from Utah into the Sonoran Region.

17. **Sceloporus torquatus**, *Peale & Green.*

Sceloporus torquatus, PEALE & GREEN, Journ. Acad. Nat. Sci. Phila., ii, ——, 131.—COPE, Check-List, 1875, 48 (quotes "Proc. Acad.' by error for Journ. Acad).

Quoted by Professor Cope from the Sonoran Region.

18. **Sceloporus yarrowi**, *Cope.*

PLATE XXIII, FIGS. 2, 2*b*, 2*c*, 2*d*.

Sceloporus jarrovii, COPE, Zoöl. Expl. W. 100th Merid., *anteà*, p .569.—COPE, Check-List, 1875, 48. (N. B. The "Check-List" is published during the passage of the present volume through the press.)

From Southern Arizona, *Mr. H. W. Henshaw.*

An account of this fine new species, appropriately dedicated to the zoölogist of the Explorations west of the one hundredth meridian, will be found on the preceding pages.

19. **Sceloporus gratiosus**, *Bd. & Gir.*

Sceloporus graciosus, BD. & GIR., Proc. Acad. Nat. Sci. Phila., 1852, 69.—*Iid.*, Stans. Rep. Exp. Great Salt Lake, 346, pl. 5, figs. 1–3.—BD., P. R. R. Rep., x, 1859, Gunnison's & Beckwith's Route, Reptiles, 17.—*Id.*, P. R. R. Rep., x, 1859, Williamson's & Abbott's Route, Reptiles, 9.—COOP. & SUCKL., Nat. Hist. Wash. Terr., 1860, 294.—HAYD., Trans. Am. Phil. Soc., xii, 1862, 177.—COPE, Proc. Acad. Nat. Sci. Phila., 1866, 303.

Sceloporus gracilis, BD. & GIR., Proc. Acad. Nat. Sci. Phila., 1852, 175.—GIR., U. S. Exp. Exped., 1858, 386, pl. 20, figs. 1–9.

Observed at Navajo Springs, near the eastern border of the Territory, and along the Colorado Chiquito River in sandy situations.

Like the last, the present species is a wide ranging one, having been obtained by Dr. Suckley at The Dalles of the Columbia, or at Steilacoom. This gentleman found it living in the crevices of basaltic rocks, and notices variations in color, according to the situation frequented, like those below mentioned in the case of *Uta symmetrica.*

20. **Uta gratiosa**, (*Hallow.*) *Bd.*

Uro-saurus gratiosus, HALLOW., Proc. Acad. Nat. Sci. Phila., vii, 1854, 92.—*Id.*, P. R. R. Rep., x, 1859, Williamson's Route, Reptiles, 4.

Uta graciosa, BD., U. S. & Mex. Bound. Surv., ii, pt. ii, 1859, Reptiles, 7.—COPE, Proc. Acad. Nat. Sci. Phila., 1866, 310.—*Id.*, Check-List, 1875, 48.

In referring this species, originally described as a *Urosaurus*, to the genus *Uta*, Professor Baird quotes specimens from the Colorado River, collected by Mr. Arthur Schott. It was not observed by me.

21. **Uta stansburiana**, *Bd. & Gir.*

Uta stansburiana, BD. & GIR, Proc. Acad. Nat. Sci. Phila., vi, 1852, 69.—*Iid.*, Stans. Rep. Exp. Great Salt Lake, 345, pl. 5, figs. 4–6.—BD., U. S. & Mex. Bound. Surv., ii, pt. ii, 1859, Reptiles, 7.—*Id.*, P. R. R. Reptiles, x, 1859, Whipple's Route, Reptiles, 37.—COPE, Check-List, 1875, 48.

Originally described from the valley of the Great Salt Lake, where specimens were secured by Capt. Howard Stansbury, this species has since been ascertained to inhabit Texas, New Mexico, and Arizona. It was taken in the desert of the Gila by Mr. Arthur Schott. No specimens are contained in my collections.

22. **Uta ornata**, *Bd. & Gir.*

Uta ornata, BD. & GIR., Proc. Acad. Nat. Sci. Phila., 1852, 126.—BD., U. S. & Mex. Bound. Surv., ii, pt. ii, 1859, Reptiles, 7.—COPE, Check.List, 1875, 48.
Uta ornata var. *linearis*, BD., *l. c.* (Los Nogales).

Obtained by the naturalists of the United States and Mexican Boundary Survey at various points along the line, and therefore properly to be included in the present connection.

?**Uta symmetrica**, *Bd.* (=*ornata?*)

Uta symmetrica, BD., Proc. Acad. Nat. Sci. Phila., 1858, p. — —*Id.*, U. S. & Mex. Bound. Surv., ii, pt. ii, 1859, Reptiles, 7.—COPE, Proc. Acad. Nat. Sci. Phila., 1866, 303.

"Larger dorsal scales in four regular series, two on either side of the median smaller ones; head short, depressed, one and a half times as wide as deep; tail one and a half times the head and body; general color light-brown above, the belly white; sides with broad transverse bands of blackish; size of *U. ornata.* Gila River."—(*Baird.*)

The above description may be supplemented with a statement of the great variation in coloration during life of different specimens, as observed by me. This variation, which I have no reason to presume to be confined to the present species of the genus, is mainly dependent upon the character of the rocks which the animals frequent. Out of great numbers of specimens procured in one locality, namely, Bero Springs, near Fort Wingate, N. Mex., and unquestionably the same species, almost the only color mark common to all was the pale yellow throat. Some were plain silvery-white below, others were bright greenish-olive on the belly. Above, the color ranges from a deep grayish-black, to a dull grayish-brown with dark lateral streak. I satisfied myself that the same individual assumed these different colors according to the kind of rocks it happened to be upon. The blackish specimens were invariably found upon dark lava rocks, the lighter ones upon yellowish sandstone. We have here an interesting case of protective assimilation. The same thing has been noted by Dr. H. C. Yarrow, as this gentleman informs me, in other cases, and it is doubtless of more general occurrence than has been fully recognized by naturalists. The common Fence

Lizard of the Eastern States (*Sceloporus undulatus*) exhibits corresponding changes in color.

This species is very abundant in certain rocky places. Like others of the genus, it is difficult to secure without mutilation, on account of its extreme agility and the fragility of the tail. Specimens may be best obtained by switching them off the rocks with a light rod drawn rapidly along the surface.

23. **Crotaphytus collaris**, (*Say*) *Holbr.*

Agama collaris, SAY, Long's Exped. Rocky Mts., ii, 1823, 252.—HARLAN, Med. & Phys. Res., 1835, 142.

Crotaphytus collaris, HOLBR., N. A. Herp., ii, 1842, 79, pl. 10.—BD. & GIR., Marcy's Rep. Exp. Red Riv., 1853, 222.—BD., U. S. & Mex. Bound. Surv., ii, pt. ii, 1859, Reptiles, 6.—*Id.*, P. R. R. Rep., x, 1859, Gunnison's & Beckwith's Route, Reptiles, 19, pl. 24, figs. 1*a*–*e*.—*Id.*, *ib.*, Whipple's Route, Reptiles, 38.—COPE, Proc. Acad. Nat. Sci. Phila., 1866, 302.—*Id.*, Check-List, 1875, 47.

A long and well known species of striking characters, type of the genus, and of very general distribution in the Southwest. It was observed by me in New Mexico and Arizona at various points from the crossing of the Rio Grande at Los Pinos to the Colorado Chiquito River, where it was particularly abundant, on dry, sandy soil among fallen timber, in brush-heaps, &c. It was not noted at Fort Whipple, and probably does not occur in the coniferous mountainous portions of the Territory.

The length of this species, as commonly observed, is eleven or twelve inches. The colors in life, when the animal is in full vigor, are strikingly rich and varied; they fade noticeably before death when the creatures fret and pine in captivity, and certainly no description taken from alcoholic specimens, even comparatively fresh, conveys an accurate idea of the richness of the tints. The throat is loose and dilatable, and the animal has a habit of puffing it out when hissing under irritation or in anger.

This lizard is one of the more agile species of its group, though not so remarkably swift-footed as some of the *Cnemidophori*, and is difficult to capture alive without injury. It is one of the boldest, fiercest, and most irascible of its kind; those that I kept in confinement proved entirely untamable. They not only defended themselves with spirit and vigor by biting when handled or irritated, but sometimes assumed the offensive, leaping to attack to the full length of the cord which confined them. Their behavior was in striking contrast to that of the Horned Frogs picketed with them. The lizards lay sullen, but not cowed, watching every movement of the persons around them with glittering eyes, ready to spring upon an intruder without warning. They clung tenaciously to a stick or the finger, in which they might fix their teeth, and suffered themselves to be suspended in this manner for some time before relinquishing hold. Now and then they seemed to have fits of ungovernable rage, during which they leaped aimlessly about, and tugged persistently at the cord. They refused to eat, apparently from pure chagrin, and all died within a few days.

24. **Crotaphytus wislizeni**, *Bd. & Gir.*

Crotaphytus wislizenii, BD. & GIR., Proc. Acad. Nat. Sci. Phila., 1852, 69.—*Iid.*, Stans. Rep. Exp. Great Salt Lake, 1852, 340, pl. 3.—BD., U. S. & Mex. Bound. Surv., ii, pt. ii, 1859, Reptiles, 7, pl. 31.—*Id.*, P. R. R. Rep., x, 1859, Gunnison & Beckwith's Route, Reptiles, 17.—*Id.*, *ib.*, x, 1859, Whipple's Route, Reptiles, 37.—COOP. & SUCKL., Nat. Hist. Wash. Terr., 1860, 294.—COPE, Proc. Acad. Nat. Sci. Phila., 1866, 303.—COPE, Check-List, 1875, 48.
Crotaphytus (*Gambelia*) *wislizenii*, BD., U. S. & Mex. Bound. Surv., *loc. cit.*, in text.
Crotaphytus gambeli, BD. & GIR., Proc. Acad. Nat. Sci. Phila., 1852, 126.
Crotaphytus fasciatus, HALLOW., Proc. Acad. Nat. Sci. Phila., 1852, 207.—*Id.*, Sitgreave's Exp. Zuñi & Col. Riv., 1853, 115, pl. 5.

This species, very distinct from the last, and only less beautiful, was found associated with it along the Colorado Chiquito River, where, however, it was less abundant. Its habits and traits appear to be much the same. Its distribution in the West is general, in suitable localities, the species being found as far north as The Dalles in Oregon.

25. **Crotaphytus reticulatus**, *Bd.*

Crotaphytus reticulatus, BD., Proc. Acad. Nat. Sci. Phila., 1858, 253 (West Texas).—COPE, Check-List, 1875, 48.

This species, originally described from Western Texas, has been recently added to the fauna of the Territory by the Explorations West of the One-hundredth Meridian.—(See Dr. Yarrow's report, *anteà*.)

26. **Dipsosaurus dorsalis**, (*Bd. & Gir.*) *Hallow.*

Crotaphytus dorsalis, BD. & GIR., Proc. Acad. Nat. Sci. Phila., vi, 1852, 126.—BD., U. S. & Mex. Bound. Surv., ii, pt. ii, 1859, Reptiles, 8, pl. 32, figs. 7–13.

Dipsosaurus dorsalis, HALLOW., Proc. Acad. Nat. Sci. Phila., vii, 1854, 92.—COPE, Proc. Acad. Nat. Sci. Phila., 1866, 310.—*Id.*, Check-List, 1875, 48.

From the Colorado Desert. Not noted by me.

27. **Callisaurus dracontoides ventralis**, (*Hallow.*) *Bd.*

b. VENTRALIS.

Homalosaurus ventralis, HALLOW., Proc. Acad. Nat. Sci. Phila., vi, 1852, 179.—*Id.*, Sitgreave's Exp. Zuñi & Col. Riv., 1854, 117, pl. 6.
Callisaurus ventralis, BD., U. S. & Mex. Bound. Surv., ii, pt. ii, 1859, Reptiles, 8.
Callisaurus dracontoides subsp. *ventralis*, COPE, Check List, 1875, 47.

Like the last, from the southern desert portions of the Territory, and not observed by me. *Callisaurus* var. *gabbi* (Cope, Check-List, 47) is a form from Owen's Valley, California, likely to be found in Arizona.

28. **Sauromalus ater**, *Duméril.*

Sauromalus ater, "DUM., Arch. du Mus."—COPE, Check-List, 1875, 47.
Euphryne obesa, BD., Proc. Acad. Nat. Sci. Phila., 1858, p. —.—*Id.*, U. S. & Mex. Bound, Surv., ii, pt. ii, 1859, 6, pl. 27.—COPE, Proc. Acad. Nat. Sci. Phila., 1866. 310.

Fort Yuma.

29. **Uma notata**, BD.

Uma notata, BD., Proc. Acad. Nat. Sci. Phila., 1858, 253 (Mojave Desert, *Heermann*).—COPE, Check-List, 1875, 47.

Ears distinct; a very long infraorbital plate; palate without teeth; outer face of upper labials plane and broadly vertical; the labials themselves much imbricated and very oblique; scales of body above equal, much smaller than ventral ones; interorbital space with two series of plates; claws very long, slender, and straight; sides with a round black spot.

Head about two-fifths the head and body; above light pea-green, spotted with darker green; beneath white; head and body about two inches long.—(*Descr. orig. gen. et sp.*)

Fort Buchanan, Ariz., *Dr. B. J. D. Irwin, U. S. A.* Originally described from the Mojave Desert.

30. **Holbrookia texana**, (*Trosch.*) *Bd. & Gir.*

Coposaurus texanus, TROSCH., Archiv f. Naturg. for 1850 (1852), 389, pl. 6.
Holbrookia texana, BD. & GIR., Proc. Acad. Nat. Sci. Phila., vi, 1852, 125.—BD., U. S. & Mex. Bound. Surv., ii, pt. ii, 1859, Reptiles, 8, pl. 30.—*Id.*, P. R. R. Rep., x, 1859, Whipple's Route, Reptiles, 38.—COPE, Proc. Acad. Nat. Sci. Phila., 1866, 303.—*Id.*, Check-List, 1875, 47.
Holbrookia approximans, BD.

Originally described from Texas, and subsequently found in Arizona by Mr. Arthur Schott and myself. The specimens do not retain indication of precise locality.

31. **Holbrookia maculata**, *Gir.*

Holbrookia maculata, GIR., Proc. Am. Assoc., iv, 1850 (1851), 201.—*Id.*, Stans. Exp. Great Salt Lake, 1852, 342, pl. 6, figs. 1–3.—*Id.*, Marcy's Rep. Exp. Red. Riv., 1852, 223.—BD., U. S. & Mex. Bound. Surv., ii, pt. ii, 1859, Reptiles, 8.—*Id.*, P. R. R. Rep., x, 1859, Gunnison's & Beckwith's Route, Reptiles, 18.—*Id.*, *ib.*, x, 1859, Whipple's Route, Reptiles, 38.—HAYD., Trans. Am. Phil. Soc., xii, 1862, 177.—COPE, Proc. Acad. Nat. Sci. Phila., 1866, 303.—*Id.*, Check-List, 1875, 47.

Holbrookia affinis, BD. & GIR., Proc. Acad. Nat. Sci. Phila., vi, 1852, 125.—BD., U. S. & Mex. Bound. Surv., ii, pt. ii, 1859, Reptiles, 8.

A species of general distribution in the southwest, along the Arkansas, Red, Canadian, and Platte Rivers, in Utah, Texas, New Mexico, and Arizona, as may be gathered from the foregoing references, and in some places extremely abundant. My specimens are from Fort Whipple, where the animal is common.

31*a*. **Holbrookia maculata propinqua**, *Bd. & Gir.*

Holbrookia propinqua, BD. & GIR., Proc. Acad. Nat. Sci. Phila., vi, 1852, 126.—BD., U. S. & Mex. Bound. Surv., ii, pt. ii, 1859, Reptiles, 8.—COPE, Proc. Acad. Nat. Sci. Phila., 1866, 303.

Numerous specimens, identified with this form by Professor Cope, are represented in my collections, from various points along my line of march, both in New Mexico and Arizona. It is perhaps the most abundant and generally distributed representative of the genus in the latter Territory. The earlier indications are from Texas.

DIPLOGLOSSA.

HELODERMATIDAE.

32. **Heloderma suspectum**, *Cope.*

Heloderma horridum, BD., U. S. & Mex. Bound. Surv., ii, pt. ii, 1859, Reptiles, 11, pl. 26 (not of WIEGM., Isis, 1829, 627, and Herp. Mex., 1834).—*Id.*, P. R. R. Rep., x, 1859, Whipple's Route, Reptiles, 38.—COPE, Proc. Acad. Nat. Sci. Phila., 1866, 303.

Heloderma suspectum, COPE, Check-List N. A. Batrach. & Rept., 1875, 47.—YARROW, Zoöl. Expl. W. 100th Merid., *anteà*.

Loss of the label from the specimen in my collection, noted by Professor Cope as above, led to an erroneous assignment of Fort Whipple as the locality. The specimen was taken in the desert in the vicinity of La Paz on the Colorado River, and I am confident it does not occur in the higher mountainous parts of the Territory.

The "Gila Monster", as this large and repulsive looking reptile is called, appears to be not uncommon in the hot, southern parts of the Territory. A poisonous property is attributed to its saliva by the Mexicans, with whom the belief is also prevalent that it has the power of spirting its supposed venom. The females of the same ignorant people have a superstitious belief in the influence that this and some other Saurians may exercise over certain periodical functions of their sex. I am informed by Dr. H. C. Yarrow that in some localities they attribute to *Amblystoma* a miraculous power of causing conception—a form of superstition doubtless found convenient at times, especially if shared by their male relatives.

GERRHONOTIDAE.

33. **Gerrhonotus nobilis**, *Bd. & Gir.*

Elgaria nobilis, BD. & GIR., Proc. Acad. Nat. Sci. Phila., 1852, 129.
Gerrhonotus nobilis, BD., U. S. & Mex. Bound. Surv., ii, pt. ii, 1859, 11, pl. 25, figs. 1–8.—COPE, Check-List, 1875, 47.

From the copper-mines of "New Mexico" (*i. e.*, Arizona).

G. infernalis is a species from Western Texas, which may yet be found to inhabit Arizona. (COPE, Proc. Nat. Sci. Acad., Phila., 1866, 322; Check-List, 1875, 47.)

LEPTOGLOSSA.

TEIIDAE.

34. **Cnemidophorus sexlineatus gularis**, (*Bd. & Gir.*).

b. GULARIS.

Cnemidophorus gularis, BD. & GIR., Proc. Acad. Nat. Sci. Phila., vi, 1852, 128.—BD. & GIR., Marcy's Rep. Exp. Exped. Red Riv., 1852, 227, pl. 10, figs. 1–4.—BD., U. S. & Mex. Bound. Surv., ii, pt. ii, 1859, Reptiles, 11, pl. 34, figs. 1–6.—*Id.*, P. R. R. Rep., x, 1859, Whipple's Route, Reptiles, 38.

Cnemidophorus sexlineatus var. *gularis*, COPE, Proc. Acad. Nat. Sci. Phila., 1866, 303.
Cnemidophorus guttatus, HALLOW., Proc. Acad. Nat. Sci. Phila., 1854, 192.—*Id.*, P. R. R. Rep., x, 1859, Parke's Route, Reptiles, 23.

Numerous specimens in my collection, from various localities nearly throughout Arizona, where it is abundant. At Fort Whipple, it is the most numerous and characteristic species. It lives in the high dry pine and oak woods, about bushes, brush-heaps, stumps, logs, &c.; but I never saw it on rocks, nor knew it to ascend the trunks of standing trees. It used to frequently enter our tents during the summer, and hunt for flies in a quiet, furtive manner, catching them with great address, and was very favorably regarded by all on this account, though its labors resulted in no sensible diminution of the pests. Though thus familiar, it is an exceedingly timorous animal, and darts out of sight upon the least alarm. It is one of the very swiftest of its agile tribe—the eye can scarcely follow it when running at its best on level ground.

35. **Cnemidophorus grahami**, *Bd. & Gir.*

Cnemidophorus grahami, BD. & GIR., Proc. Acad. Nat. Sci. Phila., 1852, 128.—BD., U. S. & Mex. Bound. Surv., ii, pt. ii, 1859, Reptiles, 10, pl. 32, figs. 1–6.—COPE, Check-List, 1875, 45.

"Los Nogales, *Kennerly*."

36. **Cnemidophorus perplexus**, *Bd. & Gir.*

Cnemidophorus perplexus, BD. & GIR., Proc. Acad. Nat. Sci. Phila., 1852, 128.—BD., U. S. & Mex. Bound. Surv., ii, pt. ii, 1859, Reptiles, 10.—COPE, Check-List, 1875, 46.

"Tucson, Arizona, *Kennerly*."

37. **Cnemidophorus tessellatus gracilis**, (*Bd. & Gir.*).

Cnemidophorus gracilis, BD. & GIR., Proc. Acad. Nat. Sci. Phila., 1852, 128.—BD., U. S. & Mex. Bound. Surv., ii, pt. ii, 1859, Reptiles, 10, pl. 34, figs. 7–14.—COPE, Proc. Acad. Nat. Sci. Phila., 1866, 310.
Cnemidophorus tessellatus subsp. *gracilis*, COPE, Check-List, 1875, 46.

Indicated as occurring in this region in Professor Cope's enumeration of species additional to those contained in my collections.

37*a*. **Cnemidophorus tessellatus melanostethus**, *Cope.*

Cnemidophorus melanostethus, COPE, Proc. Acad. Nat. Sci. Phila., 1863, 104.
Cnemidophorus tessellatus subsp. *melanostethus*, COPE, Check-List, 1875, 46.

From southeastern portions of Arizona.

37*b*. **Cnemidophorus tessellatus tigris**, (*Bd. & Gir.*).

Cnemidophorus tigris, BD. & GIR., Proc. Acad. Nat. Sci. Phila., 1852, 69.—*Iid.*, Stans. Rep. Exp. Great Salt Lake, 1853, 338.—BD., U. S. & Mex. Bound. Surv., ii, pt. ii, 1859, Reptiles, 10, pl. 33.

? *Cnemidophorus marmoratus*, BD. & GIR., Proc. Acad. Nat. Sci. Phila., 1852, 128.

? *Cnemidophorus undulatus*, HALLOW., Proc. Acad. Nat. Sci. Phila., 1854, 94.—*Id.*, P. R. R. Rep., x, 1859, Williamson's Route, Reptiles, 8.

Utah and southward in the Sonoran region. The true *tessellatus* of Say, which splits into several conspecies, is indicated by authors from Colorado.

SCINCIDAE.

38. **Eumeces obsoletus**, (*Bd. & Gir.*) *Cope*.

Plestiodon obsoletum, BD. & GIR., Proc. Acad. Nat. Sci. Phila., 1852, 129.—HALLOW., Sitgreave's Exp. Zuñi & Col. Riv., 1853, 111.

Plestiodon obsoletus, BD., U. S. & Mex. Bound. Surv., ii, pt. ii, 1859, Reptiles, pl. 25, figs. 9–16.—*Id.*, P. R. R. Rep., x, 1859, Whipple's Route, Reptiles, 39.

Plistodon obsoletus, COPE, Proc. Acad. Nat. Sci. Phila., 1866, 304.

Eumeces obsoletus, COPE, Check-List, 1875, 45.

The specimen, to which no locality is assigned in Professor Cope's article above cited, was taken at Bero Springs, N. Mex., from among some aquatic plants of a pool. It appeared sluggish, and was easily caught.

39. **Eumeces guttulatus**, (*Hallow.*) *Cope*.

Lamprosaurus guttulatus, HALLOW., Proc. Acad. Nat. Sci. Phila., 1852, 206.—*Id.*, Sitgreave's Exp. Zuñi & Col. Riv., 1853, 113, pl. 4.

Plestiodon guttulatus, HALLOW., Proc. Acad. Nat. Sci. Phila., 1857, 215.—BD., U. S. & Mex. Bound. Surv., ii, pt. ii, 1859, Reptiles, 12.—*Id.*, P. R. R. Rep., x, 1859, Whipple's Route, Reptiles, 18.

Plistodon guttulatus, COPE, Proc. Acad. Nat. Sci. Phila., 1866, 304.

Eumeces guttulatus, COPE, Check-List, 1875, 45.

From Fort Whipple, my collection.

OPHIDIA.

SOLENOGLYPHA.

CROTALIDAE.

40. **Caudisona confluenta**, (*Say*) *Cope*.

Crotalus confluentus, SAY, Long's Exped. Rocky Mts., ii, 1823, 48.—BD. & GIR., Cat. N. A. Rept., 1853, 8.—BD., P. R. R. Rep., x, 1859, Whipple's Route, Reptiles, 40, pl. 24, f. 4.—*Id.*, U. S. & Mex. Bound. Surv., ii, pt. ii, 1859, Reptiles, 14.—COOP. & SUCKL., Nat. Hist. Wash. Terr., 1860, 295, pl. 12.—COPE, Check-List, 1875, 33.

Caudisona confluenta, COPE, App. Mitchell's Researches, 1861, 122.—COPE, Proc. Acad. Nat. Sci. Phila., 1866, 307.—ALLEN, Proc. Bost. Soc. Nat. Hist., xvii, 1874, 307, 309.

Crotalus lecontei, HALLOW., Proc. Acad. Nat. Sci. Phila., vi, 1851, 180.—*Id.*, Sitgreave's Exp. Zuñi & Col. Riv., 1853, 139, pl. 18.—*Id.*, P. R. R. Rep., x, 1859, Williamson's Route, Reptiles, 18, pl. 3.—HEERM., *ibid.*, 25.

Caudisona lecontei, COPE, App. Mitchell's Res., 1861, 121.—HAYD., Trans. Am. Phil. Soc., xii, 1862, 177.

Caudisona confluenta var. *lecontei*, COPE, Proc. Acad. Nat. Sci. Phila., 1866, 307.

Crotalus cinereous (sic), LECONTE *apud* HALLOW., Sitgreave's Exp. Zuñi & Col. Riv., 1853, 140 (in text).

Examination of more material than that before Professor Cope in 1866 has failed to substantiate the distinctions sought to be maintained between *confluenta* and *lecontei;* and as the two forms do not appear to be geographically separated, there may be no necessity for their varietal recognition.

I have found this common and wide ranging species in various parts of the West, from the British possessions nearly to the Mexican border. At the North it is the only representative of its family, and one extremely abundant in some regions, as those of the Yellowstone, Upper Missouri, and Milk Rivers. In Arizona, where it is associated with several other species, it is not so abundant, nor is it the characteristic form. I found it only in the country about Prescott and on the San Francisco Mountains.

Specimens.

No. 510. San Francisco Mountains, July, 1864.

No. 572. Fort Whipple.

No. 678. Fort Whipple; length 31 inches; stomach contained an adult *Sialia mexicana* (Bluebird).

No. 801. San Francisco Mountains, at an altitude of about 10,000 feet.

41. **Caudisona molossus**, (*Bd. & Gir.*) *Cope.*

Crotalus molossus, BD. & GIR., Cat. N. A. Rept., 1853, 10 (New Mexico).—BD., U. S. & Mex. Bound. Surv., ii, pt. ii, 1859, Reptiles, 14, pl. 2.—*Id.*, P. R. R. Rep., x, 1859, Reptiles, pl. 24, f. 5 (no text).—COPE, Check-List, 1875, 45.

Caudisona molossus, COPE, Mitchell's Res., 1861, 124.—*Id.*, Proc. Acad. Nat. Sci. Phila., 1866, 307, 308.

Crotalus ornatus, HALLOW, Proc. Acad. Nat. Sci. Phila., vii, 1852, 192.—*Id.*, P. R. R. Rep., x, 1859, Parke's Route, Reptiles, 23, pl. 2.

Muzzle broad; rostral small; scales between superciliaries small, uniform, except the two anterior; two frontals; four post-frontals; two intersupercilia, all in contact; five rows of scales between labials and suborbital row; middle row not extending beyond the middle of the orbit; labials 18 above, fifth and sixth largest; 17 below; dorsal rows of scales 29; two external rows small; tail uniform black; color roll-sulphur, a series of chestnut-brown transverse lozenges, with exterior corners produced to the abdomen; centers of lozenges with one or two spots; each scale but one color; a brown patch below and behind the eye. "One of the most strongly marked of all the species" (*B. & G.*). Top of muzzle with three pairs of symmetrical shields; rattle parallelogrammic (*Cope*).

San Francisco Mountains, on dry, rocky ground, July, 1864.

42. **Caudisona lucifer**, (*Bd. & Gir.*) *Cope*.—Black Rattlesnake.

Crotalus lucifer, BD. & GIR., Proc. Acad. Nat. Sci. Phila., vi, 1852, 177.—*Iid.*, Cat. N. A. Rept., 1853, 6.—GIR., U. S. Exp. Exped., 1858, 187, pl. 15, figs. 1–6.—BD., P. R. R. Rep., x, 1859, Williamson's & Abbott's Route, Reptiles, 10, pl. 11.—COOP. & SUCKL., Nat. Hist. Wash. Terr., 1860, 295.—COPE, Check-List, 1875, 33.

Caudisona lucifer, COPE, Mitchell's Res., 1861, 121.—*Id.*, Proc. Acad. Nat. Sci. Phila., 1866, 307, 309.

Head short, broad, and deep, with much rounded angles; snout less pointed than in *C. confluenta.* Labials 15–16 above and below; dorsal rows, 25. A dorsal series of brown blotches not margined with white; two small irregular lateral series on each side. Posteriorly, 15–20 half-rings, becoming blackish in old specimens. Light stripe from orbit below superciliary to angle of jaw above labials. Light stripe before eye expanding upon the whole of the upper labials and front and sides of face below crown and in front of orbits. Single light transverse line on superciliaries often obsolete. Rostral not edged with lighter.—(*Kennicott.*) Rattle parallelogrammic; rostral elevated, narrow, cuneiform; muzzle with two marginal shields above each canthus rostralis, and numerous small plates above.—(*Cope.*)

Originally described from Oregon, as above. This species was subsequently found in California, and later its range was shown to include Arizona by Dr. B. J. D. Irwin, U. S. A., and myself.

My numerous specimens from Arizona are nearly black, especially on the head, differing so decidedly from the Oregon type as to probably warrant varietal distinction of this southern form, which may be called *C. lucifer* var. *cerberus*.

Nos. 509, 511, and others, my collection, from the San Francisco Mountains, July, 1864.

This species was found associated with *C. confluenta* in Northern and Central Arizona, where it is abundant. The great size to which it attains, the caliber of the body, and black color combine to render it peculiarly repulsive. An unusual degree of virulence is attributed by backwoodsmen to the "Black Rattlesnake", but probably without foundation.

43. **Caudisona adamantea atrox** (*Bd. & Gir*).

Crotalus atrox, BD. & GIR., Cat. N. A. Rept., 1853, 5, 156.—BD., P. R. R. Rep., x, 1859, Whipple's Route, Reptiles, 39, pl. 24, f. 3.—*Id.*, U. S. & Mex. Bound. Surv., ii, pt. ii, 1859, Reptiles, 14, pl. 1.
Caudisona atrox, COPE, Mitchell's Res., 1861, 121.—COPE, Proc. Acad. Nat. Sci. Phila., 1866, 309.
Crotalus adamanteus, subsp. *atrox*, COPE, Check-List, 1875, 33.

Head subtriangular. Plates on head; 2 anterior frontals in contact; between these and superciliaries, on side of crown, 2 imbricated plates; space inclosed occupied by smaller scales; superciliaries bordered by a row of larger scales; the anterior much largest. Three rows of scales between labials and suborbitals. Labials 16 above, first, fifth, and seventh largest; 15 below, first and third largest; dorsal rows 25–27; 2 exterior rows smooth. On the tail 3–6 half-rings. Color yellowish-brown, with a continuous succession of dorsal lozenges, sometimes truncate before and behind; intervals all narrow. A single transverse light line on superciliary. Stripe from superciliary directly to angle of mouth.—(*B. & G., descr. orig.*) Rattle and rostral plates as in *C. lucifer*.

Originally described only from Texas, and quoted with this restriction by Professor Cope in 1861 (*l. c.*). This species was ascertained by the Mexican Boundary Survey, results of which were published in 1859, to occur also in the lower portions of Arizona.

43*a*. **Caudisona adamantea scutulata**, (*Kenn.*).

Caudisona scutulata, KENN., Proc. Acad. Nat. Sci. Phila., 1861, 207.—COPE, Proc. Acad. Nat. Sci. Phila., 1866, 307, 309.
Crotalus adamanteus, subsp. *scutulatus*, COPE, Check-List, 1875, 33.

Rattle and rostral plates as in *C. lucifer.* Dorsal rows, 25; superior labials, 16; 3–4 rows of interorbital scales, bounded in front by two shields. Yellow stripe from eyebrow above rictus oris. Yellowish-brown, with a dorsal series of truncate, brown, yellow-edged rhombs; tail black-ringed.—(*Kenn.*)

Specimen in my collection from the San Francisco Mountains, July, 1864. Twenty inches long.

44. Caudisona tigris (*Kenn.*) *Cope.*

Crotalus tigris, KENN., MSS.—BD., U. S. & Mex. Bound. Surv., ii, pt. ii, 1859, Reptiles, 14, pl. 4 (Gila and Colorado, *A. Schott*).—COPE, Check-List, 1875, 33.
Caudisona tigris, COPE, Mitchell's Res., 1861, 122.—*Id.*, Proc. Acad. Nat. Sci. Phila., 1866, 309.

Body slender; head small, very much depressed, narrow behind; nose remarkably broad and obtuse; whole outline of head nearly quadrangular. Superciliaries and frontals smooth; space between superciliaries very wide; 4 frontals; 6 post-frontals; two rows of scales between suborbital chain (which is complete) and the labials. Labials 14 above; 14–15 below. Dorsal rows 21–23; very slightly carinated. Dorsal scales broad, rounded behind. Color yellowish-ash above, with rather small, indistinct, dorsal brown blotches anteriorly; two posterior thirds of body banded with brown.—(*Kennicott.*) Rattle acuminate; rostral plate equilateral, broad, depressed.—(*Cope.*)

This species, which was not met with by me, occurs in the deserts of the Gila and Colorado Rivers.

45. Caudisona pyrrha, *Cope.*

PLATE XXII.

Caudisona pyrrha, COPE, Proc. Acad. Nat. Sci. Phila., 1866, 308, 310 (type No. 6606, Mus. Smiths. Inst., Arizona, near Fort Whipple, *coll. Coues*).
Crotalus pyrrhus, COPE, Check-List, 1875, 33.

DIAG.—Scales in twenty-five series, broad and rounded; the two inferior rows smooth. Head short and very obtuse; the nostrils opening subvertically. Superior labials higher than long; three rows of temporals smooth; scales of vertex small, keeled; those more anterior, striate. Superciliaries broad, oval, striate. Canthus rostralis none. Inferior labials fifteen, the first and second margining a plate, which meets its fellow in front of the geneials, and is in other species a continuation of the first. Gastrosteges, 178; urosteges, 84; joints of rattle, 9. The general tint of this species is a bright salmon-red; the scales of the inferior rows punctulate with brown.

Nasal plates distinct; muzzle with small plates or numerous scales above [as in *cerastes* and *mitchelli*]. Upper margin of canthus rostralis with small scales like the others. Prenasal separated from rostral by scales; superciliary not prolonged. Rostral broad as long; head very obtuse, rounded. Scales 25 rows; 7 between superciliaries, 3 below orbit; labials, 14; 2 very small preorbitals and 4 loreals. Pale-vermillion varied with yellow on the sides of the belly, with numerous large reddish-bay hexagons, which become transverse bands on posterior two-thirds of length; yellow below. Rattle subacuminate.

This species is one of the most handsomely colored of the genus. Its affinities are with the *C. mitchelli*, Cope; but it exhibits an even higher degree of subdivision of the head shields.—(*Cope, descr. orig.*)

A single specimen of this richly colored species was procured at Cañon Prieto, a locality near Fort Whipple. It is not in the best order, as it was procured under the untoward circumstances of a hasty retreat from hostile Indians. The species is probably rare, as I never met with a second example, while that one procured was regarded as a curiosity by the numerous persons who came to see a "red rattlesnake".

46. Caudisona (Æchmophrys) cerastes, (*Hallow.*).

Crotalus cerastes, HALLOW., Proc. Acad. Nat. Sci. Phila., 1854, 95.—*Id.*, P. R. R. Rep., x, 1859, Williamson's Route, Reptiles, 17.—HEERM., *ibid.*, 24.—BD., U. S. & Mex. Bound. Surv., ii, pt. ii, 1859, Reptiles, 14, pl. 3.—COPE, Check-List, 1875, 33.

Caudisona cerastes, COPE, Proc. Acad. Nat. Sci. Phila., 1866, p. —.

Æchmophrys cerastes, COUES.

Head small, with rounded angles; nose depressed, obtuse; rostral as broad as high; nostril in middle of a single large plate; lateral edge of superciliary plate elongated into a horn-like process over the eye. Two rows of scales between the suborbital series and the labials; superior labials

11–13, inferior 12–13. Dorsal rows, 21; the scales slightly carinated; each scale along the middle of back with a tubercular swelling toward the center. Crown tubercular. Entire head and upper parts of a light yellowish, with a dorsal series of small indistinct blotches, below which are several irregular rows of isolated brown dots; a narrow brown stripe from orbit back over angle of mouth.—(*Kennicott.*)

The singular development of the supraorbital plate into a horn-like process, contact of prenasal with rostral, and turgidity of the dorsal and coronal scales, may be considered sufficient grounds for the establishment of a section of the genus, as above indicated.

This rather small species is found with *C. tigris* in the arid region of the Gila and Colorado, where it is common. In the position of the nostrils, it appears related to *Aploaspis*, Cope, type *C. lepida*, Kenn. (Proc. Phila. Acad., 1861, 206). This last species, described from Western Texas, will probably yet be found to inhabit Arizona.

46. **Crotalus edwardsi** (*Bd. & Gir.*) *Cope.*

Crotalophorus edwardsii, Bd. & Gir., Cat. N. A. Rept., 1853, 15.—Dum. & Bibr., Erp. Gén., vii, 1853.—Bd., P. R. R. Rep., x, 1859, pl. 24, f. 8 (no text).—*Id.*, U. S. & Mex. Bound. Surv., ii, pt. ii, 1859, Reptiles, 15, pl. 5, f. 1.
Crotalus edwardsii, Cope, Mitchell's Res., 1861, 125.
Caudisona edwardsii, Cope, Check-List, 1875, 34.

Twenty-three rows of dorsal scales; two first rows smooth. Vertical plate subpentagonal, tapering to an acute point posteriorly. Color light yellowish-brown, with chestnut blotches, lighter than in *C. tergeminus* or *C. consors*. Lateral blotches proportionally small. Yellowish line from nostril to angle of mouth as in *C. consors;* no vertebral reddish line.—(*Baird.*)

This species, with which I did not myself meet, occurs on the southern border of the Territory. Specimens have been described from Sonora on the Mexican boundary line; Tamaulipas, Mexico; and Brownsville, Texas.

Crotalus miliarius is a species probably to be added to the fauna of Arizona.

PROTEROGLYPHA.

ELAPIDAE.

48. **Elaps euryxanthus**, *Kenn.*—Harlequin Snake.

Elaps euryxanthus, KENN., Proc. Acad. Nat. Sci. Phila., 1860, 337.—COPE, Proc. Acad. Nat. Sci. Phila., 1866, 307.—*Id.*, Check-List, 1875, 34.

Head very small, narrower than the neck; entirely black as far back as the angles of the mouth. Body banded alternately with black and light brick red, separated by narrow rings of creamy white; all the bands immaculate; first broad ring behind the occiput red instead of black, as in the other species.... The plates of the head and the peculiar style of coloration in this strongly marked species cannot be mistaken. The three colors, each immaculate, glossy, and clear, form a striking contrast, and the red is probably bright carmine in life, thus affording the most beautiful coloration possessed by any North American snake.—(*Descr. orig.*)

A rather large, slender, and very beautiful species, known in Arizona as the "King" and "Ring" snake. It is apparently not abundant. One of my two specimens was taken at Fort Whipple, the other on Date Creek. Excepting the rattlesnakes, this is the only venomous reptile I observed in the Territory.

ASINEA.

COLUBRIDAE.

49. **Heterodon simus nasicus**, (*Bd. & Gir.*).

b. NASICUS.

Heterodon nasicus, BD. & GIR., Stans. Rep. Exp. Great Salt Lake, 1852, 352.—*Iid.* Marcy's Rep. Red Riv., 1852, 208.—*Iid.*, Cat. N. A. Rept., 1853, 61, 157.—HALLOW., Sitgreave's Exp. Zuñi & Col. Riv., 1853, 147.—BD., P. R. R. Rep., x, 1859, Whipple's Route, Reptiles, 41.—HALLOW., Proc. Acad. Nat. Sci. Phila., 1856, 249.—BD., P. R. R. Rep., x, 1859, Beckwith's Route, Reptiles, 19.—*Id.*, U. S. & Mex. Bound. Surv., ii, pt. ii, 1859, 18, pl. 11, f. 1.—HAYD., Trans. Am. Phil. Soc., xii, 1862, 177.—COPE, Proc. Acad. Nat. Sci. Phila., 1866, 307.—ALLEN, Proc. Bost. Soc. Nat. Hist., xvii, 1874, 69 (Yellowstone).

Heterodon simus, subsp. ***nasicus***, COPE, Check-List, 1875, 43.

This species belongs to the section of the genus in which the median plate behind the rostral is separated from the frontals by several small plates. Originally described from Utah, its range has since been ascertained to be general in the West, east of the Rocky Mountains, from the British to the Mexican boundary. I have found it along the northern boundary line, latitude 49°, the northernmost point at which any species of the genus is known to occur; and in most portions of Arizona and New Mexico it is one of the commonest serpents.

50. **Tropidonotus* sipedon couchi**, (*Kenn.*).

Nerodia couchii, KENN., Proc. Acad. Nat. Sci. Phila., 1869, 335 (Nuevo Leone).
Tropidonotus sipedon subsp. *couchii*, COPE, Check-List, 1875, 42.

Resembles *N. erythrogaster* [*i. e.*, *Trop. sipedon* var. *erythrogaster*], but the head is shorter and very broad, the muzzle broad and obtuse. Postorbitals three, much larger than in *N. erythrogaster*, the lower extending forward beyond the middle of the eye. Eight upper labials, all large; the seventh much larger than in *N. erythrogaster*. The dorsal scales are broader and less strongly keeled, and in twenty-three rows. Color uniform light dull slaty brown above, paler than in *erythrogaster*.—(*Kennicott.*)

A species of the Sonoran Region as defined by Professor Cope.

51. **Tropidonotus validus**, (*Kenn.*) *Cope.*

PLATE XXI.

Regina valida, KENN., Proc. Acad. Nat. Sci. Phila., 1860, 334.
Tropidonotus validus, COPE, Proc. Acad. Nat. Sci. Phila., 1866, 310.—*Id.*, Check-List, 1875, 42.

Body stout for the genus. Head large, short but high; broad posteriorly. Snout elongated, narrow, and pointed. Vertical very narrow, obtusely pointed posteriorly; occipitals small; nineteen rows of scales, all carinated; the exterior very feebly. Light brownish-ash above, with faint black markings upon the bases of the scales of the first, fourth, and eighth rows; abdomen entirely uniform yellowish.—(*Kennicott.*)

This species, resembling the better known *T. erythrogaster*, was originally described from Durango, Mexico, and I only know of it as an inhabitant of the Territory from Professor Cope's indication above cited.

* *Nerodia* + *Regina*, Bd. & Gir.

52. **Eutænia ornata,** *Bd. & Gir.*

Eutænia parietalis, BD. & GIR., Cat. N. A. Rept., 1853, 38 (*synon. excl. Non Coluber parietalis Say*).
Eutænia ornata, BD. & GIR., U. S. & Mex. Bound. Surv., ii, pt. ii, 1859, Reptiles, 16, pl. 9.—COPE, Proc. Acad. Nat. Sci. Phila., 1866, 305, 306.—COPE, Check-List, 1875, 41.

In this apparently well marked species, there are nineteen rows of dorsal scales, on the second and third of which are the lateral stripes. The body is rather slender, and the tail more than a fourth of the total length. The color above is olive-brown; the bands are greenish-yellow, and on the stretched skin red spots show among the black ones, which occur along the sides.

In their aquatic habits, these snakes differ from *E. cyrtopsis* and *macrostemma*, neither of which, according to my observations, are specially addicted to the water, frequenting dry herbage and bushes like *E. sirtalis* of the East. I found *E. ornata*, in June, 1864, in considerable numbers, basking about small, shallow pools in the *bosqué* of cottonwood which fringes the Rio Grande at Los Pinos, N. Mex. No specimens are contained in my collections from Arizona itself, where the aquatic species, observed at various points, is *E. vagrans*.

53. **Eutænia cyrtopsis,** *Kenn.*

PLATE XX, FIGS. 2, 2*a*.

Eutænia cyrtopsis, KENN., Proc. Acad. Nat. Sci. Phila., 1860, 333.—COPE, Proc. Acad. Nat. Sci. Phila., 1866, 306.—COPE, Check-List, 1875, 41.

Form very slender; but little stouter than that of *E. saurita;* but with shorter tail, one-fourth the total length. Head and eye large. Superior labials eight, the sixth and seventh largest. Postorbitals three; upper much the largest. Dorsal scales in nineteen rows; lateral stripe on the second and third rows. Olive-brown, with two alternating series of elongated spots between the stripes, giving the appearance of a zigzag line. Dorsal stripe narrow, distinct to the end of the tail, whitish; lateral stripe like the dorsal in color, broad, distinct from head to arms. A series of black spots on the first dorsal row. Abdomen uniform greenish-white. Orbits whitish. Occipital spots obsolete.—(*Kennicott.*)

This species, originally described from Coahuila, Durango, and the Gila River, was found to be quite common at Fort Whipple, where numerous specimens were observed besides the few transmitted to Washington. It grows to a large size, equaling or surpassing *E. sirtalis* in this respect, but preserving its very slender form, which, with the strong, clear stripes, renders it an attractive object. The larger animals, which I kept in confinement, were quite savage when first caught, biting when taken in hand, and even acting sometimes on the offensive when irritated; but they usually became gentle and submissive after a little handling.

53. **Eutænia vagrans**, *Bd. & Gir.*

Eutænia vagrans, BD. & GIR., Cat. N. A. Rept., 1853, 35.—GIR., U. S. Exp. Exped. Herp., 1858, 154, pl. 14, figs. 5–10.—BD., P. R. R. Rep., x, 1859, Beckwith's Route, Reptiles, 19, pl. 17.—COOP. & SUCKL., Nat. Hist. Wash. Terr., 1860, 297.—COPE, Proc. Acad. Nat. Sci. Phila., 305, 307.—COPE, Check-List, 1875, 41.

? *Eutænia angustirostris*, KENN., Proc. Acad. Nat. Sci. Phila., 1860, 332.

As in *E. macrostemma*, the dorsal scales are in twenty-one rows; but the lateral stripe is on the second and third, not third and fourth. There are eight superior labials, of which the sixth and seventh are very large, being higher than wide, one of them extending above the lower level of the eye. The color is a light olive-brown, or ashy-brown; the head indifferently brown or black on top. The bands are narrow, unmargined; the dorsal one strong; the lateral ones less distinct; and there are two series of small, black, lateral spots, which encroach at regular intervals upon both the longitudinal bands.

This species appears to merit its name, since, unlike several others of local distribution, it is widely dispersed in the West. My latest investigations carry its range to the northern borders of Dakota and Montana, and even a little beyond the watershed of the Missouri, into that of the Saskatchewan, as I found it in the summer of 1874 at Chief Mountain Lake, latitude 49° N., where it was associated with the form of *E. sirtalis* called *E. pickeringi*. I also found it in a corresponding latitude farther east, along with *E. radix*. It is a common species in various parts of Arizona and New Mexico. My specimens were found along the Zuñi River in New Mexico, wherever this stream spread into sluggish lagoons, basking on the floating plants, or swimming freely in the water like a *Nerodia* or *Regina.* Others were taken on the San Francisco Mountains in Arizona. They were all rather small, under two feet in length, and differed much in the coloration of the head, which, in some cases, was pitchy black, contrasting strongly with the brown of the body; in others brown. This difference, however, appears to be fortuitous, as both kinds were found together, evidently representing but one species.

?55. **Eutænia macrostemma**, *Kenn.* (*teste Cope*).

Eutænia macrostemma, KENN., Proc. Acad. Nat. Sci. Phila., 1860, 331.—(?)COPE, Proc. Acad. Nat. Sci. Phila., 1866, 306, 307 (rather var. *megalops*).

Dorsal scales in twenty-one rows, the lateral on the third and fourth. Frontal plate longer than the occipital suture. Temporal small, margining only the anterior part of penultimate labial. Post-geneials longer than pre-geneials; superior labials eight; loreal higher than long. Olivaceous, with one row of small black spots below, and two rows above the lateral stripe. Two small black nuchal spots and a short post-oral pale crescent.—(*Cope.*)

In the original description, Mr. Kennicott gives the ground color as very dull yellowish-brown; the dorsal stripe broad, covering nearly three rows of scales, light brownish, but little lighter than the ground color, indistinctly bordered with black; the lateral stripe indistinct, dull yellowish-green, and two broken series of indistinct spots along the sixth and seventh rows of scales; the abdomen immaculate, a varying shade of green; the head dark-brown above; the superior labials but little lighter, narrowly black-bordered behind. The type is from the city of Mexico.

Two specimens from Fort Whipple, identified with this species by Professor Cope. They appear, however, to rather represent the following variety, as the true *macrostemma* probably does not occur so far north.

55*a*. **Eutænia macrostemma megalops**, (*Kenn.*)

Eutænia megalops, KENN., Proc. Acad. Nat. Sci. Phila., 1860, 330.
Eutænia macrostemma subsp. *megalops*, COPE, Check-List, 1875, 41.

"Form shorter and stouter, with proportionally shorter tail than in *E. proxima*, which this species resembles. Tail one-fourth of the total length. Eye very large, greater than in *E. proxima.* First dorsal row of scales broader;

each scale as high as long, and less strongly carinate. Dorsal stripe narrow, covering one and less than two half rows of scales. . . . Color uniform dull brownish-ash or clay color, with the dorsal and lateral stripes whitish-yellow. A few of the scales have narrow black spots on their edges, but these are prominent, and never extend over a scale, appearing as indistinct mottlings of black on the ground color always on the rows next the stripes. The head above is light olive-ash. The lateral stripe is on the third and fourth rows, and is narrower than in *E. proxima*, covering rather less than two half scales. The color below the lateral stripe is a little lighter than that of the back. The exterior dorsal row is much wider than in any of the allied species, each scale being as high as long. The second row is much narrower, though a little wider than the third. The eye is strikingly large, and the superciliaries are raised, rendering the fore part of the crown an inclined plane, yet the muzzle is higher than in *E. proxima*."—(*Descr. orig.*)

With this species, described from Tucson, I am acquainted only by the above description, as no specimens were included in my collections.

56. **Bascanium* tæniatum**, (*Hallow.*) *Cope.*

Leptophis tæniata, HALLOW., Proc. Acad. Nat. Sci. Phila., vi, 1852, 181.—*Id.*, Sitgreave's Exp. Zuñi & Col. Riv., 1853, 133, 146.

Masticophis tæniatus, BD. & GIR., Cat. N. A. Rept., 1853, 103.—BD., P. R. R. Rep., x, 1859, Beckwith's Route, Reptiles, 20, pl. 23.—*Id.*, *ib.*, Abbott's Route, Reptiles, 11.—COOP. & SUCKL., Nat. Hist. Wash. Terr., 1860, 302.—COPE, Proc. Acad. Nat. Sci. Phila., 1866, 305.

Drymobius tæniatus, COPE, Proc. Acad. Nat. Sci. Phila., 1860, 561.

Bascanium tæniatum, COPE, Check-List, 1875, 40.

Masticophis schottii, BD. & GIR., Cat. N. A. Rept., 1853, 160.—BD., U. S. & Mex. Bound. Surv., ii, pt. ii, 1859, Reptiles, 20, pl. 18.

? *Leptophis lateralis*, HALLOW., Proc. Acad. Nat. Sci. Phila., 1853, 237 (a separate variety ?).

Several specimens from various parts of the Territory, including old and young; the former considered by Professor Cope to represent "*Leptophis lateralis*" of Hallowell. The *B. tæniatum*, as redescribed by Baird and Girard, has a broad, brown, dark-edged, dorsal band, six and two half scales wide, and yellow lateral bands, four scales wide, with a dark line on each scale, and another dark line along the edge of the abdomen (six dark lines

* *Bascanion* + *Masticophis*, Bd. & Gir.

in all); the belly yellowish. The characters of a supposed second species, *M. schottii*, have been found by Cope not to hold good.

Var. *ornatum* (*Masticophis ornata*, B. & G., Cat., 102), from Western Texas, is probably to be added to the list.

57. **Bascanium flagelliforme testaceum**, (*Say*) *Cope*.

Coluber testaceus, SAY, Long's Exped. Rocky Mts., ii, 1823, 48.—HARLAN, Jour. Acad. Nat. Sci. Phila., v, 1827, 348.—HOLB., N. A. Herp., iii, 1842, 63, pl. 13.—HARLAN, Med. & Phys. Res., 1835, 113.—BD. & GIR., Cat. N. A. Rept., 1853, 150.
Masticophis testaceus, BD. & GIR., U. S. Mex. Bound. Surv., ii, pt. ii, 1859, 20, pl. 16.—BD.. P. R. R. Rep., x, 1859, Whipple's Route, Reptiles, 43.—COPE, Proc. Acad. Nat. Sci. Phila., 1866, 305.
Drymobius testaceus, COPE, Proc. Acad. Nat. Sci. Phila., 1860, 561.
Psammophis flavigularis, HALLOW., Proc. Acad. Nat. Sci. Phila., vi, 1852, 178.—*Id.*, Sitgreave's Exp. Zuñi & Col. Riv., 1853, 131, 146, pl. 11.
Masticophis flavigularis, BD. & GIR., Cat. N. A. Rept., 1853, 99, 159.
Herpetodryas flavigularis, HALLOW., P. R. R. Rep., x, 1859, Williamson's Route, Reptiles, 12.
Bascanium flagelliforme subsp. *testaceum*, COPE, Check-List, 1875, 40.

The recognition, in the present species, of the *Coluber testaceus* of Say, satisfactorily completes the identification of the species of that eminent and successful naturalist.

The present is a species of general distribution in Arizona, being met with both in the mountainous and desert portions of the Territory, and is one of the common Ophidians.

57*a*. **Bascanium flagelliforme piceum**, *Cope*.

Bascanium flagelliforme subsp. *piceum*, COPE, Dr. Yarrow's Rep., *anteà*.—*Id.*, Check-List, 1875, 40.

Camp Grant, Ariz.

58. **Pityophis sayi bellona**, (*Bd. & Gir*).

a. SAYI.

Coluber melanoleucus var., SAY.—HARLAN, Jour. Acad. Nat. Sci. Phila., v, 1827, 360.—*Id.*, Med. & Phys. Res., 1835, 123.
Coluber sayi, SCHL., Ess. Physiog. Serp., 1837, 157 (not *Coronella sayi*, HOLBR., nor *Coluber sayi*, HARLAN, which are *Ophibolus*).—BD. & GIR., Cat. N. A. Rept., 1853, 151.

Pityophis sayi, BD. & GIR., Cat. N. A. Rept., 1853, 152 (in text).—KENN. *apud* COOP. & SUCKL., Nat. Hist. Wash. Terr., 1860, 300, pl. 22.—HAYD., Trans. Am. Phil. Soc., xii, 1862, 177.

b. BELLONA.

Churchillia bellona, BD. & GIR., Stans. Rep. Exp. Great Salt Lake, 1852, 350.
Pituophis bellona, BD. & GIR., Cat. N. A. Rept., 1853, 66, 157.
Pityophis bellona, KENN. *apud* BD., P. R. R. Rep., x, 1859, Williamson's Route, Reptiles, 42.—BD., P. R. R. Rep., x, 1859, Beckwith's Route, Reptiles, 19.—KENN. *apud* BD., U. S. & Mex. Bound. Surv., ii, pt. ii, 1859, Reptiles, 18.—COPE, Proc. Acad. Nat. Sci. Phila., 1866, 305.—ALLEN, Proc. Bost. Soc. Nat. Hist., xvii, 1874, 69.
Pityophis affinis, HALLOW., Proc. Acad. Nat. Sci. Phila., vi, 1852, 181.—*Id.*, Sitgreave's Exp. Zuñi & Col. Riv., 1853, 130, 146, pl. 10.
Pityophis sayi var. *bellona*, COPE, Check-List, 1875, 39.

Although perfectly harmless, this species is commonly called "Adder", like *Heterodon nasicus* and other blotched serpents, more or less resembling some of the venomous ones; it is also known as Bull Snake. *P. bellona* is one of the most abundant of all the species about Fort Whipple, where specimens could be found at any time during the summer in the grass and woods. Numbers used to be killed in the fort and surrounding buildings.

The great variability of this species, not only in color, but also in the details of the plates of the head, points not at first recognized, necessitates the reference of *bellona* to the longer known *Coluber sayi*. The use of the name "*sayi*" in this connection is to be carefully discriminated from its employ for an altogether different species of *Ophibolus*.

In addition to the foregoing, Professor Cope indicates a subspecies *mexicana* (D. & B.), as occurring in the Central and Sonoran Regions. Of this form I know nothing.

59. **Pityophis elegans** (*Kenn.*) *Cope.*

Arizona elegans, KENN., *apud* BD., U. S. & Mex. Bound. Surv., ii, pt. ii, Reptiles, 18.
Pityophis elegans, COPE, Check-List, 1875, 39.

"Sonoran Region."

60. **Ophibolus getulus boylii**, *Bd. & Gir.*

Ophibolus boylii, BD. & GIR., Cat. N. A. Rept., 1853, 82.—BD., P. R. R. Rep., x, 1859, Williamson & Abbott's Route, Reptiles, 11.—*Id.*, U. S. & Mex. Bound. Surv., ii, pt. ii, 1859, Reptiles, 20.—COPE, Proc. Acad. Nat. Sci. Phila., 1866, 305.

Lampropeltis boylii, COPE, Proc. Acad. Nat. Sci. Phila., 1860, 255.
Coronella balteata, HALLOW., Proc. Acad. Nat. Sci. Phila., 1853, 236.—*Id.*, P. R. R. Rep., x, 1859, Williamson's Route, Reptiles, 14.
Ophibolus getulus subsp. *boylii*, COPE, Check-List, 1875, 37.

Specimens from Date Creek, fifty miles south of Fort Whipple. This species, originally described from California, has an extensive dispersion in the West as far north as the Yellowstone, as I find a specimen in a collection from that region.

60*a*. **Ophibolus getulus splendidus**, *Bd. & Gir.*

Ophibolus splendidus, BD. & GIR., Cat. N. A. Rept., 1853, 83 (Sonora).—BD., U. S. & Mex. Bound. Surv., ii, pt. ii, 1859, Reptiles, 20, pl. 14 (Arizona).—COPE, Proc. Acad. Nat. Sci. Phila., 1866, 310.
Lampropeltis splendida, COPE, Proc. Acad. Nat. Sci. Phila., 1860, 255 (Fort Buchanan, Ariz).
Ophibolus getulus subsp. *splendidus*, COPE, Check-List, 1875, 37.

Not contained in my collections. It appears to be rather an inhabitant of the southern portions of the Territory and southward.

61. **Ophibolus pyromelas**, *Cope*.

PLATE XIX, FIGS. 1, 1*a*, 2.

Ophibolus pyromelanus, COPE, Proc. Acad. Nat. Sci. Phila., 1866, 305 (described from Nos. 731 and 760 of my collection from Fort Whipple, August, 1864).
Ophibolus pyromelas, COPE, Check-List, 1875, 37.

"Scales in 23 longitudinal rows; tail five and one-half times in total length. Scuta 224, 1, 66. Fifty to fifty-eight black annuli on an ochraceous white ground, on the body; each anteriorly completely, posteriorly more or less incompletely, split by a vermilion annulus; all extending with irregularities on the belly.

"Head quite distinct from body; muzzle contracted. Frontal plate broad, with prolonged apex; parietals elongate, emarginate behind; cephalic shields otherwise as in *polyzonus*, *splendidus*, &c. Postgeneials half the length of the pregeneials. Dorsal scales rather broad; outer series not abruptly enlarged. In one specimen, all the black annuli to the middle of the tail are divided by the red, thus leaving the black as a margin to it; hence the number of these annuli is fewer; they are four scales wide behind the middle of the body. In another specimen, only four anterior rings are completely divided, those on the following third of the length being divided by red on the sides; the remaining annuli black, three scales wide; white annuli one and one-half scales; anterior or nuchal red; annulus widest, its anterior black margin attaining parietals; an ochraceous band from gular region, not quite completed across parietals. Muzzle, prefontal plates, and labial margin ochraceous; remainder of top and sides of head black. Total length, 30.5 inches.

"This species has a longer body than the known red-ringed species, and is, indeed, most closely related to the *O. boylii*. It will always be distinguished for the latter by the much more numerous annuli (twenty-eight in *boylii*)."—*Cope.*

62. **Phimothyra grahamiæ**, (*Bd. & Gir.*) *Cope.*

Salvadora grahamiæ, BD. & GIR., Cat. N. A. Rept., 1853, 104, 161 (type of genus). (Sonora, Mex.).—BD., U. S. & Mex. Bound. Surv., ii, pt. ii, 1859, 21, pl. 5, f. 2.
Phimothyra grahamiæ, COPE, Proc. Acad. Nat. Sci. Phila., 1866, 310.—*Id.*, Check-List, 1875, 38.

Lower California, Sonora, and Arizona, to Utah and Texas.

62*a*. **Phimothyra grahamiæ hexalepis**, *Cope.*

Phimothyra hexalepis, COPE, Proc. Acad. Nat. Sci. Phila., 1866, 305 (described from my specimen from Fort Whipple).
Phimothyra grahamiæ subsp. *hexalepis*, COPE, Check-List, 1875, 38.

"Resembles the *P. grahamiæ* (*Salvadora grahamiæ*, B. & G.), but differs in having a shorter tail, five and one-third times in length, instead of four times; eye resting on sixth supralabial, on account of the presence of three narrow preoculars, two or three loreals—largest higher than long; nostril on suture between nasals and internasals; dorsal stripe narrow—one and two half scales, and lateral brown band wide, four and a half to five scales, whose *superior* margins are ochraceous at base. Rostral plate well developed, higher than broad; nasals elongate, much depressed, anterior extending behind first labial; postoculars two; two long narrow temporals. Width of occipitals nearly equal common suture. Nine superior labials; first pair inferior labials much dilated medially, their common suture nearly equal that of pregeneials; scales seventeen rows. Gastrosteges, 176; urosteges, 75. Tail and below uniform yellowish."—*Cope.*

Fort Whipple. The stomach contained a *Cnemidophorus sexlineatus.*

63. **Trimorphodon lyrophanes**, *Cope.*

Lycodon lyrophanes, COPE, Proc. Acad. Nat. Sci. Phila., 1860, 343.
Trimorphodon lyrophanes, COPE, Proc. Acad. Nat. Sci. Phila., 1866, 310.—*Id.*, Check-List, 1875, 38.

Scales in 21 rows, broad posteriorly and upon the middle of the body, the dorsal series not larger; body anteriorly slender; neck contracted. Tail not one-sixth the total length. Head broadest just behind the eyes; constricted at the orbits. Muzzle rather narrow and truncate. Rostral broader than high, with four sutural borders; the superior one very long; the apex apparent upon the surface of the head. Prefrontals much broader than long; one-third the size of that part of the postfrontals visible from above. Postfrontals longer than broad upon vertical view. Occipitals, superciliaries, and vertical well developed; the last right-angled posteriorly, with slightly concave and convergent lateral borders. Occipitals not longer than vertical; about as wide as long; in contact with a large scale in their posterior common emargination. Nasal plates distinctly divided, very small, higher than long. Loreals *two;* anterior one higher than long, wedged above between pre- and postfrontal; the posterior as long as high. Preoculars *three;* the superior largest, not in contact with the vertical; the inferior bounded anteriorly by the third upper labial. Postoculars three; the inferior a little the largest. Superior labials nine; fourth and fifth entering the orbit, sixth largest, higher than broad. Inferior labials twelve, third and fourth narrow, and much produced posteriorly. Geneials two pair, the anterior longest. Gastrosteges, 236; one divided anal; urosteges, 70. Length, 27 inches 10 lines; tail, 4 inches 4 lines. Ground color light-gray. Muzzle crossed by an indistinct ashy band, which extends upon the anterior part of the postfrontals. The posterior half of these plates is involved in a deep-brown band across the head between the eyes, with concave posterior border, extending upon the superciliaries, and continued posteriorly upon the inferior postorbital and sixth upper labial. A pair of broad diverging bands, commencing one on either side of the center of the vertical, crosses the superciliary and occipital, follows the expanded outline of the temporal and tympanic regions, contracts and becomes parallel and longitudinal upon the neck. A brown spot upon the posterior extremity of the verticals, with its posterior elongation, completes the resemblance to a lyre, or rather a Jew's-harp. The ground color appears upon the vertex as an anchor-shaped figure, and on the cheeks as an oblique band. The back, as far as the anus, has 21 pairs of deep-brown spots, their gemination only apparent anteriorly by the punctulate character of the scales in their intervals. These are always about 3 scales wide; the lesser ones 2½ anteriorly, 1½ posteriorly. Dorsal spots 7 scales wide; as the scales are broader posteriorly, the spots are also. An irregular series of lateral spots, one opposite each of the intervals, sometimes confluent with the dorsal spots, anteriorly forming a very narrow broken band. Another series of spots involves the tips of pairs of the gastrosteges, which are separated by 2 to 5 immaculate ones. Ten confluent pairs of spots on the tail above. Whole under surface whitish.—*From the orig. descr.*

This tropical form is one of the most interesting of the later additions to our fauna. It was first discovered in the United States by Dr. B. J. D. Irwin, U. S. A., at Fort Buchanan, Ariz.; but had been previously taken at Cape Saint Lucas. In describing it, Professor Cope at first referred it to the genus *Lycodon*, Duméril, but subsequently placed it in *Trimorphodon.*

The *Sibon annulatum* var. *septentrionalis* (*Dipsas septentrionalis*, Kenn., U. S. & Mex. Bound. Surv., ii, pt. ii, 1859, Reptiles, 16), from Southwestern Texas, will probably be found in the Territory. For the position of this form, see *Cope*, Proc. Phila. Acad., 1860, 266.

64. Hypsiglena ochrorhyncha chlorophæa, *Cope.*

a. OCHRORHYNCHA.

Hypsiglena ochrorhynchus, COPE, Proc. Acad. Nat. Sci. Phila., 1860, 246 (type of the genus), (Cape Saint Lucas).—*Id.*, Check-List, 1875, 38.

b. CHLOROPHÆA.

Hypsiglena chlorophæa, COPE, Proc. Acad. Nat. Sci. Phila., 1860, 247 (Fort Buchanan, Ariz.).

Hypsiglena ochrorhynchus var. *chlorophæa*, COPE, *op. cit.*, 1866, 304.

Hypsiglena ochrorhynchus subsp. *chlorophæa*, COPE, Check-List, 1875, 38.

My specimens from Arizona, the precise locality not indicated, are of the variety *chlorophæa*, originally described as a distinct species. The numerous small dorsal spots are mostly divided. The original form is from Lower California. The Arizonan variety is described substantially as follows:—

Rather more slender than *H. ochrorhynchus ;* scales more elongate, and rows more oblique; vertical plate rather broader, and head relatively narrower. Color greenish-ash, darker than in *H. ochrorhynchus ;* the dorsal spots black instead of brown, smaller 2 scales apart, 1½ scale long, 58 to 66 in number on the body, occupying only the space from the ninth to the thirteenth dorsal row, frequently dividing and alternating; two rows of smaller alternating spots on the sides, one on sixth and seventh rows, the other on the fourth row; crown and muzzle thickly dotted with black; beneath pale olivaceous; distribution of colors on the head and neck much as in *H. ochrorhynchus*, but the neck spots rather larger, the brown replaced by black, the ochraceous by olevaceous. Gastrosteges, 167; urosteges, 55. Length, 15½ inches; tail, 2¼.

65. Diadophis regalis, *Bd. & Gir.*

Diadophis regalis, BD. & GIR., Cat. N. A. Rept., 1853, 115, 161 (Sonora).—BD., U. S. & Mex. Bound. Surv., ii, pt. ii, 1859, 22.—COPE, Proc. Acad. Nat. Sci. Phila., 1866, 310.—COPE, Check-List, 1875, 38.

From the Sonoran border; not met with by me. According to its describers, this species lacks an occipital ring; dorsal scales 17 rows; body above uniform greenish-ash; below light-yellow, with numerous small black spots.

? Diadophis docilis, *Bd. & Gir.*

Diadophis docilis, BD. & GIR., Cat. N. A. Rept., 1853, 114 (Devil's River, Texas).—BD., U. S. & Mex. Bound. Surv., ii, pt. ii, 1859, 22, pl. 21, f. 3 (Tucson, &c.).

Not seen by me. It is described as uniform ashy-gray above, yellowish-white spotted with black below, with a broad yellowish-white occipital ring, bordered with a narrow black line; the dorsal scales in 15 rows.

66. Rhinochilus lecontii, *Bd. & Gir.*

Rhinocheilus lecontei, BD. & GIR., Cat. N. A. Rept., 1853, 120, 161 (type of the genus), (San Diego, Cal.).

Rhinochilus lecontii, BD., U. S. & Mex. Bound. Surv., ii, pt. ii, 1859, Reptiles, 21, pl. 20 (San Pedro, Tex.).

Rhinochilus lecontei, COPE, Proc. Acad. Nat. Sci. Phila., 1866, 304 (variety).—COPE, Check-List, 1875, 36.

My specimens from Arizona, without indication of particular locality, represent, according to Professor Cope, a variety, having fewer (twenty) black half-rings on the body, extending to the gastrosteges and separated by a narrow interval. Abdomen with subquadrate black spots opposite the former and their intervals.

67. Gyalopium canum, *Cope.*

PLATE XVIII, FIG. 2, 2*a*.

Gyalopion canum, COPE, Proc. Acad. Nat. Sci. Phila., 1860, 243.
Gyalopium canum, COPE, Proc. Acad. Nat. Sci. Phila., 1860, 310.—*Id.*, Check-List, 1875, 36.

Prefrontals triangular, not larger than preoculars. Postoculars of equal size. Anterior border of vertical not angulated. Occipitals as broad as long, truncate posteriorly. Superior labials 7, eye over third and fourth. Inferior labials 7, fourth largest. Geneials one pair, very short. Scales in 17 rows, nearly square. Gastrosteges, 130; one anal; urosteges, 28. Total length, 7½ inches; tail, 11 lines. Above brownish-gray, crossed by 31 irregular transverse brown bands; these are from one to three scales wide on the back, and extend to the gastrosteges; anteriorly they exhibit a tendency to divide into a dorsal and two lateral series of spots; 8 transverse spots on the tail; first spot on the neck large, produced medially to the occipitals. A brown band extends from one angle of the mouth to the other across the occipitals, involving the tip of the vertical; another brown band commences on the upper borders of the lower labial shields, passes through the eye, and crosses the anterior parts of the superciliaries, and vertical and posterior parts of postfrontals and rostral. Dirty yellowish beneath, and upon the first row of scales.—*Descr. orig.*

"It is an extraordinary serpent, resembling, at first sight, a diminutive *Heterodon*", and was discovered at Fort Buchanan, Ariz., by Dr. B. J. D. Irwin, U. S. A.

68. Sonora semiannulata, *Bd. & Gir.*

Sonora semiannulata, BD. & GIR., Cat. N. A. Rept., 1853, 117 (type of genus.)—BD., U. S. & Mex. Bound. Surv., ii, pt. ii, 1859, Reptiles, 21, pl. 19, f. 3.—COPE, Proc. Acad. Nat. Sci. Phila., 1866, 310.—*Id.*, Check-List, 1875, 36.

Sonora and Southern Arizona. Not seen by me.

69. Chionactis occipitalis, (*Hallow.*) *Cope.*

a. OCCIPITALIS.

Rhinostoma occipitale, HALLOW., Proc. Acad. Nat. Sci. Phila., vii, 1854, 95.
Lamprosoma occipitale, HALLOW., Proc. Acad. Nat. Sci. Phila., viii, 1856, 310.—KENN. *apud* BD., U. S. & Mex. Bound. Surv., ii, pt. ii, 1859, Reptiles, 21, pl. 21, f. 1.
Chionactis occipitale, COPE, Proc. Acad. Nat. Sci. Phila., 1866, 310.

b. ANNULATUS.

Lamprosoma annulatum, KENN. *apud* BD., U. S. & Mex. Bound. Surv., ii, 1859, 22, (in text: Colorado Desert, Ariz.; name proposed if distinct from true *occipitale*, HALLOW.).
Chionactis occipitalis subsp. *annulata*, COPE, Check-List, 1875, 36.

Southern and Southwestern Arizona. Not seen by me.

70. **Contia isozona**, *Cope.*

PLATE XVIII, FIGS. 1, 1*a*.

Contia isozona, COPE, Proc. Acad. Nat. Sci. Phila., 1866, 304.—*Id.*, Check-List, 1875, 36.

Two postoculars; six rows of gular scales. Rostral rounded, slightly produced backward. Scuta 158, $\frac{1}{1}$, 52. Twenty black half rings, separated by equal spaces of pinkish ground color. Eye small; its diameter contained twice in length of muzzle.

Preorbital narrower above, not extending above lower margin of superciliary; loreal twice as long as high. Prefrontals and internasals much broader than long; frontal slightly angulate in front, longer than broad; parietals rather elongate, subtruncate behind. Postorbitals subquadrate; temporals 1–2. Postgeneials minute. Superior labials seven, all higher than long; eye over third and fourth. Scales in fifteen rows, all broader than long. Tail four and two-fifths times in total length, which is 10.25 inches. Below immaculate. Tail completely six-annulate.

Resembles the *Sonora semiannulata*, B. & G., but that species has two nasals, three postoculars, the superior reaching the frontal; frontal wider behind than before, and only 149 gastroteges.—*Cope.*

Described from specimens in my collection. The same author indicates a variety of the species from Utah, with longer body, twenty-five black bars; body color vermilion above and yellow below; scuta 167, $\frac{1}{1}$, 52.

71. **Chilomeniscus ephippicus**, *Cope.*

PLATE XVIII, FIGS. 3, 3*a*.

Chilomeniscus ephippicus, COPE, Proc. Acad. Nat. Sci. Phila., 1867, 85.—*Id.*, Check-List, 1875, 35.

Scales broad, in 13 rows. Tail about one-seventh total length. Rostral plate large, entirely separating internasals, not encroaching on prefrontals;

nasal plate separating prefrontals and labials, in contact with preocular. Postoculars 2, upper only in contact with occipital. Superciliaries very narrow; occipitals as broad as long. Temporals $\frac{1}{1}$, large; labials above 7, third and fourth in orbit; these, with second, narrow, erect; first longitudinal. Inferior labials 8; first pair in contact before pregeneials; postgeneials very small. Total length, $5\frac{1}{2}$ inches. Gastrosteges, 113; separated from geneials by 4 rows of gulars; anal, 1–1; urosteges, 28–28. Above reddish or yellowish, with 21 black cross-bars to vent, which are broader than the interspaces, and do not quite reach gastrosteges. Five nearly complete rings on tail. Belly white. From occipitals to anterior part of frontal, with the labials opposite this part (except their lower edges), black.—*Cope.*

Originally described from Owen's Valley, California (*Dr. Horn*), and subsequently indicated as occurring in the Sonoran Region.

In the plate, Fig. 3*a* represents the position of the head shields as seen from above.

72. **Tantilla nigriceps,** *Kenn.*

Tantilla nigriceps, KENN., Proc. Acad. Nat. Sci. Phila., 1860, 328.—COPE, Check-List, 1875, 35.

Form more slender and head narrow than in *T. gracilis.* Vertical plate more elongate posteriorly; occipitals narrower. One anteorbital; two postorbitals. Seven upper labials. Color (in alcohol) uniform brownish-white above, lighter beneath. Crown as far back as the occipitals deep black. No indication of a postoccipital black ring, as in *T. coronata.*

Originally described from Texas and New Mexico, and since ascertained to inhabit Arizona.

SCOLECOPHIDIA.

STENOSTOMATIDAE.

73. **Stenostoma dulce,** (*Bd. & Gir.*).

Rena dulcis, BD. & GIR., Cat. N. A. Rept., 1853, 142.
Stenostoma dulce, COPE, Check-List, 1875, 44.

Indicated by Professor Cope from the Sonoran Region and Texas.

B.—BATRACHIA.

ANURA.

BUFONIFORMIA.

BUFONIDAE.

74. **Bufo microscaphus**, *Cope.*

Bufo microscaphus, COPE, Proc. Acad. Nat. Sci. Phila., 1866, 301.—*Id.*, Check-List, 1875, 29.

Head broader than long, obtuse; muzzle descending in full arc to labial border from line of orbit. Superciliary ridges well marked, but concealed by the thick skin, plane, parallel. Postorbital not prominent. Vertical gutter narrow. Eyes large, prominent; double tympanum. Parotoids broad, smooth. Skin little roughened. Toes two-thirds webbed. Shovel very small, frequently not black-edged; outer tubercle small, heel to end of muzzle. Above blackish, a black spot on each parotoid, and dark light-centered bar on femur and tibia. A yellowish bar across front and palpebræ, and spot on nape; muzzle dark. Total length, 1 inch 5.5 lines; to postorbital ridge, 7.5 lines; fore limb, 1 inch 9 lines; hind limb, 3 inches 2 lines, femur one-third included. . . . The oval well-separated parotoids and general appearance of this species ally it to the *B. speciosus*, Girard. But in that animal the supraorbital ridges are obsolete, and the metatarsal shovel is very much stronger. The *B. dorsalis*, Hallowell (*B. woodhousei*, Gird.), is also allied, but is in all proportions and details more elongate, and has a stronger shovel and head-ridges; it always has the dorsal band, which never exists in the *microscaphus*, and never the transverse face-band of the latter.—*Descr. orig.*

Numerous specimens are in my collection from Fort Whipple, where it is the characteristic species, and very abundant.

75. **Bufo lentiginosus frontosus**, *Cope.*

Bufo frontosus, COPE, Proc. Acad. Nat. Sci. Phila., 1866, 301.
Bufo lentiginosus subsp. *frontosus*, COPE, Check-List, 1875, 29.

A species most allied to the *B. americanus*, but differing in the shorter and more elevated cranium, longer and larger hind limbs, and more acuminate parotoid glands. The canthus rostrales not marked; the muzzle descending very steeply from the anterior angles of the orbits, shorter than the elevated, perpendicular muzzle. Frontal ridges higher than eyelids, rising steeply behind, terminating in two short convergent tuberosities, divergent, with interior crenations behind. Postocular ridge equally developed, sending a very small process to the anterior acuminate extremity of the parotoids. Elevation of cranium at parietal tubercle equal to length of same from the same point. Eye large; tympanum distinct, half eye; parotoid narrow, long, acuminate at both ends. Elbow to anterior margin of orbit; heel to end of muzzle. Skin everywhere with numerous small tubercles; soles rough; toes half webbed. Brown above, with pale vertebral line, and three pairs of deep-brown medium-sized spots, with paler centers. Sides and lips with small brown spots. Femur and tibia with one indistinct brown cross-bar each. Below uniform yellow. Total length, 4 inches, of which the head is 9 lines to postocular ridges; breadth between orbits, 2.5 lines; hind limb, 5 inches; sacrum, 1 inch across.—*Descr. orig.*

One specimen from Los Pinos, Rio Grande, N. Mex.; type of the species.

76. **Bufo alvarius**, *Gir.*

Bufo alvarius, GIR., *apud* BD., U. S. & Mex. Bound. Surv., ii, pt. ii, 1859, 36, pl. 41, figs. 1–6 (valley of the Gila and Colorado).—COPE, Check-List, 1875, 29.

Upper surface of head near plane upon its middle region; orbits bordered by a low and rounded-off ridge; its skin being thin and adhering to the skull. Parotoids well developed and subreniform. Eyes and tympanum rather large also. Tongue elongated, broadest posteriorly. Upper jaw emarginated. Two large carpal callosities. A membranous fold at the inner lower edge of the tarsus. Toes palmated; two metatarsal tubercles. Palms and soles coarsely granular. Upper surface of body exhibiting numerous glandular tubercles. A large pustular swelling upon the thighs. Color uniformly dark-green.—*Descr. orig.*

Originally described, as above, from the southern and western portions of the Territory; not observed by me.

77. **Bufo debilis**, *Gir.*

Bufo debilis, GIR., Proc. Acad. Nat. Sci. Phila., vii, 1854, 57.—GIR., *apud* BD., U. S. & Mex. Bound. Surv., 1859, ii, pt. ii, 27.—COPE, Check-List, 1875, 29.
Bufo insidior, GIR., Proc. Acad. Nat. Sci., 1854, 88.—GIR. *apud* BD., U. S. & Mex. Bound. Surv., ii, pt. ii, 1859, Reptiles, 26, pl. 41, figs. 13–18.

Upper surface of head plane and smooth. Snout subacute, protruding. Mouth moderate; upper jaw slightly emarginated. Tongue elongated, tapering toward both extremities. Tympanum inconspicuous. Parotoids large and elongated, obliquely situated across the shoulders. Limbs moderate. First finger equal to the second in length. A carpal disk and a tubercle. Toes slightly webbed at their base. Two metatarsal tubercles. No membranous fold at the inner lower edge of the tarsus. Skin papillous above, warty beneath. Above of a bluish slate, but with black markings. Beneath unicolor, of a dingy-yellow tint.—*Girard, l. c.*

Chihuahua and "Sonora" (*i. e.* Arizona).

78. **Bufo woodhousii**, *Gir.*

Bufo dorsalis, HALLOW., Proc. Acad. Nat. Sci. Phila., 1852, 181 (*nec Spix*).—*Id.*, Sitgreave's Rep. Exp. Zuñi & Col. Riv., 1853, 142, pl. 19.
Bufo woodhousii, GIR., Proc. Acad. Nat. Sci. Phila., 1854, 86.—*Id.*, U. S. & Mex. Bound. Surv., ii, pt. ii, 1859, Reptiles, 27.—*Id.*, P. R. R. Rep., x, 1859, Gunnison & Beckwith's Route, Reptiles, 20.—*Id.*, *ib.*, x, 1859, Whipple's Route, Reptiles, 44, pl. 25, f. 1.

Head short and thick; upper central surface but little depressed, not to say grooved; the suborbital ridge being slightly elevated. The occipito-temporal ridge is thicker, and hence a little more conspicuous. Snout rounded; nostrils terminal. Mouth wide; upper jaw emarginated. Tympanum and parotids of moderate size. Limbs rather short and stout; first finger much longer than the second; a large metacarpal disk; toes semipalmated; two metatarsal tubercles, a very large and a very small one; no membranous fold at the inner lower edge of the tarsus. Papillæ of medium size upon the back. Inferior surface with rather small, crowded, granular warts. Above dark-brown, with numerous lines of yellow. A dorsal yellowish vitta running the whole length of the body. Transverse blotches of black upon the thighs and forearms. Beneath ochraceous.—*Girard.*

A species which appears to be of very general distribution in the Southwest, and also in Sonora, hence undoubtedly occurring in Arizona, although I am not aware that as yet specimens have actually been taken within the limits of the Territory.

ARCIFERA.

SCAPHIOPIDAE.

79. **Spea hammondi,** (*Bd.*), *Cope.*

Scaphiopus hammondii, Bd., P. R. R. Rep., x, 1859, Williamson & Abbott's Route, Reptiles, 12, pl. 28, f. 2 (Fort Reading, Colo.).
Spea hammondii, Cope, Jour. Acad. Nat. Sci. Phila., 1866, 81.—Cope, Proc. Acad. Nat. Sci. Phila., 1866, 301.—Cope, Check-List, 1875, 31.

Tongue very large, orbicular, without notch behind. Spade highly developed. Color above dark olive-brown, with very indistinct blotches of darker. Summits of dorsal pustulation whitish. Beneath whitish, the chin black. Head and body, 2 inches long; hind leg, $2\frac{1}{2}$. This species is easily distinguished by its nearly uniform and very dark color, without the light lines of *S. holbrookii.*—*Descr. orig.*

Two specimens of this species were taken *in coitu* in June, 1864, near Fort Wingate, N. Mex., considerably extending the previously known range, and indicating the period at which the sexes come together. The animals were found on dry land at a distance from water, in this respect differing from the species of *Bufo*, which, at the period of sexual excitement, seek the water.

HYLIDAE.

80. **Hyla arenicolor,** *Cope.*

Hyla affinis, Bd., Proc. Acad. Nat. Sci. Phila., 1854, 61 (*nec Spix*).—*Id.*, U. S. & Mex. Bound. Surv., ii, pt. ii, 1859, Reptiles, 29, pl. 38, figs. 4–7.
Hyla arenicolor, Cope, Jour. Acad. Nat. Sci. Phila., 1866, 84.—*Id.*, Proc. Acad. Nat. Sci. Phila., 1866, 301.—*Id.*, Check-List, 1875, 31.

Body rough. Tympanum two-thirds the size of eye. Tibia not quite half the length of the body, but reaching more than half way from anus to center of eyes. Color ashy-gray or green, with numerous rounded, dorsal blotches. Three transverse bands on each thigh and leg. No vermiculation on anterior and posterior faces of hind legs, nor on lower part of sides. A light spot under the eye. Web of hand extending only to the third joint of the second finger. Arm from elbow less than tibia, but longer than hind foot; about 1½ inches long.—*Baird, l. c.*

The above description makes no mention of the bright-yellow tint of the sides of the abdomen and inner surface of the thighs, which is conspicuous in life.

My specimens are from Fort Whipple. The species was originally described (as *H. affinis*) from Northern Sonora.

RANIFORMIA.

RANIDAE.

81. **Rana halecina**, *Bosc.* (var.?)

Rana pipiens, GMEL., Syst. Nat., 13th ed., 1788, 1052 (not of authors generally).
Shad Frog, Bartram's Travels, 1790, 274.
Rana halecina, KALM.—DAUDIN, Hist. Nat. Rept., viii, 1803, 122.—HOLBR., N. A. Herp., iv, 1842, 91, pl. 91.—STORER, Mass. Rept., 2d ed., 237.—BD., P. R. R. Rep., x, 1859, Whipple's Route, Reptiles, 45.—COOP. & SUCKL., Nat. Hist. Wash. Terr., 1860, 304, pl. 29, f. 7.—HAYD., Trans. Am. Phil. Soc., xii, 1863, 177.—COPE, Proc. Acad. Nat. Sci. Phila., 1866, 301.—ALLEN, Proc. Bost. Soc. Nat. Hist., xvii, 1874, 70 (and of authors generally).
Rana utricularis, HARLAN, Am. Jour. Sci., x, 1825, 60.—*Id.*, Med. & Phys. Res., 1835, 102, 224.
Rana berlandieri, BD.

Specimens from the Rio Gallo, near Fort Wingate, N. Mex., and along the Zuñi River, where the species is common. They are probably referable to the subspecies *berlandieri*.

CADUCIBRANCHIATA.

AMBLYSTOMATIDAE.

82. **Amblystoma mavortium**, *Bd.*

a. MAVORTIUM.

Ambystoma mavortia, BD., Jour. Acad. Nat. Sci. Phila., 2d ser., i, 1849, 284, 292 (New Mexico).
Ambystoma mavortium, HALLOW., Jour. Acad. Nat. Sci. Phila., iii, 1858, 352.
Amblystoma mavortium, BD., P. R. R. Rep., x, 1859, Gunnison & Beckwith's Route, Reptiles, 20.—COPE, Proc. Acad. Nat. Sci. Phila., 1867, 184.—ALLEN, Proc. Bost. Soc. Nat. Hist., xvii, 1874, 70.
Amblystoma proserpina, BD. & GIR., Proc. Acad. Nat. Sci. Phila., 1852, 173.—BD., U. S. & Mex. Bound. Surv., ii, pt. ii, 1859, Reptiles, 29, pl. 35, figs. 7–14.

Ambystoma proserpine, HALLOW., Jour. Acad. Nat. Sci. Phila., iii, 1858, 354.
Ambystoma maculatum, HALLOW., Jour. Acad. Nat. Sci. Phila., iii, 1858, 355.—*Id.*, Proc. Acad. Nat. Sci. Phila., 1857, 215.
Desmiostoma maculatum, SAGER, Penins. Jour. Med., 1858, 428.
Camarataxis maculata, COPE, Proc. Acad. Nat. Sci. Phila., 1859, 123.
Ambystoma nebulosum, HALLOW., Proc. Acad. Nat. Sci. Phila., 1852, 209.—*Id.*, Jour. Acad. Nat. Sci. Phila., iii, 1858, 352.—*Id.*, Sitgreave's Exp. Zuñi & Col. Riv., 1853, 143, pl. 20.
Amblystoma? nebulosum, COPE, Proc. Acad. Nat. Sci. Phila., 1866, 300.

b. CALIFORNIENSE.

Ambyostoma californiense, GRAY, Proc. Zoöl. Soc. Lond., 1853, 11, pl. 7 (Monterey).—HALLOW., Jour. Acad. Nat. Sci. Phila., iii, 1858, 355.

In the *Siredon* stage differing from the *S. gracilis*, Baird, in its oblique branchial arches, with finer pectinations, and in coloration. On the anterior side of the third arch, twenty pectinations, or rakers, may be counted; in the *S. pisciformis* (or *mexicanus*) there are but twelve. . . . Male about seven inches long. Branchiæ well developed. Gular derm free half way to symphysis mandibuli. Twelve costal folds. Muzzle slightly narrowed; jaws equal. Lateral and dorsal pertonæum black. The lungs extend to opposite the inguinal region. Corpus adiposum extending on testes to their anterior extremity. Testes undivided, broad; length equal to half that from axilla to anus; efferent vessels numerous, not entering directly the vas uro-spermaticus. The latter is very slender, lying along the outer margin, but not in contact with the narrow kidney; opposite the latter recurrently convolute, anterior to it straight, and extending to opposite axilla with decreasing diameter. It empties into the rectum near the cloaca. Cloaca protected on each side by a large, vertical, compressed gland, which is fringed on its inferior border (which is received into the lip of the cloaca), and also on its superior margin, which lies next the caudal vertebræ; it is continuous in front of anus; behind the two edges are pressed together. Integument of cloaca thrown into numerous appressed vertical plicæ, as in other Siredons. Stomach straight, extending to the left groin, filled with larvæ of *Diptera Nematocera*. Intestines long; rectum large. Female smaller; many of the ova black.

In these animals, the tarsal and carpal bones are fully formed, but cartilaginous. The palatine and pterygoid teeth in continuous series; the latter slightly separated medially, and concentric with maxillary series. On this character, preserved in a stage of all allied species without branchiæ, I proposed the genus *Camarataxis*, the validity of which can only be established when the development of all our Amblystomas is known. It is a stage nearer the larval condition than the transverse series of *A. opacum*, while the Λ-shaped series of *A. luridum* is intermediate.—*Cope.*

Several specimens of this interesting animal are contained in my collections; Nos. 455, 456, ♀, from Jacob's Well, nearly on the boundary between Arizona and New Mexico, and No. 491, ♂, from water holes in the San Francisco Mountains. The foregoing is Professor Cope's notice of these specimens. They are still in the *Siredon* stage, though they have attained a length of six or eight inches, and were confined to the water; on being exposed to the air, the skin dried with remarkable rapidity, but would soon regain its natural condition upon re-immersion. They lived for a half hour or so in the air. The color in life was shining-green above and silvery greenish-white beneath, shading into yellowish about the head and gills, with a few obsolete black spots on the head and back. The now well-known variations in color of this wide-ranging species have necessitated the reduction to synonyms of several species at one time current as valid. Specimens which I lately procured in the northern portions of Dakota and Montana, where they were common in suitable situations, were darker in color than these, more boldly blotched with black, and showed no yellowish hue. They had all completed the metamorphosis at a length of three or four inches, and were often found in damp places at a distance from water.

Amblystoma trisruptum, Cope, Proc. Acad. Nat. Sci., Phila., 1867, 194, from the Ocate River, New Mexico, is a species which may be expected to occur in Arizona.

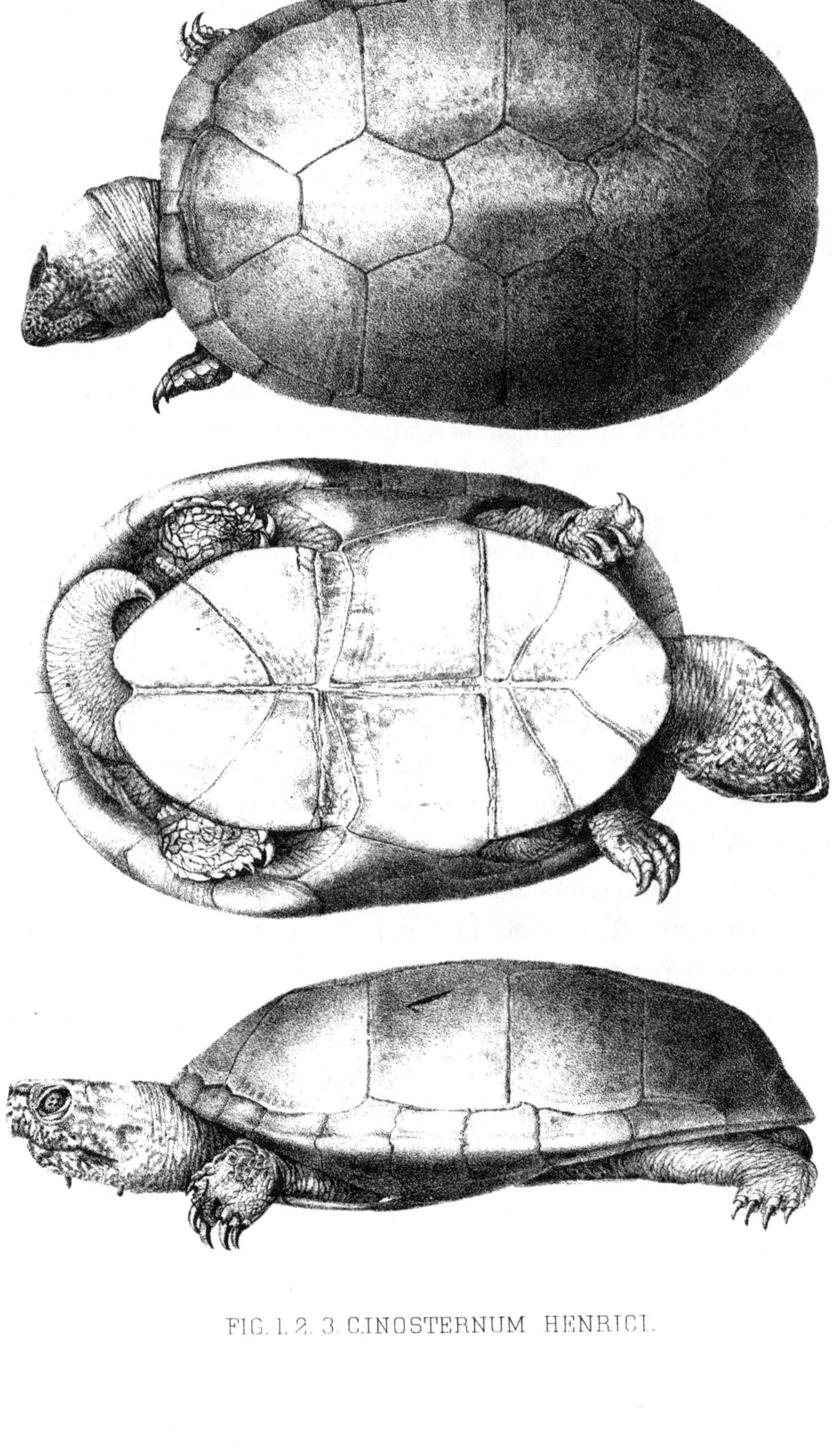

FIG. 1. 2. 3. CINOSTERNUM HENRICI.

FIG. 1. 2. 3. CINOSTERNUM FLAVESCENS.

T. Sinclair & Son lith. Phila.

FIG. 1 and 1a. CONTIA ISOZONA. FIG. 2 and 2a. GYALOPIUM CANUM.
FIG. 3 and 3a. CHILOMENISCUS EPHIPPICUS. COPE.

T. Sinclair & Son lith. Phila.

FIG. 1 and 1a. 2. OPHIBOLUS PYRRHOMELAS. COPE.

2

2a.

1

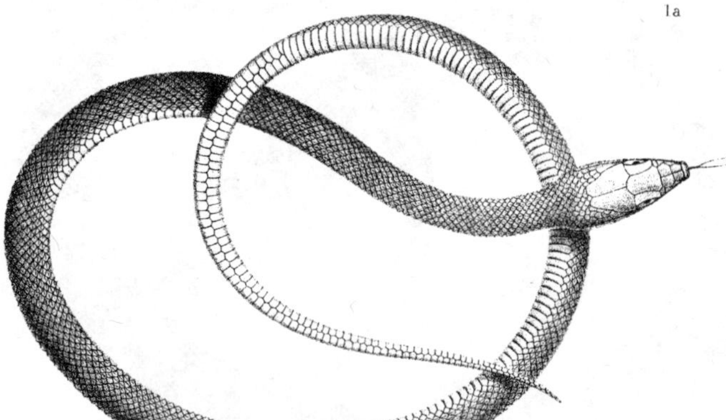

1a

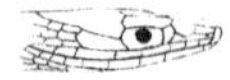

T. Sinclair & Son lith. Phila.

FIG. 1 and 1a. CHILOPOMA RUFIPUNCTATUM.
FIG. 2 and 2a. EUTÆNIA VAGRANS ANGUSTIROSTRIS.

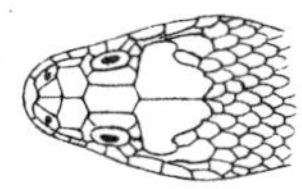

1a.

1

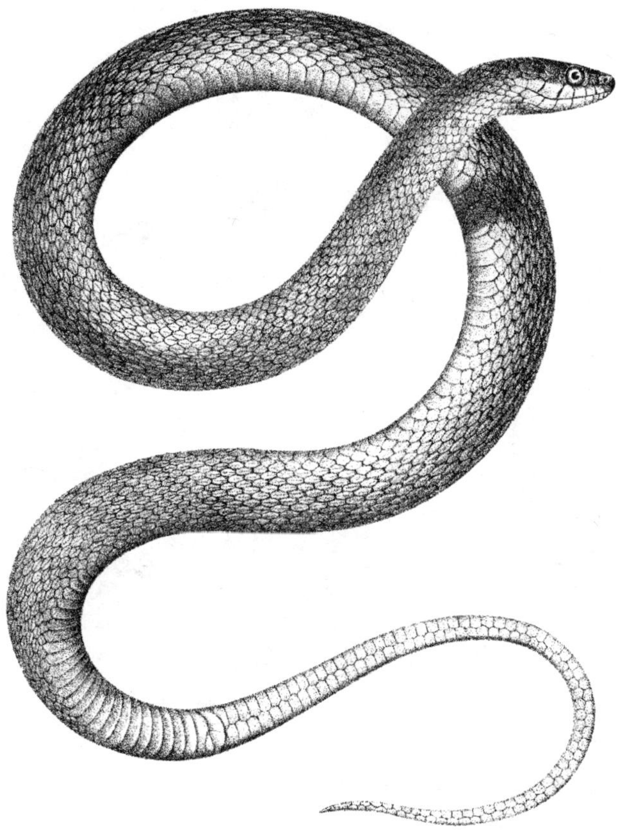

FIG. 1 and 1a. TROPIDONOTUS VALIDUS

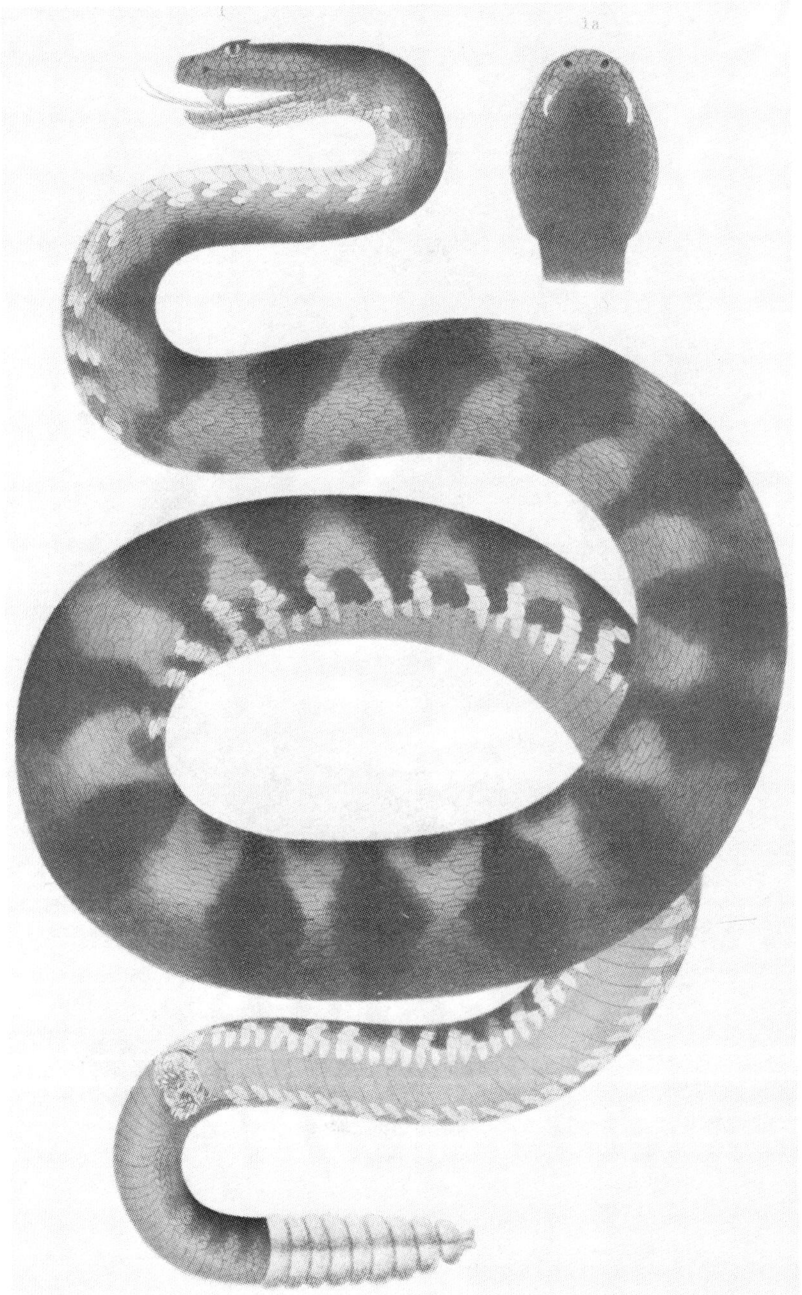

T. Sinclair & Son lith.

FIG. 1 and 1a. CROTALUS PYRRHUS COPE.

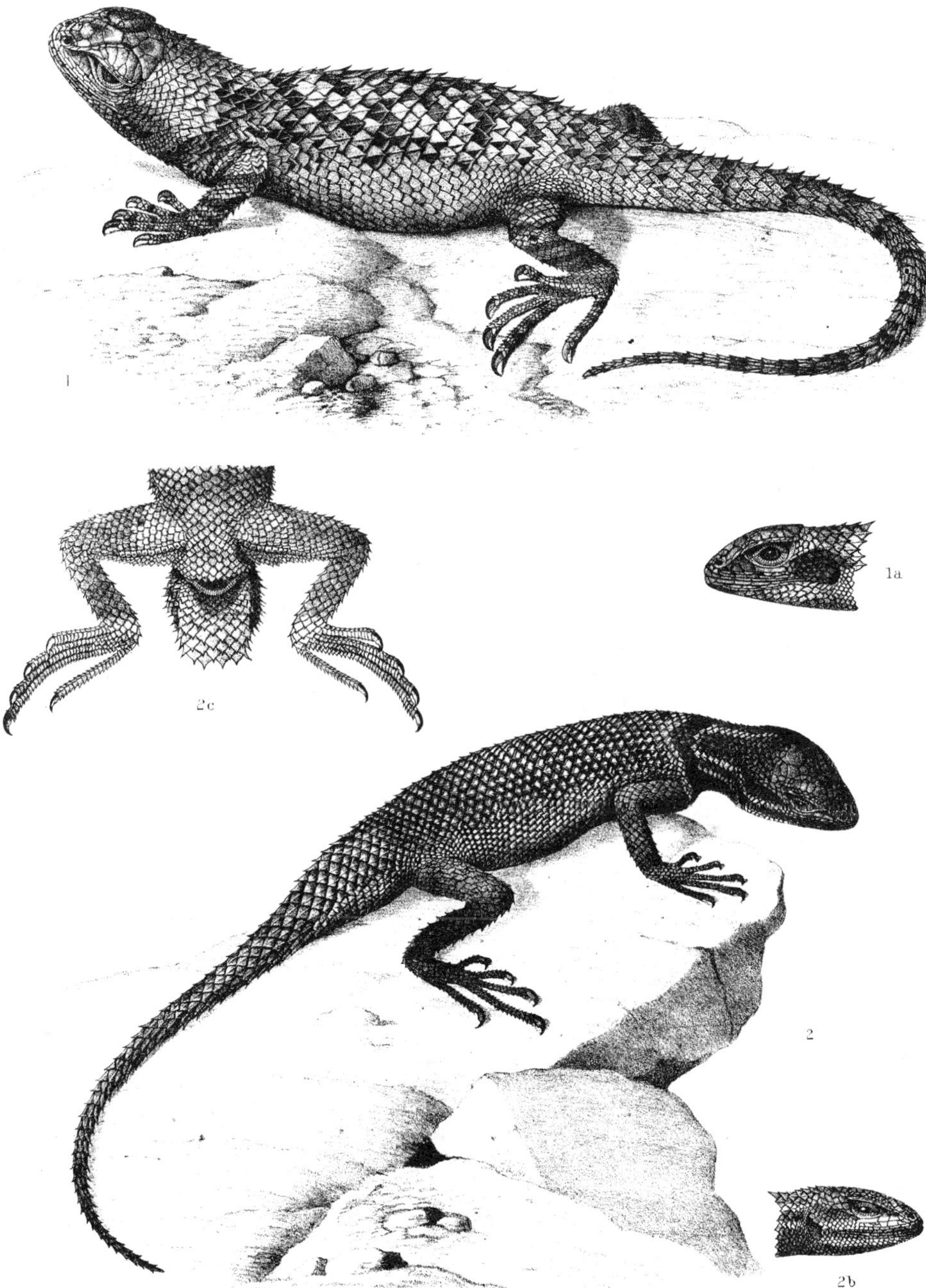

FIG. 1 and 1a. SCELOPORUS CLARKII, VAR.
FIG. 2. 2b. and 2c. SCELOPORUS JARROVII.

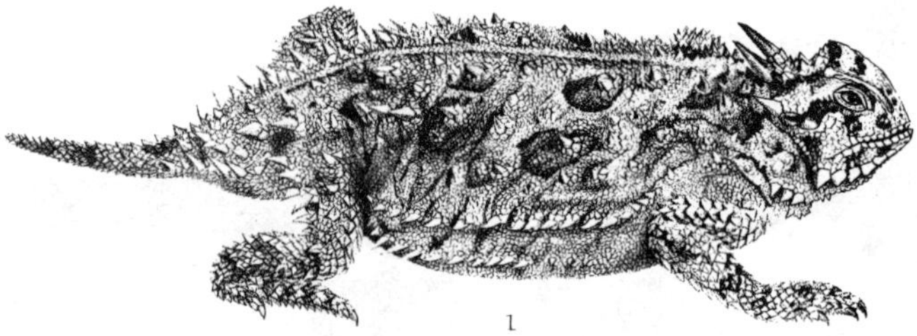

1

1b

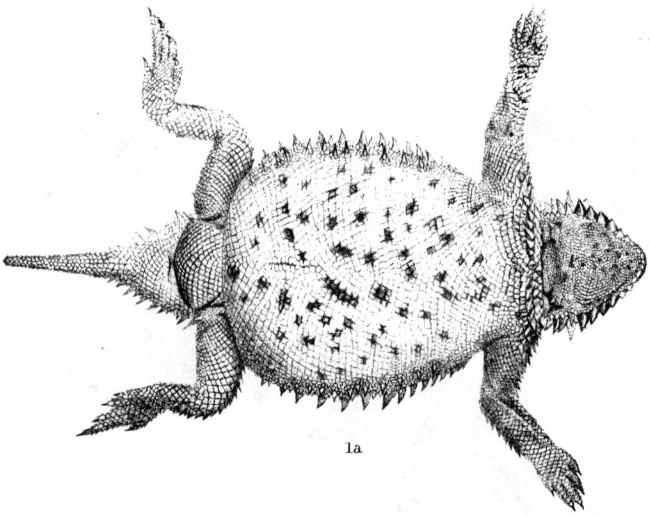

1a

2a

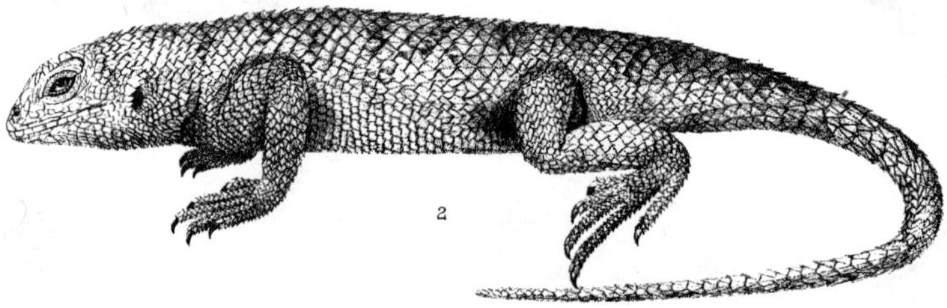

2

T. Sinclair & Son, lith. Phila.

FIG. 1—1a. and 1 b. PHRYNOSOMA. PLANICEPS.
FIG. 2. and 2 a. SCELOPORUS. SMARAGDINUS.

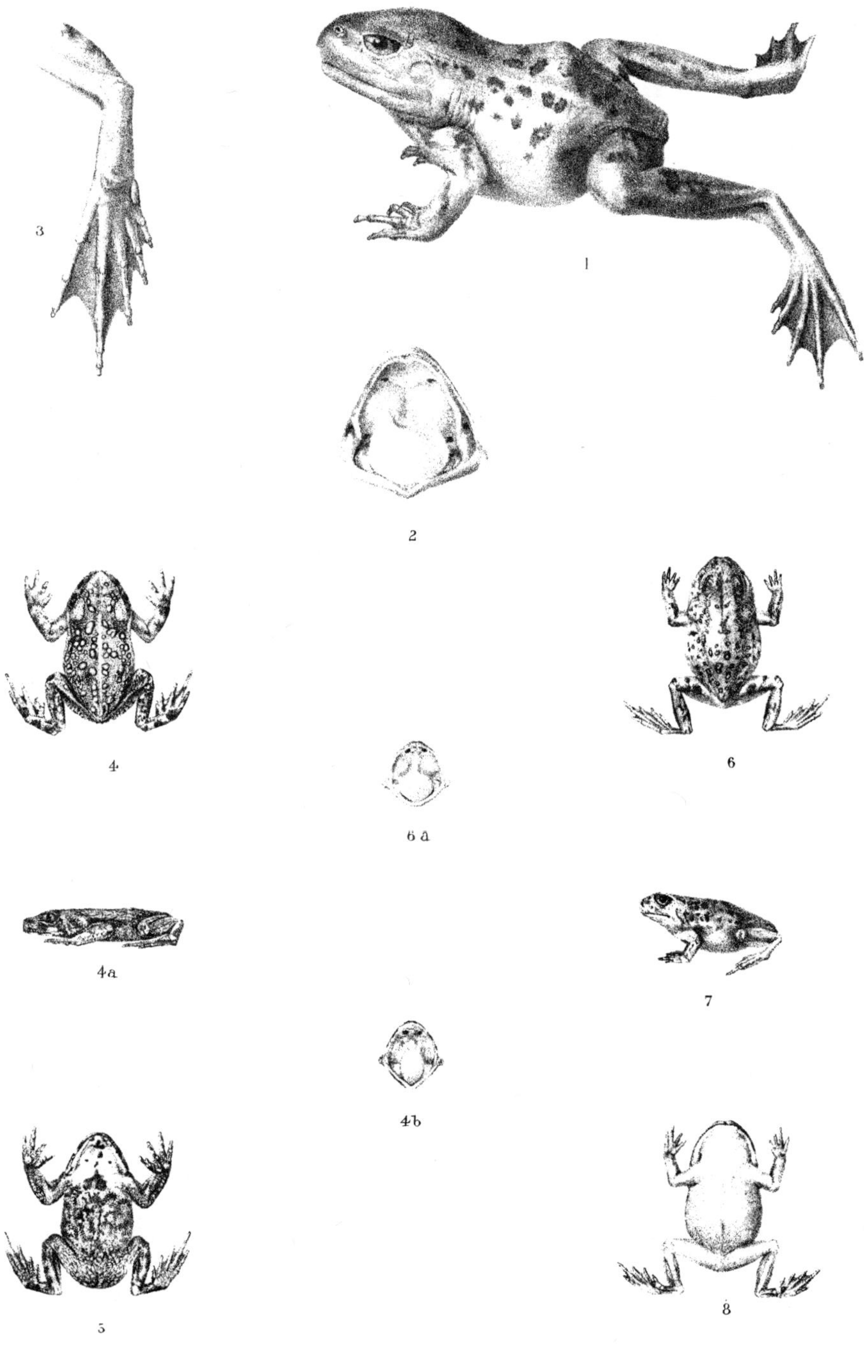

T. Sinclair & Son lith. Phila.

FIG. 1-2 and 3. RANA ONCA. FIG. 4 and 5. BUFO PICTUS.
FIG. 6, 7 and 8. SPEA STAGNALIS.

NOTES ON THE HERPETOLOGY OF DAKOTA AND MONTANA

Elliott Coues and H[enry] C[rècy] Yarrow

ART. XI.—NOTES ON THE HERPETOLOGY OF DAKOTA AND MONTANA.

BY DRS. ELLIOTT COUES AND H. C. YARROW.

The present article is based primarily upon a collection of Reptiles and Batrachians made in Dakota and Montana in 1873–74 by Dr. Coues, as Naturalist of the United States Northern Boundary Commission.

In identifying these specimens, the authors have diligently compared them with other material from the same geographical area in the National Museum, and have added to the species collected by Dr. Coues others known to occur in the region under consideration, thus presenting a tolerably complete list of the Reptiles and Batrachians of the two Territories. No species is introduced that is not fully identified and determined to inhabit this portion of the United States. Care has been taken with the synonymy to exclude doubtful references, except in one or two instances. The descriptions are drawn directly from the specimens, and considerable matter of popular interest has been introduced. The nomenclature and classification are mainly according to Professor Cope's recent Check List, though the authors have not hesitated to differ from this authority on occasion.

A.—REPTILIA.

CHELONIA.

Family EMYDIDÆ.

Genus CHRYSEMYS. *Gray.*

CHRYSEMYS OREGONENSIS. (*Harlan*) *Ag.*

Oregon Golden Turtle.

Emys oregonensis, HARL. Am. Journ. Sci. xxxi. 382, pl. 31.—HOLBR. N. Am. Herpet. i. ——, 167, pl. 16.—DEKAY, N. Y. Fn. iii. 1842, 20.

Chrysemys oregonensis, AG. Contr. Nat. Hist. U. S. i. 1857, 440, pl. 3, f. 1-3.—BD. U. S. Mex. B. Surv. ii. pt. ii. 1859, Reptiles, 4 (Texas).—ALLEN, Proc. Bost. Soc. N. H. xvii. 1874, 68 (Fort Rice, Dakota).

Specimen.

1096. Mouse River, Dakota. Aug. 30, 1873.

Shield Reptiles are not well represented in the region surveyed by the commission, where the present, the only one observed, appears to be the most characteristic species. Professor Agassiz notices specimens from different localities in Minnesota and from the Yellowstone, where it was also observed by the Prince Maximilian and Mr. J. A. Allen. The former naturalist expresses great doubts respecting the accuracy of Nuttall's statement that it is found in Oregon, as it has never been seen in that Territory by any of the recent explorers, the only true Turtle of the Pacific slopes being the *Chelopus marmoratus* Bd. & Grd. (*Emys nigra* of Hallowell). It is, however, a species of wide distribution in the central region, having been observed southward nearly to the Mexican border in Texas.

The following additional species of this order are indicated by authors as occurring on or near the northern boundary:—

Genus PSEUDEMYS. *Gray.*

PSEUDEMYS ELEGANS. (*Maxim.*)

Elegant Terrapin.

Emys elegans, MAXIM. Reise Nord-Amer. i. 1839, 213 (Upper Missouri).—HAYD. Trans. Am. Phil. Soc. xii. 1862, 177 (Yellowstone).

Trachemys elegans, AGASS. Contrib. Nat. Hist. U. S. i. 1857, 435.—BD. U. S. Mex. B. Surv. ii. pt. ii. 1859, Reptiles, 3 (Texas.)

Pseudemys elegans, GRAY.—COPE, Check List Bat. Rept. N. A. 1875, 53.

Emys cumberlandensis, HOLBR. N. Am. Herpet. i. 115, pl. 118 (Tennessee).—DEKAY, N. Y. Fauna, iii. 1842, 20.

Emys holbrookii, GRAY, Cat. Brit. Mus. 1844, 23.

Emys terrapin, WAILES, Geol. Rep. Mississippi, 1854, p. — (*fide* Agass.).

A species originally described from the Upper Missouri by Prinz Maximilian von Neu Wied, and subsequently ascertained to occur throughout the Central region, east to the Ohio, and south to Texas.

Genus CISTUDO. *Fleming.*

CISTUDO ORNATA. *Agass.*

Ornate Box-turtle.

Cistudo ornata, AGASSIZ, Contrib. Nat. Hist. U. S. i. 1857, 445, pl. 3, f. 12, 13.—COPE, Check List Bat. and Rep. N. A. 1875, 53.

The Northwestern type of *Cistudo*, Professor Agassiz remarked, in proposing *C. ornata*, is of all the forms the most likely to be distinct, and such has proven to be the case. "It is round, broad, and flat, without keel, even when young, while the young of *Cistudo virginea* are always strongly keeled." The species is based upon specimens from the Upper Missouri and from Iowa.

Family TRIONYCHIDÆ.

Genus ASPIDONECTES. *Wagler.*

ASPIDONECTES SPINIFER. (*Les.*) *Ag.*

Trionyx spiniferus, LE SUEUR, Mém. Mus. d'Hist. Nat. xv. 258, pl. 6.
Aspidonectes spinifer, AGASS. Contrib. Nat. Hist. U. S. i. 1857, 403.—COPE, Check List N. A. Bat. and Rep. 1875, 51.
Trionyx ocellatus, LE SUEUR (young ♀, *fide* Agass.; not of DeKay, which is *Amyda mutica*).
Trionyx ferox, partim, ALIQ.

The Northern and Northwestern *Aspidonectes*, the characters and synonymy of which were first satisfactorily distinguished from those of the Southern *A. ferox* by Professor Agassiz in the work above cited, is represented as a common species from New York and Pennsylvania to the Rocky Mountains, where it is mentioned as occurring by Lewis and Clarke. According to Say and Allen, it is frequently found in the tributaries of the Missouri; the last-named naturalist took it in the Musselshell and Yellowstone. (*See* ALLEN, Proc. Bost. Soc. Nat. Hist. 1874, p. 69.)

Family CHELYDRIDÆ.

Genus CHELYDRA. *Schw.*

CHELYDRA SERPENTINA. (*L.*) *Harl.*

Snapping Turtle.

Testudo serpentina, LINN. Syst. Nat. ed. 12, i. 1766, 354 (localities erroneously assigned as Algiers and China). Also of other older authors.—LEC. Ann. Lyc. Nat. Hist. N. Y. iii. 127.
Chelonura serpentina, SAY, Journ. Acad. Nat. Sci. Phila. iv. 217.—HOLBR. N. Am. Herpet. 1st ed. iv. 21, pl. 3; 2d ed. i. 139, pl. 23.—DEKAY, N. Y. Fn. iii. 1842, 8, pl. 3, f. 6.
Emys serpentina, GRAY, Syn. Rept. in Griffith's An. Kingd. ix. 14.
Chelydra serpentina, HARL. Med. & Phys. Res. 1835, 157.—AGASS. Contrib. Nat. Hist. U. S. i. 1857, 417. And of most late authors.—COPE, Check List N. A. Bat. and Rep. 1875, 51.
Emysaurus serpentina, DUMÉR. & BIBR. Erp. Gén. ii. 350.—STORER, Rep. Mass. ——, 212.
Chelydra emarginata, AGASS. *op. cit.* in text.
"*Chelydra lacertina*, SCHW." (young).
"*Testudo serrata*, PENN."
"*Testudo longicauda*, SHAW."

A species of remarkably extended distribution, from the Northern border of the United States to South America; not, however, in the Pacific region.

OPHIDIA.

Family CROTALIDÆ.

Genus CROTALUS. *Linn.*

CROTALUS CONFLUENTUS. *Say.*

Missouri Rattlesnake.

Crotalus confluentus, SAY, Long's Exped. R. Mts. ii. 1823, 48.—BD. & GIR. Cat. N. Am. Reptiles, 1853, 8.—BD. Pac. R. R. Rep. x. 1859, Reptiles of Whipple's Route, 40; pl. 24, f. 4.—BD. U. S. and Mex. B. Surv. ii. pt. ii. Reptiles, 14.—COOP. & SUCKL. Nat. Hist. Wash. Terr. 1860, 295, pl. 12.—COPE, Check List N. A. Bat. and Rep. 1875, 33.

Caudisona confluenta, COPE, App. Mitchell's Researches, 1861, 122.—COPE, Proc. Acad. Nat. Sci. Phila. 1866, 307, 309.—ALLEN, Proc. Bost. Soc. N. H. xvii. 1874, 69.

Crotalus lecontei, HALLOW. Proc. Acad. Nat. Sci. Phila. vi. 1851, 180.—HALLOW. Sitgreaves's Rep. Expl. Zuñi and Colorado, 1853, 139, 147, pl. 18.—HALLOW. Pac. R. R. Rep. x. 1859, Williamson's Route, Reptiles, 18, pl. 3.

Caudisona lecontei, COPE, App. Mitchell's Researches, 1861, 121.—HAYD. Trans. Am. Phila. Soc. xii. 1862, 177.—COPE, Proc. Acad. Nat. Sci. Phila. 1866, 307.

Crotalus cinereus, LECONTE, *apud* HALLOW. Sitgreaves's Rep. Expl. Zuñi and Colorado, 1853, 140 (in text).

Specimens.

1149.	Sweet Grass Hills, Montana.	July 29, 1874.
1150.	Sweet Grass Hills, Montana.	July 29, 1874.
1197.	Black Coulé, near Teton River, Montana.	Sept. 6, 1874.
1198.	Black Coulé, near Teton River, Montana.	Sept. 6, 1874.
1199.	Black Coulé, near Teton River, Montana.	Sept. 6, 1874.

The ascribed characters of *C. lecontei*, as compared with *C. confluentus*, are found not to hold good when sufficient series are examined. The number of superior labials in our specimens, and in others from the Yellowstone, ranges from thirteen to sixteen, while in others eighteen are described. Certain ascribed features of coloration are altogether uncertain, as specimens vary interminably in the distinctness of the dorsal blotches and in the details of the light markings about the head. This appears to be due in part to age, as the smaller specimens are usually the most boldly blotched, while on some of the largest examined the markings are nearly obsolete. These statements are fully borne out by our experience, we having noticed in this particular species that the blotching cannot be relied upon as a distinctive character, as in some individuals, notably from localities where the color of the soil is light, the blotches in some instances are barely perceptible; moreover, the forms of the blotches vary indefinitely, as some are serrated on their borders, others present an unbroken line. The distinctness of the pattern of coloration also depends somewhat upon season, the markings being clearest just after the shedding of the skin. There is certainly no specific difference between the two supposed species, and varietal distinction can hardly be predicated, at least upon the characters as yet adduced. In any event, the specimens above enumerated are the true *C. confluentus* of Say. Two of these are of unusual size, being both over four feet in length, a dimension near the known maximum of this species, though less than that of some others. The average length is less than three feet, and the calibre of the body, even in the largest examples, is relatively inferior to that of several Southern species.

The number of rattles in this and other species, though of course increasing with age, is not an infallible clue to the age of a specimen; for, accidental variation aside, it is far from proven that an annual increase by one is regular. On the contrary, the growth of the organ must depend largely, as in all parallel cases, upon the vigor of the individual, which is not the same at all periods of life, granting even a continuous state of perfect health. The purpose subserved in the economy of the animal by this singular organ has been the subject of much speculation and discussion. It is difficult to perceive of what use the rattle can be, either in procuring prey or avoiding enemies. We do not know that it comes into play at all in the pursuit of prey, while the actual result of its use as a menace in self-defence is the reverse of beneficial to the serpent, since the sound serves to direct and provoke attack from all enemies which the animal has occasion to fear. The theory that the rattle is a part of the serpent's means of terrifying its intended victim, used as an adjunct of other supposed powers of fascination, may be safely held in check until it is proven that this peculiar influence is ever exerted to the extent of preventing its prey from seeking safety in flight. The notion that the rattle is intended to serve as a warning, and thus offset the venomous nature and highly dangerous powers of the serpent, is contrary to all analogy, since animals are endowed with attributes for their own good, irrespective of the result upon others, and would require a faith in the intervention, for the benefit of the dominant species of the *Mammalia*, of special Providences, a belief now held by few thoughtful persons. It has been suggested that the rattle may be used to call the sexes together, and thus serve a useful purpose in the perpetuation of the species,—a hypothsis less untenable than some of the others which have been advanced. Another supposition, made irrespective of "final causes", is, that the rattle has resulted, in the course of time, from the continual agitation of the caudal extremity of these highly nervous and irritable creatures, and that it has no special function. This seems not unreasonable, although, in view particularly of the fact that rattlesnakes alone, of the many equally or more venomous reptiles, have such appendage, it is not entirely satisfactory. One thoroughly established fact concerning the rattle is that its practical operation is injurious to its possessor by provoking attack from those who can cope with it successfully. It may be suggested, that inasmuch as to an unpracticed ear the rattle of the *Crotalus* cannot be distinguished from the crepitation of the large Western grasshopper, it may serve the purpose of attracting within reach of the fangs of the snake the many birds who greedily devour these insects. The rattle has been heard at times when no apparent cause of irritation to the snake existed, and a case has been reported in which a biped was drawn within reach of a rattler, thinking it a grasshopper.

The principal enemies of the rattlesnake, besides man, are wild hogs, peccaries, and deer. The latter kill the serpent when coiled by striking with the hoofs; the former attack it successfully with hoofs and teeth, and in some regions derive no small part of their subsistence from this source. The popular belief that the venom of the rattlesnake is innocuous to hogs is merely a partial statement of the fact that the fluid usually fails to enter the circulation through the layer of adipose tissue with which these animals are commonly covered. The venom is conceded to be innocuous when introduced to the stomach, and the flesh of the rattlesnake is as edible as that of other serpents. The fatality of the rattlesnake's bite is by no means the constant element generally supposed, but the result may vary from the slightest amount of poisoning to one rapidly fatal. This depends altogether upon the amount of venom absorbed in the system, and the rapidity of its diffusion through the circulation, matters which turn upon the amount of venom in store at the moment of striking, the vigor of the animal at the time, the penetration of the tooth, the part of the body struck, and, finally, the state of health of the person attacked. No positive specific antidote is known. Surgical means of preventing dispersion of the poison through the system, and alcoholic stimulation to the highest pitch, are the usual resorts.

It may not be out of place to refer in this connection to the interesting mechanism of the poison apparatus, as it is a matter not very generally known as yet, though clearly set forth by the researches of specialists, notably Dr. S. Weir Mitchell. The venomous fluid to be injected into a wound made by the teeth has nothing to do with the ordinary saliva, as popularly supposed; nor does the forked tongue or any of the numerous small teeth of the mouth take part in the infliction of the wound. The tongue and smaller teeth are essentially the same as in any harmless serpent. The active instruments are a pair of fangs,* one on each side of the upper jaw, rooted in the maxillary bones, which bear no other teeth. The fangs vary in size, being sometimes half an inch long. They are somewhat conical and scythe-shaped, with an extremely fine point; the convexity looks forward, the point downward and backward. The fang is hollow, for transmission of the venom; but

* It may be mentioned, as a fact of some interest, that, while in *C. confluentus* the fangs are generally shed or pushed out of place at variable periods of time (probably in twelve months), in *C. adamanteus atrox*, a species common in the Sonoran region, this shedding, or loss, frequently fails to take place, and it is common to find generally in the right side of the jaw of this species two or more fangs in position. In one specimen lately examined, three were found in position, and behind them three or four others were advanced in growth.

the construction of the tube is not as if a hole had been bored through a solid tooth. It is in effect a flat tooth, with the edges rolled over together till they meet, converting an exterior surface, first into a groove, finally into a tube. This is shown, on microscopic examination of a section of the tooth, by the arrangement of the dentine. Unlike an ordinary tooth, the fang is movable, and was formerly supposed to be hinged in its socket, since it is susceptible of erection and depression. But the tooth is firmly socketed, and the source of this movement is the maxillary bone itself, which rocks to and fro by a singular contrivance. The maxillary is a small, stout, triangular bone, movably articulated above with a smaller one, the lachrymal, which is itself hinged upon the frontal. Behind, the maxillary articulates with the palatal and pterygoid, both of which are of rod-like shape, and are acted upon by the spheno-pterygoid muscle, the contraction of which pushes them forward. This forward impulse of the palatal and pterygoid is communicated to the maxillary, against which they abut, causing the latter to rotate upon the lachrymal. In this rocking forward of the maxillary, the socket of the fang, and with it the tooth itself, rotates in such manner that the apex of the tooth describes the arc of a circle, and finally points downward instead of backward. This protrusion of the fang is not an automatic motion, consequent upon mere opening of the mouth, as formerly supposed, but a volitional act, as the reverse motion, namely, the folding back of the tooth, also is; so that, in simply feeding, the fangs are not erected. The folding back is accomplished by the ecto-pterygoid and spheno-palatine muscles, which, arising from the skull behind as a fixed point of action, in contracting draw upon the jaw-bones in such a way that the maxillary, and with it of course the fang, are retracted, when the tooth is folded back with an action comparable to the shutting of the blade of a pocket-knife. All the motions of the fangs are controlled by these two sets of antagonistic muscles, one of which prepares the fangs for action, while the other stows them away when not wanted.

The fangs, when not in use, are further protected by a contrivance for sheathing them, so that they rest like a sword in its scabbard. This is a fold of mucous membrane, the *vagina dentis*, which envelopes the tooth like a hood, enwrapping its base, and slipping down over its length, partly as a consequence of its elastic texture, partly on account of its connections. Erection of the fang causes the sheath to slip off, like the finger of a glove, and gather in folds around the base of the tooth. This arrangement can be readily examined without dissection.

The poisonous fluid is secreted in a gland which lies against the side of the skull, below and behind the eye, of a flattened oval shape, obtuse behind, tapering in front to a duct that runs to the base of the tooth. Without going into the minute anatomy of the gland, it may be described as a sac, or reservoir, in the walls of which the numerous secretory follicles are imbedded; it is invested with two layers of dense, white, fibrous tissue, the outer of which gives off three strong ligaments that hold it in place. In a large snake, the entire gland may be nearly an inch long and one-fourth as wide, weighing, empty, ten or twelve grains, and having a capacity of ten or fifteen drops of fluid. There is no special reservoir for the venom, other than the central cavity of the gland. A certain dilatation of one portion of the duct, formerly supposed to be such storehouse, is due to thickening of its walls, without corresponding increase of capacity, resulting from muscular fibres which serve as a sphincter to compress the canal and prevent wasteful flow of the contents. There is further provision to this same end. When the tooth is folded back, the duct attached to its root is submitted to some strain, which pushes it against a shoulder of the maxillary bone, and tends to shut off the communication.

The injection of the venom, though to all appearance instantaneous, is a complicated process of several rapidly consecutive steps. Forcible voluntary closure of the jaws may always be, if desired, accompanied by a gush of the venom, owing to the arrangements of the muscles which effect such movement of the under jaw. These are the *temporales*, one of the three of which is situated in such relation to the poison-sac that its swelling in contraction presses upon the receptacle and squeezes out the fluid. The force of ejection is seen when the serpent, striking wildly, misses its aim; under such circumstances, the stream has been seen to spirt five or six feet. A blow given in anger is always accompanied by the spirt of venom, even when the fang fails to engage, from whatever cause. But since this result does not follow upon mere closure of the mouth, it is probable that the two posterior temporals ordinarily effect this end, the more powerful action of the anterior temporal (the one which presses upon the poison-sac) being reserved for its special purpose. There is one very curious piece of mechanism to be noted here. Since the serpent always snaps its jaws together in delivering a blow, the points of the erected fangs would penetrate the under jaw itself in case they failed to engage with the object aimed at, were there no contrivance for preventing such disaster to the snake. But there is a certain movement among the loose bones of the skull, perhaps not well made out, the result of which is to spread the points of the fangs apart in closure of the mouth, so that they clear the sides of the under jaw, instead of impinging upon it.

The complicated mechanism of the act of striking may be thus described:—The snake prepares for action by throwing itself into a number of superimposed coils, upon the mass of which the neck and a few inches more lie loosely curved, the head elevated, the tail projecting and rapidly vibrating. At the approach of the intended victim, the serpent, by sudden contraction of the muscles upon the convexity of the curves, straightens out the anterior portion of the body, and thus darts forward the head. At this instant, the jaws are widely separated, and the back of the head fixed firmly upon the neck. With the opening of the mouth, the spheno-palatines contract, and the fangs spring into position, throwing off the sheath as they leap forward. With delivery of the blow and penetration of the fangs, the lower jaw closes forcibly, the muscle that executes this movement causing simultaneously a gush of venom through the tubular tooth into the wound. There are also some secondary actions, though all occur at nearly the same instant. The mouth fixed at the wound drags upon it with the whole weight of the snake's body. This dragging motion is accompanied by contraction of the ectopterygoid and spheno-palatine muscles, which ordinarily fold back the tooth; but the fang being at this moment engaged in the flesh, the action of the muscles only causes it to bury itself deeper, and thus enlarge the puncture. The train of action seems to be, the reaching of the object, the blow, the penetration, the injection of the poison, and the enlargement of the wound. These actions completed, the serpent loosens its hold by opening the jaws, and disengages itself, sometimes not without difficulty, especially when the bitten part is small and the numerous small teeth have caught. The head is withdrawn, the fangs folded, the mouth closed, and the former coiled attitude of passive defense is resumed.

These remarks apply in substance to other species as well as to the one now under special consideration. Upward of eighteen species, not counting *Ancistrodon*, are described as inhabitants of the United States, nearly all of which occur in the West and Southwest. Our rattlesnakes fall in two genera, *Crotalus* and *Caudisona*, readily distinguished by the scutellation of the head. In the former, the top of the head is covered with a large number of small asymmetrical scales like those on the body; in *Caudisona*, the same region is shielded by a definite small number of large flat plates symmetrically disposed. *Crotalus horridus* is a most widely dispersed species of Eastern North America, the only other species of the same portion of the continent being *C. adamanteus*, the "Diamond" rattlesnake of the South Atlantic and Gulf States. With the latter is found associated a species of *Caudisona*, namely, *C. miliarius*, the small spotted rattlesnake; but the best known species of the latter genus is *Caudisona tergeminus*, the common "Massasauga" of the interior States and of the Plains. Texas, New Mexico, and Arizona furnish the largest number of species.

Crotalus confluentus is a species of more and general distribution in the West, from the Mexican to the British boundary, and on both sides of the Rocky Mountains. It is associated in some parts with the Massasauga, but in other regions, as in the one now under consideration, it is the only known representative of its family. It appears to be particularly numerous in the region of the Yellowstone, where, according to Mr. Allen, it was estimated that two thousand were killed during the expedition of 1872. Farther northward, it is less abundant, though fairly to be considered common in the region of the Upper Missouri and Milk River and some of their northern tributaries. Along the northern boundary line, its distribution appears to be determined very nearly by the Missouri watershed, as limited by the Coteau. This carries its range somewhat into the British Possessions, so that it is probably the most northern species of the genus. It does not appear to exist along that portion of the line represented by the watershed of the Red River of the North, where none were observed by the commission. I shall have frequent occasion to allude to the great difference in the fauna and flora of these two portions of the line sharply divided by the Coteau of the Missouri. The presence among plants of the *Opuntia missouriensis* and the prevalence of *Artemisia*, the occurrence of *Centrocercus urophasianus* among birds, of *Cynomys ludovicianus* among mammals, and of *Phrynosoma douglasi* and *Crotalus confluentus* among reptiles, are some of the prominent features of the Missouri watershed as contrasted with that of the Red River. *C. confluentus* is also found extending to New Mexico, Utah, Colorado, Arizona, California, Nevada, and even to the islands of Santa Barbara Channel, California; but the typical *Crotalus* of the Sonoran region is *C. adamanteus atrox*, a subspecies of our Southern serpent, of the Pacific. *C. lucifer*, a well-marked and distinct form, is found in Arizona and the Pacific region.

The pairing season of these serpents is in midsummer, when they have several times been observed *in coitu*. Little, however, has been ascertained respecting their reproduction. About half the year, in most latitudes, they hibernate in holes in the ground. They have, however, been observed abroad after severe frosts in the Yellowstone region. At Fort Randall, on the Missouri, they were stated to reappear in May with the loosening of the ground from frost. As in the case of other species, there is a regular *annual*, or perhaps, as in some other Ophidians and as in Saurians, a more frequent casting of the skin. During the moult, they are reported to be specially venomous, but probably upon no other foundation than that at this time their sluggishness results in the accumulation of a large supply of the poisonous fluid. In one of the specimens secured were found the remains of a Prairie Squirrel (*Spermophilus richardsoni*), the most abundant mammal of the Milk River region; and it is probable that these animals, together with the allied species, form a large part of their subsistence.

While the venomous properties of these reptiles, not easily overrated, should suffice to ensure due caution in capturing or killing them, it is as well to remember that the utmost range of a rattlesnake's blow is less than its own length. They may readily be captured alive by pinning down the neck with a forked stick, and may be handled with impunity, when not too large and powerful, if seized immediately behind the head. In case of a strong snake, however, the power of constriction is sufficient to paralyze the muscles of both arms, as in the case of a person we knew who had seized two of these reptiles by the back of the neck. He had to be relieved by a bystander. A method employed in the South to capture the *C. adamanteus* is perhaps worthy of mention. A silk handkerchief is fastened to the end of a pole, which is held toward the reptile, which strikes fiercely at it, the fangs and teeth become engaged in the fibre of the silk, and a dexterous movement of the stick readily pulls out the fangs, and the reptile can be approached with safety.

There seems to be a special and peculiar enmity existing between the Rattlesnake and Moccasin and the Blacksnake (*Bascanium*) and "King Snake" (*Ophibolus getulus sayi*); these two latter species waging a constant warfare against the former, and invariably conquering, according to information received from reliable parties. After the conflict, the vanquished is eaten by the victor. In one case reported, a large Blacksnake (*Bascanium constrictor*) had seized a Rattlesnake (*Crotalus adamanteus*), and entwined two or more folds behind his head and several six or eight inches farther back; then by muscular effort had torn the body. It is a well-known fact that both Rattlesnakes and Moccasins will endeavor to get away from the "King Snake" (*Ophibolus getulus sayi*); and in the South this beautiful and harmless species is protected in view of this fact.

Genus CAUDISONA. *Laur.*

CAUDISONA TERGEMINA. (*Say*) *Cope.*

The Massasauga, or Prairie Rattlesnake.

Crotalus tergeminus, SAY, Long's Exped. R. Mts. i. 1823, 499.—HARLAN, Jour. Acad. Nat. Sci. Phila. 1827, 372.—COPE, Mitchell's Researches, App. 1861, 125.—HAYD. Trans. Amer. Phil. Soc. xii. 1862, 177 (Yellowstone).

Crotalophorus tergeminus, GRAY, Synop. Rept. 78; Cat. Rept. Br. Mus. 18.—HOLBR. N. Am. Herpet. 2d ed. iii. 1842, 29, pl. 5.—DEKAY, N. Y. Fn. ill. 1842, 57.—BD. & GIR. Cat. N. A. Rept. 1853, 14.—BD. P. R. R. Rep. x. 1851, pl. 25, f. 9 (no text).

Caudisona tergemina, COPE, Check List N. A. Bat. and Rep. 1875, 34.

Crotalophorus ——, AGASS. L. Superior, 1850, 381, pl. 6, f. 6, 7, 8.

Var ? (*Black Massasauga.*)

Crotalophorus kirtlandii, HOLBR. N. Am. Herpet. 2d ed. iii. 1842, 31, pl. 6.—GRAY, Cat. Br. Mus. 18.—BD. & GIR. Cat. N. A. Rept. 1853, 16.—BD. P. R. R. Rep. x. 1859, pl. 251, f. 11, 11 *bis* (no text).

Crotalophorus massasauga, KIRTL. *apud* BD. Serpents N. Y. 11, pl. 1, f. 2.

This species is distributed in prairie countries from Ohio and Michigan westward, finding its most western limit in the region of the Yellowstone. It is readily distinguished from any species of *Crotalus* by the presence of few (9) large symmetrical plates on the head, as in serpents generally, instead of numerous small scales, like those on the body. The rattle is much smaller than in *Crotalus*. The size varies from one to three feet. The ground-color above is brown, marked with blotches of deep chestnut-brown, blackish on the periphery, and margined with yellowish-white.

Family COLUBRIDÆ.

Genus HETERODON. *Beauv.*

HETERODON SIMUS NASICUS. (*B. & G.*) *Cope.*

Hog-nosed Snake; Sand Viper; Puffing Viper; Blowing Adder.

Heterodon nasicus, BAIRD & GIRARD, Stansbury's Exp. Great Salt Lake, 1852, 352.—BAIRD & GIR. Cat. N. A. Reptiles, 1853, 61, 157.—HALLOW. Sitgr. Rep. Expl. Zuñi and Colorado R. 1853, 147.—BD. P. R. R. Rep. 4, 1859, Whipple's Route, Reptiles, 41.—BD. P. R. R. Rep. x. 1859, Beckwith's Route, Reptiles, 19.—BD. U. S. Mex. B. Survey, ii. pt. ii. 1859, 18, pl. 11, f. 1.—HAYD. Trans. Amer. Phil. Soc. xii. 1862, 177.—COPE, Proc. Acad. Nat. Sci. Phila. 1866, 307.—ALLEN, Proc. Bost. Soc. Nat. Hist. xvii. 1874, 69.

Heterodon simus subsp. *nasicus*, COPE, Check List Bat. and Rep. N. A. 1875, 43.

Specimen.

No. 1101. Big Muddy River, Montana. June 25, 1874.

The serpents of the genus *Heterodon* are medium-sized or rather small species, thick-set in form and sluggish, of repulsive aspect, not distantly resembling some of the venomous species, especially the Copperhead (*Ancistrodon contortrix*). The similarity to poisonous species is heightened by the flat, broad, triangular shape of the head and the habit of hissing when irritated. They are commonly called "Adders" and "Vipers", and are reputed venomous; nevertheless, they are perfectly harmless. They cannot be provoked to bite. The belief in the poisonous qualities is further heightened by the presence of two tolerably large teeth in each upper jaw, resembling fangs, these teeth being the ninth (?) of the series in some individuals, their bases being below the fifth upper labial. There is no groove present, nor is the tooth movable. We do not know that this fact has ever been before mentioned, although the post-palatine teeth are spoken of as being larger than others. These large teeth have, however, a sort of sheath over them, similar to the fang-sheath of *Crotalus*. There is an interval between the small anterior teeth, and these are not contained in the same sheath as the fang-like tooth, which in some cases is found to have in its sheath one or more smaller fang-like teeth. They may be distinguished from any other serpents of this country by the sharp-pointed and elevated end of the muzzle, the rostral plate being prolonged into a spur.

The present species finds its nearest ally in the *Heterodon simus* of the Southern States, sharing with this species the separation of the median plate behind the rostral from the frontals by the interposition of several small plates. From *H. simus* it is distinguished by the slaty-black, which occupies all or most of the under surface. The color above is an obscure grayish-brown, with very numerous darker blotches along the dorsal line, and usually other smaller ones on the sides. But the markings are never bold, sometimes nearly obsolete.

This is the most abundant and wide-ranging species of the genus, occurring throughout the West east of the Rocky Mountains. The specimen above noted is particularly interesting in the fact that it is the northernmost one hitherto recorded, demonstrating a wider range, not only of the species, but of the genus, than was before known. Mr. Allen procured it on the Yellowstone, and it appears to increase in numbers southward, being one of the more common serpents of New Mexico and Arizona. I am under the impression that I saw the same species beyond the Missouri watershed, at Chief Mountain Lake; but the individual was unfortunately not secured.*

---

* Professor Cope, in his recent Check List, has seen fit to reduce the numbers of species of this genus to four, which are readily separated into two groups as follows:—

A. *Heterodon platyrhinus.*

Heterodon platyrhinus subspecies *atmodes.*

With the azygos behind the rostral plate in contact with the frontal plates.

B. *Heterodon simus* subspecies *simus.*

Heterodon simus subspecies *nascius.*

With the azygos behind the rostral plate separated by a varying number of small plates.

In this connection, it may be mentioned that if color should be taken into consideration as a specific marking, it seems that Baird and Girard's *H. niger* should be admitted as a subspecies of *H. platyrhinus*, for not only is there a very marked difference of coloration (some species of *H. niger* being entirely black), but as a rule the rostral of *H. niger* is much more developed than that of *H. platyrhinus*, and the dorsal carina are acute and very well marked, and there are obvious differences in the size of the scales. It is true that in examining a number of specimens of *H. platyrhinus*, *H. niger*, and *H. atmodes*, it will be found that a regular intergradation of color exists; but if *atmodes* is to be admitted as a good and valid subspecies of *H. platyrhinus*, it would seem that *H niger* is entitled to the same respect. In an examination of the different specimens of *Heterodon* in the National Museum, Smithsonian Institu.ion, the authors were fortunate enough to discover a species called *Heterodon kennerlyi* by Kennicott, in the Proceedings of the Acad. of Nat. Sci. Phila., 1860, pp. 336 and 337; and as after a careful examination of several specimens, the specific characters of them compare entirely with the type, the entire description is here given. The species naturally falls near the *B.* or *simus* group, in which the azygos is separated from the frontals, not by a varying number of plates, but by exactly two plates in five specimens and by three in one specimen. These specimens are from the following localities:—

1282. Matamoras, Texas.
7290. Lower Rio Grande.
5185. Fort Stockton, Texas.
8878. Southern Arizona.
8413. Southern Arizona.

A comparison of these specimens with eighteen well-marked species of *H. simus nasicus* shows that although these latter vary as to the number of scales separating the azygos and frontals, in no respect does it approach the regularity and systematic arrangement of the scales in *H. simus kennerlyi.*

HETERODON SIMUS KENNERLYI. (*Kennic.*) *C. & Y.*

H. kennerlyi, Kennicott.

SPEC. CHAR.—Head broad, very short anteriorly. Rostral plate very large. Loral plate very small, sometimes absent. Only two supplemental plates behind the azygos;

Genus EUTÆNIA. *Baird & Girard.*

Coluber, Tropidonotus, sp., AUCT.
Eutainia, BAIRD & GIRARD, Cat. N. Am. Reptiles, 1853.
Eutænia, emend.

To the few species of this genus known to the older authors under the names of *Coluber* or *Tropidonotus*, many were added, in 1853, by Baird and Girard, at the date of establishment of the genus *Eutænia*, a majority of the fifteen species described in the catalogue of the authors

---

the latter is sometimes replaced by two symmetrical contiguous plates, and without any supplemental. The prenasal and prefrontal in contact with the posterior process of the rostral. Dorsal row of scales twenty-three, all carinated except the first and second, which are perfectly smooth. Ground-color light yellowish gray; a dorsal series of rather indistinct, rounded or subquadrate, brown blotches; a second series of smaller, circular spots, much darker and more distinct; below this a third and more indistinct series.

DESCR.—In its general form and appearance, this resembles the *H. nasicus*, with which it is sometimes found associated. The body, however, is rather shorter and thicker than in *H. nasicus*, and the head is broader, with the part of the head anterior to the eye decidedly shorter. The nasals are not as well developed longitudinally as in *H. nasicus*, but the result of this shortness of the anterior part of the head is seen in the very small loral, which is frequently wanting entirely. There is never more than one loral, while frequently two are seen in *H. nasicus*, in which the loral is in every case strikingly larger than in *kennerlyi*. The most striking difference between these species is in the number of small plates surrounding the azygos or postrostral. While in *H. nasicus* there are always at least ten of these, one or two of which margin the inner edges of the prenasals and prefrontals, there are never more than two, and frequently but one additional plate, in *H. kennerlyi*, and the prenasal and prefrontal are always in contact with the posterior process of the rostral. The azygos is short, nearly as broad as long, and usually there are just behind it two contiguous plates of about the same size, separating it from the postfrontal, but not from the prefrontal. Frequently, however, the azygos is longitudinally divided, and without any additional plates, but in contact with the rostral anteriorly, and the vertical posteriorly, and not separated from the postfrontals. The vertical, occipitals, superciliaries, and labials are much as in *H. nasicus*, though generally less developed longitudinally. The rostral is as large as in *H. nasicus*. The two outer dorsal rows are both perfectly smooth; in *H. nasicus*, the second is distinctly though delicately carinate.

The ground-color is light yellowish-gray, with a dorsal series of rather indistinct subquadrate or rounded blotches, two to two and a half scales long, and separated by intervals of one or two scales, rather wider anteriorly. Below this is a series of very distinct, purplish-black, circular blotches, covering four scales transversely and two longitudinally; below this one or more indistinct series of spots. This pattern of coloration is very similar to that of *H. nasicus*, but the ground-color is always lighter, and the dorsal spots are lighter and less distinct. The upper lateral series is of a purplish-black, and much more distinct, forming a prominent character.

Abdomen nearly entirely black, except a few yellow scuta. The head is marked as in *H. nasicus*, except that the nasals, prefrontals, and rostral are all yellowish; while, in the latter species, they are dark in front of the light transverse line which crosses the crown behind the rostral; and, in *H. kennerlyi*, the light line across the superciliaries and vertical is much broader than in *H. nasicus*. This species differs from *H. simus* in many of the same features as does *H. nasicus*. These, together with the small or absent loral and small number of supplemental plates, will readily distinguish it.

Rio Grande (*Dr. Kennerly*). Sonora.

just named being new. To these several more have since been added, chiefly by Mr. Kennicott and Professor Cope, from various parts of the West, the present number of current United States species being over twenty.

That the species of this large and difficult genus require thorough critical revision, with a reduction of the number of accredited species, will be evident to any one who undertakes the identification of any considerable series of specimens. A certain proportion of the specimens cannot be referred without hesitation to the described species they are supposed to represent, or, rather, may be referred, with about equal propriety, to more than one such species. This indicates either that the descriptions drawn from particular type-specimens are too exclusive to fairly afford specific diagnoses, or that the supposed species they represent are not valid, but blend with each other through intermediate specimens. There is unquestionably a gentle and complete intergradation between several of the accredited species.

Too much stress altogether has been laid, in the preparation of specific diagnoses, upon points which should properly be only adduced in illustration of the normal inherent range of variation of the individual, and have no value whatever as functions of the actual specific equation. For instance, "superior labials seven" and "superior labials eight" are expressions found in the diagnosis of certain species as distinguishing marks. Whereas the fact is, as any one may satisfy himself by examination of the first dozen specimens of *Eutænia* that come to hand, that the superior labials may be either seven or eight in different specimens of indubitably the same species, or that there may be eight of them on one side of the mouth, and seven on the other, in the same specimen. Subdivisions of the genus have been based upon the number of dorsal rows, whether 17, 19, or 21. Whereas it is a fact that different specimens vary a pair or two of scales in this respect, and that different parts of the body of the same specimen show a different number of rows of scales. Other matters, such as the width and sharpness of definition of the characteristic stripes, and the special tinge of coloration of these and other parts of the body, might be mentioned in similar terms.

As far as we have seen, the position of the lateral stripe may be a means of grouping the species. Though this varies within certain limits, mainly according to the width of the band, yet its position on the third and fourth, or on the second and third, dorsal rows, affords a ready means of distinguishing certain sets of species or varieties.

Along the northern boundary, *Eutænia* is the best represented genus of *Ophidia*, and indeed of *Reptilia*, not only in numbers of individuals, but of species as well. They occur in all situations, excepting, as a rule, the most arid regions, and are particularly numerous about the prairie pools and sloughs and along the banks of the various streams. All the species represented are more or less aquatic, particularly during the season of reproduction. Several hundred individuals fell under Dr. Coues's observation, of which a sufficiently large series of upward of fifty specimens, representing all the species known to occur in this region, was preserved.

These specimens include three perfectly distinct species. One of these is the wide-ranging *E. vagrans*, almost universally distributed in the West. Another is of the *sirtalis* type, and the third, more abundant and characteristic than either of the others, belongs to the *radix* group, and occurs under two varieties geographically distinguished. Without reference to other species of the genus, the three just indicated may be readily distinguished by the following analysis:—

A. Lateral stripe on the second and third rows of dorsal scales.
  a. Dorsal rows commonly 21; normally 8 superior labials.
    Body brown, with numerous small dark spots in two rows, nicking into the narrow inconspicuous stripes; no red; no bands on head; belly variably plumbeous .... *vagrans.*
  b. Dorsal rows commonly 19; normally 7 superior labials.
    Body pitchy-black, without spots, but mixed with small vermilion-red spaces; the stripes broad, firm, and perfectly continuous .... *sirtalis parietalis.*
B. Lateral stripe on the third and fourth rows of dorsal scales; superior labials normally 7; normally 21 rows of scales.
  a. Pitchy-black and equally so below and above the lateral stripe; dorsal and lateral stripes narrow; both gamboge-yellow, not contrasted with each other .... *radix.*
  a''. Olivaceous-black; lighter or interrupted below the lateral stripe; dorsal stripe broad, rich chrome-yellow, contrasted with the pale gamboge-yellow lateral stripe .... *radix twiningi.*

EUTÆNIA VAGRANS. *B. & G.*

Wandering Garter Snake.

Eutainia vagrans, BD. & GIR., Cat. N. A. Reptiles, 1863, 35 (Texas and California to Puget Sound).—GIR. U. S. Expl. Exped. Herpet. 1858, 154, pl. 14, f. 5-10.—BD. P. R. R. Rep. x. 1859, Beckwith's Route, Reptiles, 19, pl. 17.—COOP. & SUCKL. N. H. W. Terr. 1860, 297.—COPE, Proc. Acad. Nat. Sci. Phila. 1866, 305, 307.
Eutænia vagrans subsp. *vagrans*, COPE, Check List N. A. Bat. and Rep. 1875, 41.

Specimens.

1151.	Sweetgrass Hills.	August 3, 1874.
1157.	Sweetgrass Hills.	August 3, 1874.
1157 *bis.*	Sweetgrass Hills.	August 3, 1874.
1157 *ter.*	Sweetgrass Hills.	August 3, 1874.
1184.	Chief Mountain Lake.	August 23, 1874.

DESCRIPTION (from Nos. 1157, *bis*, *ter*).—This is a rather small species, the largest specimens seen being little over two feet in length, of slender form, and inconspicuous coloration, by reason of the narrowness of the bands, their dull color, and their indentation by the series of dark spots. On a general view, these spots are nearly as evident as the bands themselves, being quite blackish and set off upon the general dull grayish-brown ground-color. The under parts are dull slaty-gray, variously speckled and blotched with slaty-black, which in some cases prevails over the gray, especially on the hinder part of the body. The dorsal and lateral bands are alike pale dull yellowish. The dorsal stripe at its broadest points is one scale and two half scales wide; where encroached upon by the black spots, it is reduced to a single scale, or even interrupted altogether. These spots are generally opposite, giving a beaded character to the dorsal stripe; sometimes alternate, when the band appears zigzag; and both these conditions may be found at different points on the same specimen. The lateral stripe is less firm than the dorsal, since it is not only beaded along its upper edge by the lower one of the two series of lateral spots, but also blended to a degree with the color of the first row of scales along which it lies, as usual in those species in which this stripe is on the second and third rows. The first row of dorsal scales is colored like the belly, not like the back. The lateral dark spots, very numerous, and, as already said, quite conspicuous, are mostly alternate with each other, in some places opposite. The plates of the head are light brown, excepting the labials, which are colored like the body.

Twenty-one rows of scales is normal in this species, and the lateral stripe occupies the second and third. The head is large and especially wide, and the muzzle blunt. The superior labials are eight in two and a half of the three specimens under examination, the other half of the third specimen having seven. The discrepancy occurs, as usual, among the smaller anterior ones, the eye being in all situate over the fourth and fifth, counting from behind. The third from behind is the largest of the series. The length of the tail is contained $4\frac{2}{3}$ times in the total length; $3\frac{2}{3}$ times in the length of the body alone.

E. vagrans exhibits in a marked degree the variation in number of labials, also of the anterior and postorbital plates. The species is peculiarly characteristic of the Central region, but it is found exceedingly numerous in Utah, Colorado, and Arizona, generally in mountains.

The Wandering Garter Snake does not appear to be generally distributed along the northern boundary line. It was not met with during the first year of my connection with the Survey in any part of the Red River watershed, nor was it seen the second season except to the westward from the outliers of the Rocky Mountains to the main chain itself. We may conclude that its northwestern limits of distribution are indicated in these points. The species was originally described from the Pacific slope, Puget's Sound, California, and New Mexico, and has since been shown to be of very general dispersion in the West, on both sides of the mountains.

EUTÆNIA SIRTALIS PARIETALIS. (*Say*) *Cope*.

Parietal Garter Snake.

Coluber parietalis, SAY, Long's Exp. R. Mts. i, 1823, 186.—HARL. Journ. Phila. Acad. v. 1827, 349.
Eutainia parietalis, BD. & GIR. Cat. N. A. Rep. 1853, 28.
Eutænia sirtalis subsp. *parietalis*, COPE, Check List N. A. Bat. and Rep. 1875, 41.

Specimens.

1180. Chief Mountain Lake. August 19, 1874.
1193. St. Mary's River. August 28, 1874.

DESCRIPTION (No. 1180, Chief Mountain Lake).—This is one of the larger species of the genus, frequently attaining a length of three feet, although, at the same time, the average dimension is less than this. It belongs strictly to the *sirtalis* group, and in fact is not specifically separable from that species. It is a rather slender and elegant snake; and, when found inhabiting the clear cold mountain streams or lakes, one of the most beautifully colored representatives of the genus, the stripes being firm and bold, and the dark body-color being relieved with rich red in marked contrast. The coloration as observed in life in the Rocky Mountain specimens in August is as follows:—

The dorsal band, which is one scale and two half-scales broad, firm and perfectly continuous from head to end of tail, without indentation for the dark body-color, is pure yellow, fading to pale naples-yellow in alcohol. The tint is clearer than that of the lateral bands, which are rather of a heavier golden-yellow from some suffusion with the red that beautifully mottles the sides. The lateral stripe is as firm and continuous as the dorsal one, and broader, occupying two whole scales (of the second and third rows). The body-color is black, without obvious shade of brown or olivaceous, speckled between the scales with rich vermilion-red, which is very conspicuous on stretching the skin, forming an incomplete zigzag annulation. This red does not reach as high as the dorsal stripe, but extends through the lateral stripe, and occurs on the first dorsal row as a speck on the lower corner of each scale, and on the corresponding angle of the gastrosteges. The color is chiefly on the skin itself between the scales, but also suffuses the edges of many scales themselves. The first dorsal row of scales, which are much wider than the others, are colored like the belly; this lighter inferior boundary of the lateral stripe causes the stripe itself to appear less firm in outline below than above. The belly is not blackish or even slaty, but of a peculiar pale glaucous-greenish, much as in *sirtalis;* but black appears as a pair of small, round, lateral spots on each scute at its front border. The head is olivaceous-blackish, the rostral, lorals, and labials being like the belly. Length 31 inches; tail about $\frac{1}{5}$ of the total length—$\frac{1}{4}$ of the length of body alone. Gastrosteges 157; urosteges 64, all bifid. Superior labials seven on both sides; the fifth largest in this and a second specimen examined. In No. 1180, the labials of the left side have a small intercalated scale between the second and third. No. 1193, from the Saint Mary's River, August, is larger than the other, measuring about thirty-six inches; it is similar in general coloration, but less richly marked, the red suffusion being of less extent and intensity.

This slender and elegant species was only observed at and near the end of the Line, at the eastern base of the Rocky Mountains, in August. It was common in the clear cold waters of the lakes and streams, and appeared to be one of the most thoroughly aquatic species of the genus, being often seen swimming freely in deep water at some distance from the shore. At this season, all the female individuals observed were gravid with nearly matured embryos. Like other of the genus, the species is ovo-viviparous, the young being some six inches in length when born. Newly-born individuals are of an indefinite dark color, with pale bands and under parts, without red, but with two rows on each side of very evident blackish specks—markings like those that persist in the adults of *E. vagrans*, for example. In two young specimens found *in utero*, the genital or intromittent organs are external to the anus, and extremely large when compared with the size of the individual. The placental cord is attached a short distance in advance of the anus.

EUTÆNIA RADIX. (B. & G.)

Racine Garter Snake.

Eutainia radix, BD. & GIR. Cat. N. Am. Reptiles, 1853, 34 (Racine, Wisconsin).—KENN. *apud* COOP. & SUCKL. N. H. Wash. Terr. 1860, 299 (Minnesota).
Eutænia radix, COPE, Check List N. A. Bat. and Rep. 1875, 40.
Eutænia haydeni, KENN. *apud* COOP. & SUCKL. Nat. Hist. Wash. Terr. 1860, 298, pl. 14 (Fort Pierre, Dakota).
Thamnophis haydeni, COPE.—HAYD. Trans. Amer. Assoc. xii. 1862, 177.

DESCRIPTION (No. 1027, Pembina, June 5, 1873).—This is one of the stoutest species of the genus. A specimen two feet long equals or rather exceeds in calibre a three feet long individual of *sirtalis* for example. The rapidly tapering tail, in a specimen 30 inches long, is 6½ inches, or contained about 4⅔ times in the total length—nearly 4 times in the length of body alone. The head is very short and thick, with a broad obtuse muzzle. The dorsal stripe, one scale and two half-scales broad throughout, is firm and continuous along the body, but less evident (sometimes extinguished altogether) on the tail. The lateral stripe is fairly two scales wide along most of the body, but only a scale and a half posteriorly, and but one scale on most of the tail; on the body it occupies the third and, for the most part, the fourth row of dorsal scales; on the tail it descends at once to the first row. Both dorsals and lateral bands are alike clear pale yellow, the former only occasionally deepening anteriorly into a more golden- or chrome-yellow. The body is olivaceous-blackish or obscure brownish-black, and of much the same tint above and below the lateral stripes. In the darkest and most "pitchy" black individuals, no markings are evident; in some lighter ones, there are indications of obscure dark spots, scarcely or not, however, traceable in definite rows. In all the specimens, the series of spots below the lateral line are well marked and distinguishable, and in many cases the line of spots just above the lateral line is fairly perceptible. The belly is pale glaucous-olivaceous, touched with blackish on the sides of the scutes, and sometimes this color mingles with the dark of the sides below the lateral stripe. Unlike those species in which the lateral band is lower down, there is not so much difference in the size or shape of the first and second dorsal rows. The superior labials are normally seven, sometimes eight on one or both sides; they are light-colored, like the belly, but each has a touch of blackish along the posterior border. The dorsal rows are normally 21, and all quite broad. All the dorsal scales are strongly carinated, giving the animal a roughened, scabrous appearance. The length is oftener 2 to 2½ feet than more, but at all ages the bulk of the snake, as already indicated, is considerable.

The specimen here described, a gravid female, and others of the numerous ones collected, agree perfectly with the original diagnosis of the type from Racine, Wis. (whence the name "radix" is derived), and equally well with Mr. Kennicott's subsequent description of a specimen from Fort Snelling, Minn. Throughout the Red River region, from Pembina to where the Coteau de Missouri crosses the line, it is the characteristic Ophidian, the principal and almost the only representative of its order, outnumbering all the others put together. Indistinguishable specimens also occur in the eastern portions of the Missouri region at the same latitude, though there the greater number are of the *twiningi* type, which farther westward prevails altogether.

In the more fertile portions of the Red River Valley itself, this snake may be found almost anywhere in the brush and herbage. Out on the dryer prairie beyond, it is chiefly confined to the pools and streams, or their immediate vicinity. Numbers are found basking together on the muddy borders of the sloughs, or among masses of aquatic vegetation, where they find ample subsistence during the summer months in the tadpoles, young frogs, and various water insects. They are themselves preyed upon by hawks, especially the Marsh Harrier (*Circus cyaneus hudsonius*) and Swainson's Buzzard (*Buteo swainsoni*). They are less active than some of the slenderer species, are readily caught, and when captured make little or no resistance. Only the largest individuals assume for the moment a defensive attitude and attempt to bite; most may be at once handled with impunity. The greater part of the females observed in July and August will be found pregnant, the young numbering sometimes as many as thirty or forty. Individuals were taken *in coitu* in September and part of October. These observations together indicate a period of gestation protracted for the greater part of a year. The snakes become much less numerous in the latter part of September, but Dr. Coues occasionally saw them abroad on warm days up to the middle of October, even after there had been snow, sleet, and freezing of the more shallow waters.

EUTÆNIA RADIX TWININGI. *Coues & Yarrow.*

Twining's Garter Snake.

DESCRIPTION (No. 1135, Two Forks of Milk River, July 15, 1874).—From the Coteau de Missouri westward, in the arid region of the Upper Missouri and Milk Rivers, the characters of *E. radix* undergo considerable modification. The difference is easily recognized in life by an observer familiar with both kinds. The principal character is seen in the increased breadth and intensity of coloration of the dorsal band, especially on the anterior portion. This band is of a rich chrome-yellow or reddish-golden, contrasting strongly with the clear pale yellow of the lateral stripe. This richly-colored cadmium-yellow band commences as a minute linear trace on the middle borders of the two scales just posterior to the occipitals; it then covers a single scale in rear of these, gradually increasing until three or even four scales are covered, finally settling down to thin scales which continue down two-thirds of body, then covers one whole and two halves; opposite the anus, and to its termination, it is confined to two half-scales. At its broadest part, near the head, it is full three scales broad, and sometimes even three and two half-scales in width. There are slight or no indications of darker mottling, even in the lighter-colored specimens. Below the lateral band, the dark color is usually much broken up with mottling of the color of the belly. With much the same general form as in *E. radix*, the head appears decidedly narrower and less obtuse. In the specimen 1135 there are eight upper labials on right side, seven on left; it is a gravid female.

This form corresponds perfectly with certain geographical faunal areas which are represented in the region under consideration, a fact in further evidence of the propriety of distinguishing it. It is abundant about the prairie pools of the Upper Missouri and Milk Rivers; its habits are the same as those of *E. radix*. It does not appear to extend into the Saskatchewan watershed.

Dedicated to Maj. W. J. Twining, United States Engineers, in recognition of his cordial coöperation in the scientific interests of the Boundary Commission, and in expression of our personal consideration.

List of specimens (of both forms).

1013.	Pembina.	June 5, 1873.
1019.	Pembina.	June 7, 1873.
1020.	Pembina.	June 7, 1873.
1027.	Pembina.	June 7, 1873.
1047.	Pembina.	June 24, 1873.
1065.	Turtle Mountain	July 22, 1873.
1068.	Turtle Mountain.	July 23, 1873.
1089.	Mouse River.	Aug. 20, 1873.
1090.	Mouse River.	Aug. 20, 1873.
1091.	Mouse River.	Aug. 20, 1873.

1093.	Mouse River.	Aug. 25, 1873.
1100.	Big Muddy River.	June 25, 1874.
1117.	Porcupine River.	June 27, 1874.
1119.	Big Porcupine River.	June 29, 1874.
1130.	Frenchman's River.	July 8, 1874.
1132.	Near Frenchman's River.	July 12, 1874.
1132 *bis*.	Near Frenchman's River.	July 12, 1874.
1132 *ter*.	Near Frenchman's River.	July 12, 1874.
1135.	Two Forks of Milk River.	July 15, 1874.

To the foregoing species of Ophidians observed by the Boundary Commission may be added short notices of the following, known to occur in the Yellowstone region, and very probably extending farther north:—

EUTÆNIA PROXIMA. (*Say*) *B. & G.*

Say's Garter Snake.

Coluber proximus, SAY, Long's Exped. R. Mts. i. 1823, 187.—HARL. Journ. Acad. Nat. Sci. Phila. v. 1827, 353.

Tropidonotus proximus, HALLOW. Sitgreaves's Rep. Expl. Zuñi and Colorado R. 1853, 134, 146.

Eutainia proxima, BD. & GIR. Cat. N. A. Reptiles, 1853, 25.—ALLEN, Proc. Bost. Soc. N. H. xvii. 1874, 69 (Yellowstone).—KENN. *apud* BD. U. S. Mex. B. Surv. ii. pt. ii. 1859, Reptiles, 16.

Eutænia proxima, COPE, Check List N. A. Bat. and Rep. 1875, 40.

This is a stout species, like *E. radix* and *E. twiningi*, the total length only about 3½ times that of the tail. The dorsal stripe is ochraceous-yellow; the lateral greenish-white or yellow on the third and fourth rows of scales; the dorsal rows are 19 in number. Carinæ of scales of dorsal region are of a whitish hue, which gives the species a streaked appearance, and the upper anterior border of the last row of scales is lined with white. There are also irregular white spots near the row of black ones above lateral line; these are more profuse in some places than others. The belly is greenish-white, more yellowish anteriorly. Its known range is east of the Rocky Mountains, from the region of the Yellowstone to New Mexico and Texas. Specimens are in the National Museum from Texas, California, Wisconsin, North Carolina, Mexico, &c. A specimen from Tomales Bay, California, is the type of *E. imperialis*.

EUTÆNIA SIRTALIS PICKERINGI. (*B. & G.*)

Pickering's Garter Snake.

a. SIRTALIS.

Coluber sirtalis, LINN. Syst. Nat. i. ed. 12, 1766, 383.—GM. Syst. Nat. i. pt. iii. ed. 13, 1788, 1107.—HARL. Journ. Acad. Nat. Sci. Phila. v. 1827, 352.—HARL. Med. and Phys. Res. 116.—STORER, Rep. Reptiles Massach. 1839, 221.

Tropidonotus sirtalis, HOLBR. N. Am. Herpet. iii. 1842, 41, pl. 11.

Eutainia sirtalis, BD. & GIR. Cat. N. A. Rept. 1853, 30.

Tropidonotus tænia, DEKAY, N. Y. Fauna, iii. 1842, 43.
Tropidonotus bipunctatus, SCHL. Ess. Physiogn. Serp. 1837, 320.
Tropidonotus tænia, DEKAY, N. Y. Zool. 1842, 43, pt. 13, f. 27.

b. PICKERINGI.

Eutainia pickeringii, BD. & GIR. Cat. N. A. Rept. 1853, 27 (Puget Sound).—GIR. U. S. Expl. Exped. Herpet. 1858, 150, pl. 13, f. 14–20.—COOP. & SUCKL. Nat. Hist. Wash. Terr. 1860, 296.
Eutænia sirtalis subsp. *pickeringii*, COPE, Check List Bat. and Rep. N. A. 1875, 41.

This species was not procured by Dr. Coues, but is introduced on the strength of specimens from Fort Benton, Montana, collected by Lieutenant Mullan.

There are two well-marked forms of *E. sirtalis* subspecies *pickeringi*.

The spots confluent into a dark band.
a. With a lateral band.
b. No lateral band.

Baird and Girard say of the species,—"Body slender; black above, slate-color beneath; lateral stripe irregular, confluent with the light-colored intervals between the dark spots. This species exhibits great variation of color, principally in regard to black of abdomen." In the reserve collection of reptiles in the National Museum are quite a number of specimens of the two differently marked subspecies all from one locality, viz, Fort Benton, Missouri, collected by Lieutenant Mullan of the Army. In this subspecies, the differences in number of labials may be frequently seen.

TROPIDONOTUS SIPEDON. (*L*.)

Water Snake.

Coluber sipedon, LINN. Syst. Nat. ed. 12, 1766, i. 379.—GMEL. Syst. Nat. ed. 13, 1788, pt. iii. 1098.—HARL. Journ. Phila. Acad. v. 1827, 351; Med. and Phys. Res. 114.—THOMPS. Hist. Vermont, 1842, 118.
Tropidonotus sipedon, HOLBR. N. Am. Herpet. iii. 1842, 29, pl. 6.—DEKAY, N. Y. Fauna, iii. 1842, 42, pl. 14, f. 31.—HAYD. Trans. Amer. Phil. Soc. xii. 1862, 177.
Nerodia sipedon, BD. & GIR. Cat. N. Am. Reptiles, 1853, 38.
Coluber pœcilogaster, MAXIM. Reise Nord-Amer. i. 1839, 106.

This serpent appears to have been first found in the Upper Missouri region by the Prinz Maximilian von Neu Wied, who described it under the name of *Coluber pœcilogaster;* and it was subsequently observed in the Yellowstone country by Dr. F. V. Hayden. It is one of the commonest and best known species of the Eastern United States. The serpent of this region, however, may not be typical *sipedon*, but rather *woodhousii* or *erythrogaster*. Reptiles of this genus (comprising *Nerodia* and *Regina* of Baird and Girard) are the most completely aquatic ones of this country. The species of *Nerodia* proper are dark-colored, more or less evidently blotched, stout and rather repulsive, quite pugnacious when full grown, and commonly regarded as venomous under the name of Water Adder, or Water Moccasin: needless to say, like other true *Colubrines*, they are perfectly harmless. Those of the other section, *Regina*, are slenderer, and banded lengthwise, much like *Eutænia*.

PITYOPHIS SAYI BELLONA. (*B. & G.*) *Cope*.

Say's Pine Snake.

a. SAYI.

Coluber sayi, SCHL. Ess. Physiogn. Serp. 1837, 157. (Not *Coronella sayi* of Holbrook or *Coluber sayi* of DeKay, which is *Ophibolus*.)
Pituophis sayi, BD. & GIR. App. Cat. N. A. Rept. 1853, 152 (in text under *Coluber sayi*, p. 151).—KENN. *apud* COOP. & SUCKL. Nat. Hist. Wash. Terr. 1860, 300, pl. 22.—HAYD. Trans. Amer. Phil. Soc. xii. 1862, 177.

b. BELLONA.

Churchillia bellona, BD. & GIR. Stansbury's Rep. Great Salt Lake, 1852, 350.
Pituophis bellona, BD. & GIR. Cat. N. Am. Rept. 1853, 66, 157.
Pityophis bellona, KENN. *apud* BD. P. R. R. Rep. x. 1859, Williamson's Route, Reptiles, 42.—KENN. *apud* BD. U. S. Mex. B. Surv. ii. pt. ii. 1859, Reptiles, 18.—BD. U. S. P. R. R. Rep. x. 1859, Beckwith's Route, Reptiles, 19.—COPE, Proc. Acad. Nat. Sci. Phila. 1866, 305.—ALLEN, Proc. Bost. Soc. Nat. Hist. xvii. 1874, 69.
Pityophis sayi var. *bellona*, COPE, Check List Bat. and Rep. N. A. 1875, 39.
Pituophis affinis, HALLOW, Proc. Acad. Nat. Sci. Phila. vi. 1852, 181.—HALLOW. Sitgr. Rep. Expl. Zuñi and Colorado R. 1853, 130, 146.

The species of this genus, known as "Pine" and "Bull" Snakes, are of large size, sometimes attaining a length of six feet or more. They are perfectly harmless, and appear of a rather sluggish and inoffensive disposition. They are light-colored (whitish, yellowish, or even reddish), but thickly blotched above with a dorsal series of numerous large brown or brown black-bordered spots, and other smaller lateral ones; on each side of the belly is usually found (as in the case of the present species) a row of black spots, one on each scutellum. Several upper dorsal series are lightly carinated; the rest are smooth. The tail is very short, about one-twelfth of the whole length, half-ringed above with black, and having lateral black spots. There is a dark stripe across the head from one eye to the other, continued behind each eye to the angle of the mouth. The head is very small, and the neck contracted. The general blotched character of the upper parts is somewhat in superficial appearance like that of *Crotalus confluentus* or *Heterodon nasicus*; but very little further observation is required to recognize the decided distinctions.

The best known species of this genus is the *P. melanoleuca*, the Common Pine or Bull Snake of the Eastern United States. An excellent and interesting account of the habits of this species, by the Rev. S. Lockwood, will be found in the American Naturalist for January, 1875.

Serpents of this genus vary notably in the construction of the plates of the head. A specimen of *P. bellona*, from the Yellowstone, collected by Mr. Allen in the expedition of 1873, presents the following case:—A large subpentagonal, shield-shaped vertical. Single large, triangular superorbital. A pair of moderate occipital and numerous small temporals. In advance of the vertical is a small azygos plate, wedged in between the connivent postfrontals. Two pairs of postfrontals. One pair of prefrontals. Two nasals, the nostril between them. A small loral. Two anteorbitals; the lower very small; the large, upper one bounding nearly all of the orbit anteriorly. Three small postorbitals. A large obtuse rostral. Eight superior labials, the eye over the fourth and fifth, the penultimate one largest.

After a careful examination of many specimens of *P. sayi bellona* and *P. sayi mexicana* in the National Museum, we find no absolute diagnostic value in the entire number of superior and inferior labials and number of dorsal scales, and are rather of the opinion that these two species should be grouped together under Baird and Girard's original name of *bellona*. Further investigation may show that *catenifer* Blainv. should be brought under the same head.

OPHIBOLUS GETULUS BOYLII. (*B. & G.*) *Cope.*

Ophibolus boylii, BD. & GIR. Catal. N. Am. Reptiles, 1853, 82.—BD. P. R. R. Rep. x. 1859, Williamson's and Abbott's Route, Reptiles, 11.—BD. U. S. Mex. B. Surv. ii. pt. ii. 1859, Reptiles, 20.—COPE, Proc. Acad. Nat. Sci. Phila. 1866, 305.

Lampropeltis boylii, COPE, Proc. Acad. Nat. Sci. Phila. 1860, 255.

Ophibolus getulus subsp. *boylii*, COPE, Check List N. A. Bat. and Rep. 1875, 37.

Coronella balteata, HALLOW. "Pr. Acad. Nat. Sci. Phila. vi. 1853, 236".—HALLOW. P. R. R. Rep. x. 1859, Williamson's Route, Reptiles, 14.

A specimen of this species, contained in a collection from the Yellowstone, offers the following characters:—The color is lustrous browni h-black, crossed at intervals of about an inch by narrow rings of pure white, which gradually widen on the sides to a breadth greater than that of the black interspaces. On the belly, these white rings are sometimes opposite, and then are continuous with the white coming down from the other side, and sometimes alternate, when they abruptly meet the black, producing a checkered pattern. These points are wholly irregular, both being observable in different parts of the same specimen. In this specimen, which is about 3½ feet long, there are in all forty-four rings, including some which are incomplete, that is, existing only on one side; for the rings on the back, as on the belly, are not always continued all around, some broken ones finding no fellow on the opposite side. In other specimens, there is also the greatest variety in all these details of pattern. The fore part and sides of the head are irregularly blotched with black and yellowish, and there are yellowish specks on the occiput.

This species is found abundantly in Pacific and Sonoran districts, and grows to a large size. The dark bands in Californian specimens in life are of a lustrous blackish-green bronze. Its discovery in Montana is an interesting fact, as there is but one specimen in the National Museum from that region.

The genus *Ophibolus* comprises a considerable number of species of very handsomely marked serpents, in all of which a black, brown, or red ground is crossed by light markings. The *Ophibolus getulus* is a common Eastern species, black like the present, and ringed with yellow, but the rings bifurcate on the sides.

OPHIBOLUS MULTISTRATA. (*Kenn.*)

Lampropeltis multistriata, KENN. Proc. Acad. Nat. Sci. Phila. 1860, 328. (By err. typog. for *multistrata*. "Fort Lookout, Nebraska", by err. for Fort Benton, Montana.)—HAYDEN, Trans. Amer. Phil. Soc. xii. 1862, 177 (Fort Benton).
Ophibolus multistratus, COPE, Check List N. A. Bat. and Rep. 1875, 37.

The locality of the original specimen is stated by its discoverer to be Fort Benton, Montana, not "Fort Lookout, Nebraska". Another error occurred in the original notice of the species, the name being printed *multistriata* for *multistrata*, in allusion to the number of rows of scales.

BASCANIUM FLAVIVENTRE. (*B. & G.*)

Yellow-bellied Black Snake.

Coluber flaviventris, SAY, Long's Exped. R. Mts. ii. 1823, 185.
Bascanion flaviventris, BD. & GIR. Cat. N. A. Reptiles, 1853, 96.—BD. U. S. Mex. B. Surv. ii. pt. ii. 1859, Reptiles, 20.—HAYD. Trans. Amer. Phil. Soc. xii. 1862, 177.—ALLEN, Proc. Bost. Soc. N. H. xvii. 1874, 69 (Yellowstone).

This species is dark olive-green above and bright yellow beneath, being distinguished from the usual style of *B. constrictor* by these characters, the last-named species being lustrous pitch-black above and ordinarily greenish-black below. These characters, however, are not diagnostic, as more or less yellow-bellied Eastern *constrictor* often occurs.

Inasmuch as individuals of *B. constrictor*, which have not attained their adult state, resemble greatly in coloration *B. vetustum*, color cannot be relied upon as a specific point in diagnosis. The position of certain of the upper labial and their relation to the eye and that of the lower postorbital afford the most reliable means of distinguishing the species. In *Bascanium constrictor*, a line drawn slightly obliquely backward from the junction of the third and fourth upper labials will pass directly through the centre of the pupil of the eye. The same line drawn in *B. vetustum* would pass slightly anterior to the centre of the pupil, and in the latter species the lower postorbital lies in a notch between the fourth and fifth upper labials. In *B. constrictor*, the lower postorbital rests on the upper border of the fourth upper labial. In some cases, the position of the lower postorbital in *B. vetustum* differs on different sides of the same individual. The young of *B. vetustum* can hardly be distinguished from the young of *B. constrictor* except by the position of the lower postorbital. The description of the young by Baird and Girard, p. 94 of their Catalogue, is excellent, and should be relied upon, as young specimens differ so materially in coloration from adults.

The typical and best known species of this genus is the common Black Snake of the United States (*B. constrictor*). All the species agree in their slender form and perfectly smooth, lustrous scales and uniform coloration while adult, though the young are somewhat particolored. They grow to a large size, and are noted for their powers of constriction. They are among the most active and agile of our serpents, possessing eminent scansorial powers, and are persistent enemies of numerous small birds, whose nests they rob of the eggs or young.

CYCLOPHIS VERNALIS. (*DeK.*) *Günth.*

Coluber vernalis, DEKAY, MS.—HARL. Journ. Acad. Nat. Sci. Phila. v. 1827, 361; Med. and Phys. Res. 1835, 124.—STORER, Rep. Mass. Rept. 1839, 224.—HOLBR. N. Am. Herpet. iii. 1842, 79, pl. 17.—DEKAY, N. Y. Fauna, iii. 1842, 40, pl. 11. f. 22.—Thomps. Nat. Hist. Vermont, 1842, 117.

Chlorosoma vernalis, BD. & GIR. Catal. N. Am. Rept. 1853, 108.

Herpetodryas vernalis, HALLOW. Proc. Acad. Nat. Sci. Phila. 1856, 243.

Liopeltis vernalis, COPE, Proc. Acad. Nat. Sci. Phila. 1860, 560.—HAYD. Trans. Amer. Phil. Soc. xii. 1862, 177.

Cyclophis vernalis, GÜNTHER, Cat. Col. Snakes Brit. Mus. 1853, 119.

Observed by Dr. Hayden on the Yellowstone.

In vol. v, Zoology, of Ex. for Expl. West of One Hundredth Meridian, mention is made by Dr. Yarrow, p. 539, of the discovery of this species at Abiquiu, N. Mex., and in the Am. Nat. vol. —, p. —, the same author relates that it was found by Lieutenant Carpenter in Colorado. These facts extend greatly its southern limit. *C. æstivus*, the Eastern and Southern species, has been found in New Mexico, Texas, and Western Missouri.

The two species are readily distinguishable: *C. vernalis* having 15 rows of smooth scales, 7 upper labials, 8 lower; and *C. æstivus* having 17 rows of strongly carinated scales, except the outer row, which is smooth, and the second slightly keeled; upper labials 7, lower labials 8.

SAURIA.

Family IGUANIDÆ.

Genus PHRYNOSOMA. *Wieg.*

PHRYNOSOMA DOUGLASSI. (*Bell*) *Gray.*

Horned Lizard; Horned "Toad"; Horned "Frog".

a. DOUGLASSI.

Agama douglassii, BELL, Trans. Linn. Soc. xvi. 1829, 105, pl. 10.—BELL, Isis, Bd. xxiii. 1830, 910.—HARL. Med. and Phys. Res. 1835, 141, f. 3.

Phrynosoma douglassii, GRAY, Griffith's An. King. ix. 1831, 44.—WAGL. Nat. Syst. Amph 1830, 146.—WIEGM. Herp. Mex. 1834, 54.—HOLB. N. Am. Herpet. i. 1842, 101, pl. 14.—GRAY, Cat. Br. Mus. 1845, 227.—GIRARD, Stansbury's Rep. Expl. Great Salt Lake, 1852, 362, pl. 7, f. 6-9 (monographic).—COPE, Proc. Acad. Nat. Sci. Phila., 1866, 302.—ALLEN, Proc. Bost. Soc. Nat. Hist. xvii. 1874, 69.

Phrynosoma douglassii subsp. *douglassii*, COPE, Check List N. A. Bat. and Rep. 1875, 49.
Tapaya douglassii, GIRARD, U. S. Ex. Ex. Herpet. 1858, 398, pl. 21, f. 1-5.—BD. P. R. R. Rep. x. 1859, Gunnison's and Beckwith's Route, Reptiles, 18.—BD. P. R. R. Rep. x. 1859, Williamson's and Abbott's Route, Reptlles, 9.—COOP. & SUCKL. N. H. Wash. Terr. 1860, 294.

b. ORNATISSIMA.

Phrynosoma orbiculare, HALLOW. Sitgreaves's Rep. Expl. Zuñi and Colorado Rivers, 125, pls. 8, 9. (*Nec* Wiegm.)
Tapaya ornatissima, GIR. U. S. Expl. Exped. Herpet. 1858, 396.—BD. P. R. R. Rep. x. 1859, Whipple's Route, Reptiles, 38.—BD. U. S. Mex. B. Surv. ii. pt. ii. 1859, Reptiles, 9.
Phrynosoma douglassii subsp. *ornatissimum*, COPE, Check List N. A. Bat. and Rep. 1875, 49.

Specimens.

1148.	Milk River at 49°.	July 26, 1874.
1153.	Sweetgrass Hills.	Aug. 3, 1874.
1154.	Sweetgrass Hills.	Aug. 3, 1874.

Representatives of the order *Sauria* are even fewer in number than those of the *Ophidia* in this latitude (49° N.), and the present is the only species which was observed by the Commission. Two others, however, are given beyond as probably occurring on the line east of the Rocky Mountains. Six, including the present species, are described from Washington Territory by Drs. Cooper and Suckley.

Douglass's Horned Lizard is the most abundant and widely diffused of the six or eight known United States species of the genus *Phrynosoma*. It ranges in fact throughout nearly all parts of the West, from the Mexican to the British Boundary, reaching the latter in the region of the Milk River. Its northernmost extension east of the Rocky Mountains at any rate appears to be only in the Missouri watershed. Dr. Coues obtained no indication of its presence in any part of the Red River region. It was found quite commonly on the Milk River, where this stream crosses the Line, and thence westward to the Rocky Mountains. In these latitudes, its range appears to coincide with that of *Crotalus confluentus*.

The present species may be known among the congeners by the orbicular shape of the body in connection with the very slight development of the cephalic spines, which are, in fact, no more than pointed scales, little different from those on other parts of the body. In some other species, the body is more elongated or oval, and certain plates upon the head are developed into long spines. The coloration is variagated and diffuse, and differs greatly in different individuals, especially farther south, where there is greater latitude in this respect than at the north. In this region of northernmost extension, the colors are almost uniform and quite pale, and the size is usually small. The coloration of the animal depends greatly upon the color of the ground where found, and it is our impression that this species possesses to a limited degree a modified power of chameleonization seen in other Saurians.

Although commonly called "Horned Toads", or "Horned Frogs", from hasty consideration of a superficial resemblance in shape to some of the

Batrachia, these animals are true Lizards, belonging to a large group which contains the Iguana, and numerous well known smaller species of this country. The presence of a tail, which is not found in adult life in the typical Batrachians; the scaly armor of the body, as contrasted with the naked skin of Batrachians; the mode of development, in not passing through a larval Tadpole stage, in which the species breathe in the water by means of gills; and the ambulatory, not saltatorial, mode of progression, are some of the prominent characteristics by which the *Phrynosomæ* show their true affinities.

Although of rather repulsive aspect, the Horned Lizards are inoffensive and perfectly harmless animals. They are rather sluggish, easily captured, make no resistance, and are readily tamed, when they make rather amusing pets for those who may desire to watch their sly and furtive ways. They feed principally upon flies, ants, and other small insects, which they catch by rapid protrusion of their viscid tongue, leaping, or rather running, at their prey sometimes. One that had been tamed would eat from the fingers, and also take a drop of milk, appearing greatly to relish it. Like other cold-blooded animals, they sustain long fasts without apparent inconvenience, and may be safely *mailed* alive, as curiosities, to almost any part of the country. We have never been able to keep them alive over four months.

The subspecies of this same *P. douglassii* (*ornatissimum*) is principally found in the Sonoran region, but specimens are in the National Museum from Utah and Upper Colorado and Pole Creek (wherever that may be).

SCELOPORUS CONSOBRINUS. *B. & G.*

Sceloporus consobrinus, BD. & GIR. Marcy's Rep. Expl. Red R. 1853, 224, pl. 10, f. 5-12.—BD. P. R. R. Rep. x. 1859, Whipple's Route, Reptiles, 37.—BD. U. S. Mex. B. Surv. ii. pt. ii. 1859, Reptiles, 5.—HAYD. Trans. Amer. Phil. Soc. xii. 1862, 177.—COPE, Proc. Acad. Nat. Sci. Phila. 1866, 303.—COPE, Check List N. A. Bat. and Rep. 1875, 49.—ALLEN, Proc. Bost. Soc. Nat. Hist. xvii. 1874, 69 (from the Yellowstone).

Lately obtained by Mr. J. A. Allen on the Yellowstone, and has been found in Utah and Nevada. Belongs to the Sonoran and Central region, Oregon?, and Texas.

Family SCINCIDÆ.

Genus EUMECES. *Wieg.*

EUMECES SEPTENTRIONALIS. (*Bd.*)

Northern Skink.

Plestiodon septentrionalis, BD. Proc. Acad. Nat. Sci. Phila. 1858, —.—BD. P. R. R. Rep. x. 1859, Whipple's Route, Reptiles, 38, pl. 24, f. 2.—HAYD. Trans. Amer. Phil. Soc. xii. 1862, 177.

Eumeces septentrionalis, COPE, Check List N. A. Bat. and Rep. 1875, 44.

Originally described from Minnesota, and also known to occur in Nebraska and Kansas.

B.—BATRACHIA.

ANURA.

Family BUFONIDÆ.

Genus BUFO.

BUFO LENTIGINOSUS FOWLERI. (*Putnam, MSS.*)

Fowler's Toad.

Bufo americanus var. *fowleri*, "PUTN. MSS."
Bufo lentiginosus subsp. *fowlerii*, COPE, Check List N. A. Bat. and Rep. 1875, 29.

Specimens.

1006.	Pembina.	June 3, 1873.
1012.	Pembina.	June 5, 1873.
1028.	Pembina.	June 7, 1873.
1031.	Pembina.	June 7, 1873.
1066.	Turtle Mountain.	July 23, 1873.
1070.	Turtle Mountain.	July 24, 1873.
? 1086.	Mouse River.	Aug. 17, 1873.
? 1092.	Mouse River.	Aug. 25, 1873.

Abundant along the line throughout the Red River watershed, where it was the only species observed. Westward, in the Missouri and Milk River region, it appears to be entirely replaced by the following species:—

BUFO COLUMBIENSIS. *Bd. & Gir.*

Columbia Toad.

Bufo columbiensis, BD. & GIR. Proc. Acad. Nat. Sci. Phila. 1853, 378.—GIR. Proc. Acad. Nat. Sci. Phila. 1854, 87.—GIR. U. S. Expl. Exped. Herpet. 77, pl. 5, f. 4-9.—BD. P. R. R. Rep. x. 1859, Williamson's and Abbott's Route, Reptiles, 12.—COOP. & SUCKL. N. H. Wash. Terr. 1860, 304.—ALLEN, Proc. Bost. Soc. Nat. Hist. xvii. 1874, 70.—COPE, Check List N. A. Bat. and Rep. 1875, 29.

Specimens.

? 1177.	Kootanie River.	Aug. 17, 1874.
1181.	Chief Mountain Lake.	Aug. 19, 1874.
1183.	Chief Mountain Lake.	Aug. 23, 1874.
1190.	Chief Mountain Lake.	Aug. 24, 1874.
1191.	Chief Mountain Lake.	Aug. 24, 1874.

This species, originally described from the Pacific slopes, and not generally recognized as occurring east of the Rocky Mountains, was found to be very common in the above-named localities. It was also procured on the Yellowstone by Mr. Allen. The palmation of the feet renders it much more decidedly aquatic in habit than is usual in this genus. I found it swimming freely in the lake, as well as in various streams and pools about the eastern base of the Mountains. Specimens were taken from the stomach of *Salmo namaycush* and other fish of the same genus in this locality, further indicating its aquatic nature. The colors of the specimens inhabiting these clear cold waters are notably fresh and firm.

Family RANIDÆ.

Genus RANA.

RANA HALECINA. *Kalm.*

Leopard Frog.

"*Rana pipiens*, GM.", Syst. Nat. 13th ed. 1788, 1052 (*nec auct.*).

Rana halecina, KALM.—DAUD. Hist. Nat. Rept. viii. 1803, 122.—HARL. Journ. Acad. Nat. Sci. Phila. v. 337; Med. and Phys. Res. 102, 224.—DEKAY, N. Y. Fauna, iii. 1842, 63, pl. 20, f. 19.—HOLBR. N. Am. Herpet. iv. 1842, 91, pl. 13.—BD. P. R. R. Rep. x. 1859; Whipple's Route, Reptiles, 45.—COOP. & SUCKL. N. H., Wash. Terr. 1860, 304, pl. 29, f. 7.—HAYD. Trans. Amer. Phil. Soc. xii. 1862. 177.—COPE, Proc. Acad. Nat. Sci. Phila. 1866, 301.—ALLEN, Proc. Bost. Soc, Nat. Hist. xvii. 1874, 70.

Specimens.

1064, *bis*, *ter*, etc.	Near Pembina Mountains.	July 9, 1873.	Numerous specimens.
?1081.	Souris River.	Aug. 16, 1873.	
?1081, *bis*.	Souris River.	Aug. 16, 1873.	
1118.	Wolf Creek.	June 28, 1874.	
1118, *bis*.	Wolf Creek.	June 28, 1874.	
?1167.	Head of Milk River.	Aug. 14, 1874.	

A series of specimens, demonstrating the general dispersion of the species in the permanent waters of the region explored. It occurs in Washington Territory, and I have found it in New Mexico and Arizona, as well as in various localities in the Eastern United States, where it is one of the most abundant and well-known species, conspicuous in its size, rich coloration, and agility.

The common Western form is *Rana halecina berlandieri*, which is only distinguished from *R. halecina* by its larger size and generally coarser and more pustulated skin. The specimens represent "*berlandieri*", but this we are disinclined to adopt without further investigation of its alleged distinctness.

Rana septentrionalis of Baird (Proc. Phila. Acad. 1854, p. 61) is accredited with a range from "Canada to Montana", but was not observed.

Family HYLIDÆ.

Genus CHOROPHILUS.

CHOROPHILUS TRISERIATUS. (*Maxim.*)

Helecetes triseriatus, MAXIM. Reise Nord-Am. i. 1839, —.—HAYD. Trans. Amer. Phil. Soc. xii. 1862, 177.
Chorophilus triseriatus, COPE, Check List N. A. Bat. and Rep. 1875, 30.

Specimens.

1123, *bis*, *ter*, etc. Frenchman's River. July 6, 1874.

This small species was found in the greatest abundance in prairie pools* and streams at various points along the line, especially at Frenchman's River, where numerous specimens were secured. It forms a considerable portion of the food of the *Eutæniæ* of this region. Specimens are also in the Natural Museum from Colorado, New Mexico, Nebraska, and Kansas.

Family AMBLYSTOMATIDÆ.

Genus AMBLYSTOMA. *Tsch.*

AMBLYSTOMA MAVORTIUM. *Bd.*

a. MAVORTIUM.

Ambystoma mavortia, BD. Journ. Acad. Nat. Sci. Phila. 2d ser. i. 1849, 292, 284 (New Mexico).
Ambystoma mavortium, HALLOW. Journ. Acad. Nat. Sci. Phila. iii. 1858, 352.
Amblystoma mavortium, BD. P. R. R. Rep. x. 1859, Gunnison's and Beckwith's Route, Reptiles, 20.—COPE, Proc. Acad. Nat. Sci. Phila. 1867, 184.—ALLEN, Proc. Bost. Soc. Nat. Hist. xvii. 1874, 70.—COPE, Check List N. A. Bat. and Rep. 1875, 25.
Amblystoma proserpina, BD. & GIR. Proc. Acad. Nat. Sci. Phila. 1852, 173.—BD. U. S. Mex. B. Surv. ii. pt. ii. 1869, Reptiles, 29, pl. 35, f. 7–14.
Ambystoma proserpine, HALLOW. Journ. Acad. Nat. Sci. Phila. iii. 1858, 354.
Ambystoma maculatum, HALLOW. Journ. Acad. Nat. Sci. Phila. iii. 1858, 355.—HALLOW. Proc. Acad. Nat. Sci. Phila. 1857, 215.
Desmiostoma maculatum, "SAGER, Penins. Journ. Med. 1858, 428".
Camarataxis maculata, COPE, Proc. Acad. Nat. Sci. Phila. 1859, 123.
Ambystoma nebulosum, HALLOW. Proc. Acad. Nat. Sci. Phila. 1852, 209 (Arizona).—HALLOW. Sitgreaves's Rep. Expl. Zuñi and Colorado R. 1853, 143, pl. 20.—HALLOW. Journ. Acad. Nat. Sci. Phila. iii. 1858, 352.
Amblystoma? nebulosum, COPE, Proc. Acad. Nat. Sci. Phila. 1866, 300.

b. CALIFORNIENSE.

Ambystoma californiense, GRAY, Proc. Zool. Soc. Lond. 1853, 11, pl. 7 (Monterey).—HALLOW. Journ. Phila. Acad. Nat. Sci. iii. 1858, 355.
Amblystoma mavortium subsp. *californiense*, COPE, Check List N. A. Bat. and Rep. 1875, 25.

* These pools also furnished great numbers of an interesting Phyllopod, *Lepidurus couesi* of Packard.

Specimens.

1045.	Pembina.	June 24, 1873.
1057.	Pembina.	June 28, 1873.
1071.	Turtle Mountain.	July 28, 1873.
1074.	Turtle Mountain.	Aug. 11, 1873.
1074 *bis.*	Turtle Mountain.	Aug. 1, 1873.
1137.	Two Forks of Milk River.	July 15, 1874.

Common in suitable situations all along the line. In all the specimens observed, the metamorphosis from the *Siredon* stage was completed at a length of four or five inches. In other regions, I have procured the same species, still in the *Siredon* stage, but nearly twice as large. Individuals were found in damp places about the buildings at Fort Pembina and vicinity, and still more numerously around the pools at the western base of Turtle Mountain. They wandered freely away from the water, and in some instances entered our tents.

In life, the coloration of the specimens examined was clear olive above, more glaucous or greenish-white below, everywhere variegated in bold pattern with black.

In addition to the foregoing, the only species of the genus observed by the commission, a second is described as inhabiting the region about the eastern portions of the line. This is the *Ambystoma laterale* of Hallowell (Journ. Acad. Nat. Sci. Phila. iii. 1858, 352), now regarded as a variety of *Amblystoma jeffersonianum* Bd. (*op. cit.* i. 1849, 283) (*Xiphonura jeffersoniana* Tschudi, Class. Batrach. 1838).

Amblystoma aterrimum Cope (Proc. Acad. Nat. Sci. Phila. 1867, 201) is a species described from the Northern Rocky Mountains, in the region explored by Lieutenant Mullan.

BIOLOGISTS AND THEIR WORLD

An Arno Press Collection

Adler, Kraig, editor. **Early Herpetological Studies and Surveys in the Eastern United States.** 1978

Adler, Kraig, editor. **Herpetological Explorations of the Great American West.** Two vols. 1978

Agassiz, [Jean] Louis [Rodolphe]. **Contributions to the Natural History of the United States of America.** Four vols. in two. 1857/1860/1862

Allard, Dean Conrad, Jr. **Spencer Fullerton Baird and the U.S. Fish Commission.** 1978

Altum, Bernard. **Der Vogel Und Sein Leben.** 1868

Azara, Don Felix de. **Apuntamientos Para La Historia Natural De Los Quadrúpedos Del Paragüay Y Rio De La Plata.** Two vols. in one. 1802

Baer, Karl Ernst v[on]. **Reden Gehalten in Wissenschaftlichen Versammlungen und Kleinere Aufsaetze Vermischten Inhalts.** Three vols. in two. 1864/1876/1873

Barrett-Hamilton, Gerald E. H. and Martin A. C. Hinton. **A History of British Mammals.** Two vols. 1910-1921

Boulenger, G[eorge] A[lbert]. **The Tailless Batrachians of Europe.** Two parts in one. 1897/1898

Brocchi, [Paul]. **Mission Scientifique Au Mexique Et Dans L'Amérique Centrale,...** Étude Des Batraciens De L'Amérique Centrale. 1882

Buffon, [Georges L. L.]. **The History of Singing Birds.** Translated from the French. 1791

Buffon, [Georges L. L.]. **The Natural History of Oviparous Quadrupeds and Serpents.** Translated by Robert Kerr. Four vols. in two. 1802

The Cabinet of Natural History and American Rural Sports. Three vols. in one. 1830/1832/1833

Candolle, A[ugustin] P[yramus] de and K[urt] Sprengel. **Elements of the Philosophy of Plants.** Translated from the German. 1821

Cassin, John. **United States Exploring Expedition During the Years 1838, 1839, 1840, 1841, 1842 Under the Command of Charles Wilkes, U. S. N.:** Mammalogy and Ornithology. Two vols. 1858

Chapman, Frank M. **Essays in South American Ornithogeography.** Edited by Keir B. Sterling. 1978

Cope, Edward D[rinker]. **The Vertebrata of the Tertiary Formations of the West.** One vol. in two. 1883

Cuvier, [Georges]. **The Class Mammalia.** Five vols. 1827

Donovan, E[dward]. **The Natural History of British Fishes.** Five vols. in two. 1802/1803/1804/1806/1808

Duméril, Auguste [Henri André], [Marie-Firmin] Bocourt and [François] Mocquard. **Mission Scientifique Au Mexique Et Dans L'Amérique Centrale,...** Étude Sur Les Reptiles. Two vols. 1870-1909

Flower, William Henry and Richard Lydekker. **An Introduction to the Study of Mammals, Living and Extinct.** 1891

Forbush, Edward Howe. **Birds of Massachusetts and Other New England States.** Three vols. 1925/1927/1929

Girard, Charles. **United States Exploring Expedition During the Years 1838, 1839, 1840, 1841, 1842 Under the Command of Charles Wilkes, U. S. N.:** Herpetology. Two vols. 1858

Grinnell, Joseph. An Account of the Mammals and Birds of the Lower Colorado Valley. 1914

Howard, H[enry] Eliot. **Territory in Bird Life.** 1920

Hume, Edgar Erskine. **Ornithologists of the United States Army Medical Corps.** 1942

Huxley, Leonard. **Life and Letters of Sir Joseph Dalton Hooker.** Two vols. 1918

LeConte, John L. and George H. Horn. **Classification of the Coleoptera of North America.** 1883

Linsley, E. Gorton, editor. **Beetles From the Early Russian Explorations of the West Coast of North America, 1815-1857.** 1978

Linsley, E. Gorton, editor. **The Principal Contributions of Henry Walter Bates to a Knowledge of the Butterflies and Longicorn Beetles of the Amazon Valley.** 1978

Murray, Andrew. **The Geographical Distribution of Mammals.** 1866

Osborn, Henry Fairfield. **Cope:** Master Naturalist. 1931

Peale, Titian R. **United States Exploring Expedition During the Years 1838, 1839, 1840, 1841, 1842 Under the Command of Charles Wilkes, U. S. N.:** Mammalogy and Ornithology. 1848

Phillips, John C. **American Game Mammals and Birds.** 1930

Rafinesque: Autobiography and Lives. 1978

Ray, John. **Synopsis Methodica Animalium Quadrupedum et Serpentini Generis.** 1693

Ray, John. **Synopsis Methodica Avium & Piscium.** [Edited by William Derham]. 1713

Richardson, John. **Fauna Boreali-Americana;** Or the Zoology of the Northern Parts of British America, Part Third: The Fish. 1836

Richardson, John, William Swainson and William Kirby. **Fauna Boreali-Americana;** Or the Zoology of the Northern Parts of British America, Part Four: Insecta, 1837

Riley, Charles V[alentine]. **[Nine] Annual Report[s] on the Noxious, Beneficial and Other Insects of the State of Missouri, 1869-1877** *and* **General Index and Supplement.** Ten vols. in three. 1869-1877/1881

Say, Thomas. **The Complete Writings of Thomas Say on the Entomology of North America.** Edited by John L. LeConte. Two vols. 1859

Schuchert, Charles and Clara Mae LeVene. **O. C. Marsh:** Pioneer in Paleontology. 1940

Sclater, William Lutley and Philip Lutley Sclater. **The Geography of Mammals.** 1899

Seton, Ernest Thompson. **Trail of an Artist-Naturalist:** The Autobiography of Ernest Thompson Seton. 1940

Smith, James Edward. **A Selection of the Correspondence of Linnaeus and Other Naturalists, From the Original Manuscripts.** Two vols. 1821

Stuckey, Ronald L., editor. **Development of Botany in Selected Regions of North America Before 1900.** 1978

Stuckey, Ronald L., editor. **Essays on North American Plant Geography From the Nineteenth Century.** 1978

Stuckey, Ronald L., editor. **Scientific Publications of Charles Wilkins Short.** 1978

Swammerdam, John. **The Book of Nature.** 1758

Wadland, John Henry. **Ernest Thompson Seton:** Man in Nature and the Progressive Era, 1880-1915. 1978

Waterhouse, G[eorge] R[obert]. **A Natural History of the Mammalia.** Two vols. 1846/1848

Weiss, Harry B. and Grace M. Ziegler. **Thomas Say:** Early American Naturalist. 1931

Wheeler, Geo[rge] M., [editor]. **[Reports Upon Insects Collected During Geographical and Geological Explorations and Surveys West of the One Hundredth Meridian During the Years 1872, 1873, and 1874].** 1875

Willughby, Francis. **De Historia Piscium** *and* **Icthyographia Ad Amplisimum Virum Dnum., Samuelem Pepys, Presidem Soc. Reg.** Two vols. in one. 1686/1685

Youmans, William Jay, editor. **Pioneers of Science in America.** Revised edition. 1896